Lectures on the Theory of Phase Transformations

2nd Edition

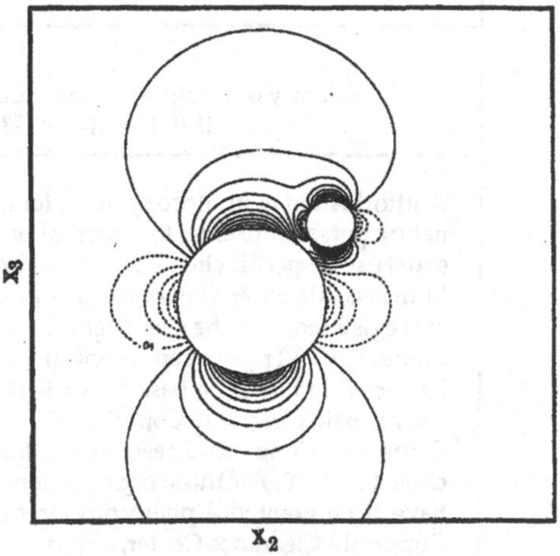

Edited by Hubert I. Aaronson

T0156039

A Publication of
The Minerals, Metals & Materials Society
184 Thorn Hill Road
Warrendale, Pennsylvania 15086-7528

Visit the TMS web site at
http://www.tms.org

The Minerals, Metals & Materials Society is not responsible for statements or opinions and is absolved of liability due to misuse of information contained in this publication.

Library of Congress Catalog Card Number 00-130293
ISBN 978-0-87339-476-5

Authorization to photocopy items for internal or personal use, or the internal or personal use of specific clients, is granted by The Minerals, Metals & Materials Society for users registered with the Copyright Clearance Center (CCC) Transactional Reporting Service, provided that the base fee of $7.00 per copy is paid directly to Copyright Clearance Center, 27 Congress Street, Salem, Massachusetts 01970. For those organizations that have been granted a photocopy license by Copyright Clearance Center, a separate system of payment has been arranged.

TMS

© 1999

If you are interested in purchasing a copy of this book, or if you would like to receive the latest TMS publications catalog, please telephone 1-800-759-4867 (U.S. only) or 724-776-9000, Ext. 270.

Preface to the Second Edition

The first edition of Lectures on the Theory of Phase Transformations was published by TMS in 1975. This edition was twice reprinted; the number of xeroxed reproductions is, of course, unknown but appears to have been rather large. Particularly in the U. S., this book found extensive use as an ancillary textbook in graduate courses on both phase transformations and on broader topics of which transformations was an important component.

Pressure recently developed, particularly amongst academicians, for a second edition. All of the original principal authors agreed to update their contributions. Additionally, Prof. William C. Johnson of the University of Virginia agreed to prepare a tutorially oriented overview of the effects of coherency upon phase equilibria and phase transformations; this is a topic which assumed major importance in phase transformations theory since publication of the first edition.

Perhaps the most important feature of the first edition was its emphasis upon more detailed presentations of the mathematics involved than is allowable in either research or review papers or is customary in monographs and textbooks. This feature has been further emphasized in the second edition.

The Phase Transformations Committee of ASM-International served as the informal sponsor of the second edition--and also as the primary source of vigorous reminders to the editor to spur the authors toward completion of their papers!

Hubert I. Aaronson, Editor
Department of Materials Science and Engineering
Carnegie Mellon University
Pittsburgh, PA 15213-3890, USA

Table of Contents

Chapter I

Chapter II

Chapter III

Theory of Capillarity

Rohit K. Trivedi, Ames Laboratory, Department of Science and Engineering, Iowa State University, Ames Iowa, U.S.A.

APPLICATIONS OF GIBBS ENERGY–COMPOSITION DIAGRAMS

Mats Hillert
Royal Institute of Technology
SE-10044 Stockholm, Sweden

1. Introduction

The discussion of equilibria and of the driving force for reactions in alloys is usually based upon Gibbs energy, G, because conditions of constant temperature and pressure are usually considered. (G is often called Gibbs free energy but according to recommendations by IUPAC one should use the shorter term Gibbs energy.) The purpose of this chapter is to present the basis for the use of molar Gibbs energy diagrams for such purposes. However, molar diagrams for all extensive quantities have many properties in common. This will be emphasized by using the molar volume V_m as the illustrative example when such properties are discussed in the first part. However, in order to make it easier to interpret such diagrams as G_m diagrams, the shape of the V_m curves will resemble the shape of typical G_m curves.

Under conditions of constant temperature, pressure and composition the Gibbs energy is particularly interesting because equilibrium is then characterized by a minimum in this quantity and the decrease in Gibbs energy during a spontaneous reaction can be regarded as the driving force. These facts make the application of molar diagrams of Gibbs energy versus composition very useful and the main part of the present chapter will be concerned with such diagrams, usually called G_m diagrams. It should be emphasized, however, that the same properties hold for the Helmholtz energy F at given temperature and volume, for the enthalpy H at given entropy and pressure, and for the internal energy U at given entropy and volume.

The treatment is based on the fundamentals of thermodynamics as given in textbooks like E. A. Guggenheim, "Thermodynamics, An Advanced Treatment for Chemists and Physicists", North-Holland Publ. Comp., Amsterdam 1949. A more thorough account of the use of diagrams is given in the textbook M. Hillert, "Phase Equilibria, Phase Diagrams and Phase Transformations - Their Thermodynamic Basis", Cambridge University Press, Cambridge 1998.

2. Notations

The two components in a binary system will be denoted by A and B. Any property, the value of which for the whole system is equal to the sum of its values for the separate parts of the system, is an extensive quantity. Examples are the volume V and the Gibbs energy G but also the total content of matter N and the content of a particular component as N_B. It is often convenient to divide an extensive quantity by a measure of the size of the system under consideration. Gibbs divided by the total content of matter, N, expressed as the mass of the system. The resulting quantities are called specific, e.g. the specific volume. Today it is more common to express N as the number of moles of atoms and the resulting quantity is then called a molar quantity and it is denoted by a subscript m, for example molar volume $V_m = V/N$. With this rule of notation the molar content of B, i.e. N_B/N, should be denoted by $(N_B)_m$ but is usually denoted by x_B and called mole fraction. Sometimes it is convenient to express the size of the system through the content of a particular component e.g. N_A. The word "concentration" will be avoided because it really means the amount per volume.

The phase to which the quantity refers is given as a superscript to the right of the symbol as in V^α. The symbol $x_B^{\alpha/\beta}$ denotes the B content of the α phase in contact with the β phase. The relative amount of a phase, expressed as $N^\alpha/\Sigma N^i$, will be denoted by f^α and may be regarded as the mole fraction of the α phase. The superscript position to the left will be used to denote a standard state of a pure component as in ${}^oV_B^\alpha$ or the change on mixing as in ${}^MV_B^\alpha$. For mole fractions this position will be used to characterize particular compositions like ${}^1x_B^\alpha$ or ${}^2x_B^\alpha$ or the equilibrium content like ${}^ex_B^{\alpha/\beta}$. The composition of a binary α phase will usually be identified by giving x_B^α, but x_A^α can immediately be obtained as $1-x_B^\alpha$.

Following international recommendations, the partial quantities will be denoted as V_B instead of the older notation $\bar{V}_B$. The partial molar Gibbs energy for B, G_B, will simply be called partial Gibbs energy. It is identical to the chemical potential which is usually denoted by μ_B. For clarity, G_m is often called the integral molar Gibbs energy but will here simply be called molar Gibbs energy.

3. Derivation of the Tie-Line Rule

Interesting diagrams can often be obtained by combining two molar quantities but it is important to realize that both quantities must be defined by the use of the same definition of the size of the system in order to obtain a diagram that can be interpreted easily. This chapter is concerned with the properties of such molar diagrams and particularly with the properties of diagrams obtained by plotting any molar quantity of a binary system versus the mole fraction of B. All molar quantities will here be defined by dividing with N. Diagrams for molar quantities obtained by dividing with N_A have somewhat different properties but they will not be discussed here.

The definition of molar quantities can also be applied to a single phase α and we shall write $V_m^\alpha = V^\alpha/N^\alpha$ and $x_B^\alpha = N_B^\alpha/N^\alpha$. Consider the position of two such phases in a diagram of V_m versus x_B, Fig. 1. Any point (x_B,V_m) on the straight line between α and β is subject to a simple condition

$$\frac{V_m^\beta - V_m}{V_m - V_m^\alpha} = \frac{x_B^\beta - x_B}{x_B - x_B^\alpha} \tag{1}$$

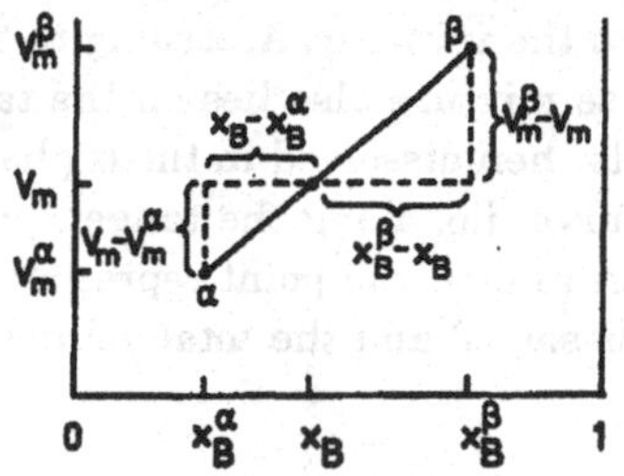

Fig. 1. Definition of tie-line between points representing two different phases in a molar diagram.

We shall now show that any point representing a two-phase alloy, α+β, fulfills this condition and thus falls on the straight line. Since V, N and N_B are extensive quantities, we have $V = V^\alpha + V^\beta$, $N = N^\alpha + N^\beta$ and $N_B = N_B^\alpha + N_B^\beta$. By dividing V and N_B by N we obtain for an alloy

$$V_m^{alloy} = \frac{V^{alloy}}{N^{alloy}} = \frac{V^\alpha + V^\beta}{N^\alpha + N^\beta} = \frac{N^\alpha V_m^\alpha + N^\beta V_m^\beta}{N^\alpha + N^\beta} = f^\alpha \cdot V_m^\alpha + f^\beta \cdot V_m^\beta \tag{2}$$

$$x_B^{alloy} = \frac{N_B^{alloy}}{N^{alloy}} = \frac{N_B^\alpha + N_B^\beta}{N^\alpha + N^\beta} = \frac{N^\alpha x_B^\alpha + N^\beta x_B^\beta}{N^\alpha + N^\beta} = f^\alpha \cdot x_B^\alpha + f^\beta \cdot x_B^\beta \tag{3}$$

Each one of these equations may be regarded as an application of the lever rule. By combination we can eliminate f^α and f^β because $f^\alpha + f^\beta = 1$:

$$\frac{V_m^\beta - V_m^{alloy}}{V_m^{alloy} - V_m^\alpha} = \frac{f^\alpha}{f^\beta} = \frac{x_B^\beta - x_B^{alloy}}{x_B^{alloy} - x_B^\alpha} \tag{4}$$

This relationship is identical to Eq. 1 with $V_m = V_m^{alloy}$ and $x_B = x_B^{alloy}$ and the point $(V_m^{alloy}, x_B^{alloy})$ for any alloy which is a mixture of α and β thus falls on the straight line between α and β. This we shall call the tie-line rule because the straight line between two points representing one phase each is called a tie-line. The rule holds independently of what measure of the size has been used as long as the same measure has been used for both axes. The tie-line rule will now be used to derive a number of important relationships.

4. Significance of Intersections with Component Axes.

Consider a system with two phases, α and β. Suppose the α phase initially has such a composition that the tangent to the α curve at the point representing the initial α composition goes through the β point. For the sake of simplicity, we shall assume that the β phase has a well defined composition and it is thus represented by a single point in Fig. 2. Let us increase the total B content of the system by adding a very small amount of β to the initial α. According to the tie-line rule the point representing the two-phase mixture also lies on the tangent, as is demonstrated in Fig. 2. If the β phase is then dissolved in the α phase, the molar volume of α is changed along the α curve, i.e. along the tangent, as long as the amount of β was very small. As a consequence, the point representing the alloy will not move as the β phase is being dissolved and the total volume of the system will not change.

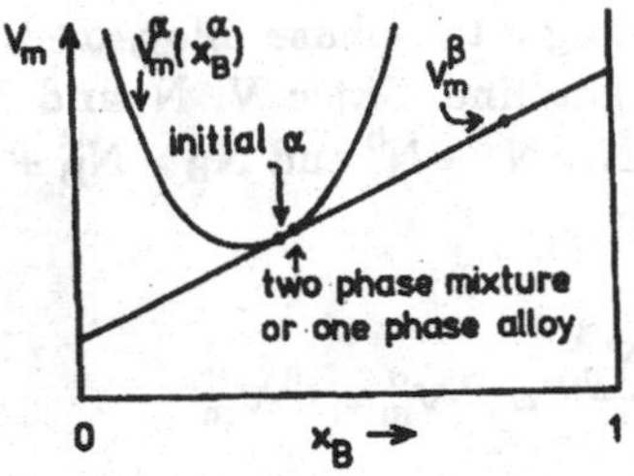

Fig. 2. An application of the tie-line rule. The value of the molar volume of the system is not changed when the small amount of β phase in an $\alpha+\beta$ two-phase mixture is dissolved in the α phase.

As a special case we shall now assume that the β phase is pure B in a hypothetical state situated on the α tangent under discussion. See Fig. 3. Again we can start with α of the initial composition and add some small amount of β. The two-phase mixture will fall on the tangent and the volume of the system has increased by V_m^β per mole of added β. We can then dissolve the β phase in the initial α phase without changing the total volume of the system, but the α phase has changed its volume by V_m^β per mole of added β, i.e., in this case B. We can express this mathematically by

$$(\partial V^\alpha/\partial N_B)_{N_A} = V_m^\beta \tag{5}$$

However, this derivative is the thermodynamic definition of the partial molar volume of B in the α phase, usually denoted by V_B^α. Fig. 3 immediately yields the well-known relation

$$V_B^\alpha = V_m^\alpha + x_A^\alpha \cdot dV_m^\alpha/dx_B \tag{6}$$

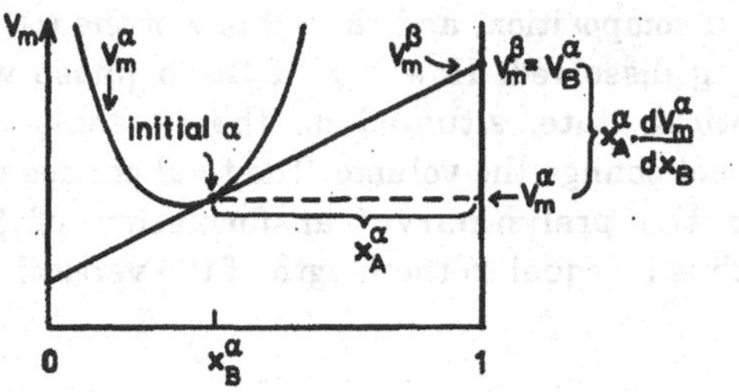

Fig. 3. Derivation of a relation between partial molar volume, V_B, and molar volume, V_m.

The intersection of the same tangent on the left-hand side of the diagram defines the corresponding quantity for the other component in the same α phase, V_A^α, Fig. 4. Two relations are obtained directly from this diagram,

$$V_m^\alpha = x_A^\alpha V_A^\alpha + x_B^\alpha V_B^\alpha \tag{7}$$

$$dV_m^\alpha/dx_B = V_B^\alpha - V_A^\alpha \tag{8}$$

Of course, all the quantities in these equations must be evaluated for the actual alloy composition, x_B^α.

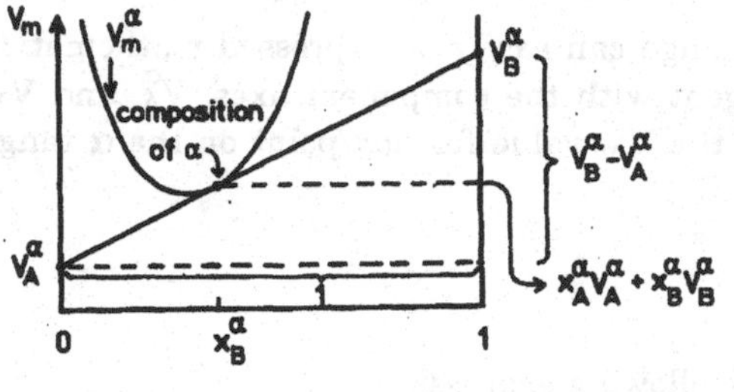

Fig. 4. Evaluation of molar volume, V_m^α, from the two partial molar volumes, V_A^α and V_B^α.

The constructions in Figs. 3 and 4 are particularly important when applied to the molar Gibbs energy. It should thus be noted that the intersections of a tangent with the two component axes will give the partial Gibbs energies of the two components, i.e. their chemical potentials. The intersections thus give the chemical potentials of reservoirs from which small amounts of A or B can be dissolved into the α phase without changing the Gibbs energy of the whole system. The α phase is thus in equilibrium with such reservoirs and it may itself be regarded as a reservoir from which a small amount of A or B can be transferred to another phase of the same chemical potential without changing the total Gibbs energy.

5. Change in a Molar Quantity when a Second Phase is Dissolved

In the general case, the β phase does not lie on the tangent to the α curve at the point representing the initial α composition, and the volume of the α+β system will then change as β is being dissolved. However, if the β phase was first transformed into a hypothetical state, situated on the tangent, then the subsequent dissolution would not change the volume. The total change would be the change connected with the preliminary transformation of β to the hypothetical state. It would thus be equal to the length of the vertical arrow in Fig. 5.

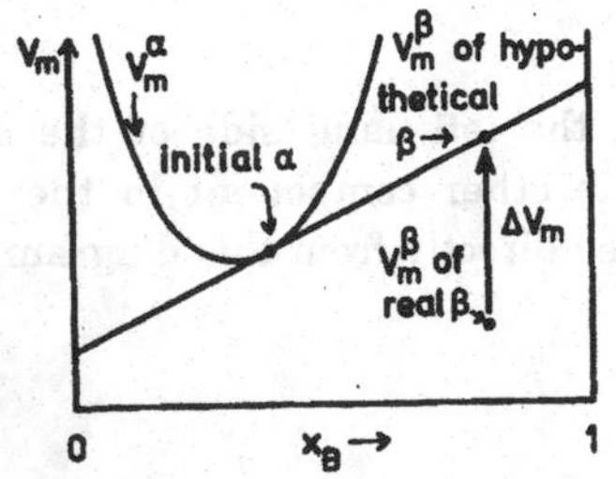

Fig. 5. Evaluation of the change in volume when a small amount of a β phase is dissolved in an α phase. The change ΔV_m is counted per mole of the dissolved β.

The vertical arrow, ΔV_m, represents the increase in volume per mole of β phase being dissolved because the diagram is a molar diagram. It should again be emphasized that this construction is valid only as long as the change in composition of α is small.

The magnitude of the above change can easily be expressed mathematically if the intersections of the α tangent with the component axes, V_A^α and V_B^α, are known. Fig. 6 illustrates that the V_m value for any point on the α tangent is given by

$$V_m^{\alpha tg} = x_A V_A^\alpha + x_B V_B^\alpha, \tag{9}$$

From Fig. 7 we thus obtain the following expression

$$\Delta V_m = x_A^\beta V_A^\alpha + x_B^\beta V_B^\alpha - V_m^\beta \tag{10}$$

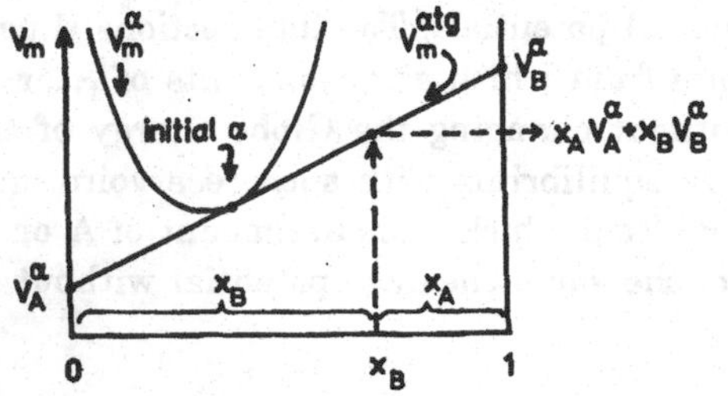

Fig. 6. Derivation of the equation for a tangent in a molar diagram.

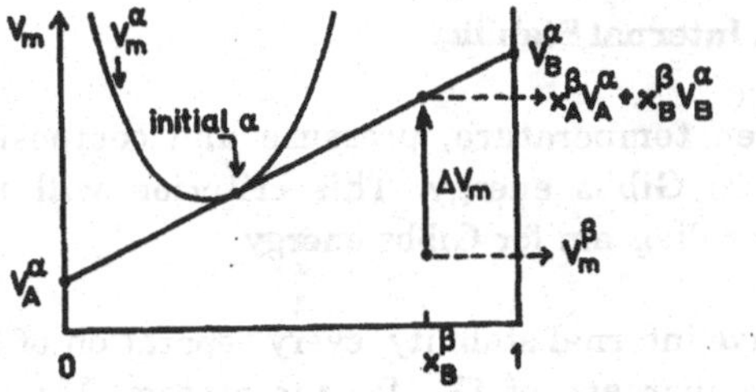

Fig. 7. Derivation of an expression for the change in volume accompanying the dissolution of a very small amount of β in α.

6. Use of the Curvature

So far, we have considered small changes in composition of the α phase. For larger changes we can no longer approximate the new α phase with a point on the initial tangent. In order to treat such changes one needs to know the rate of change of V_A^α and V_B^α when the composition of the phase is changed. The following relations can be obtained directly from the molar diagram in Fig. 8 by comparing triangles.

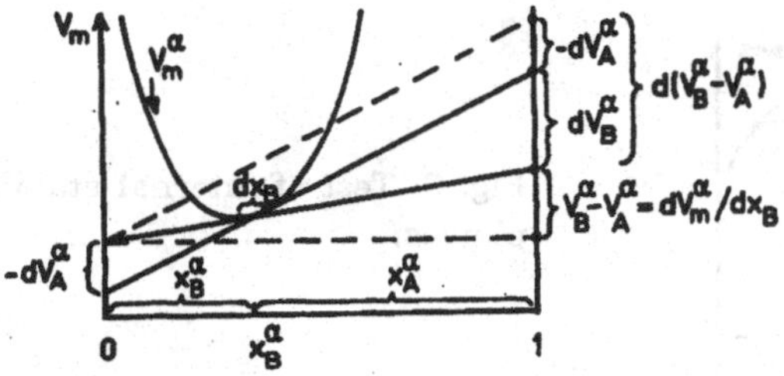

Fig. 8. Evaluation of the change in partial molar volume due to a change in composition.

$$-\,dV_A^\alpha/x_B^\alpha = dV_B^\alpha/x_A^\alpha = d(V_B^\alpha - V_A^\alpha)/1 \tag{11}$$

As already shown by Eq.8, we can substitute dV_m^α/dx_B for $V_B^\alpha - V_A^\alpha$. Multiplying Eq. 11 by $x_A^\alpha x_B^\alpha$ we obtain

$$-\,x_A^\alpha dV_A^\alpha = x_B^\alpha dV_B^\alpha = x_A^\alpha x_B^\alpha \cdot d(V_B^\alpha - V_A^\alpha) = x_A^\alpha x_B^\alpha \cdot d^2V_m^\alpha/dx_B^2 \cdot dx_B \tag{12}$$

The left-hand equality yields

$$x_A^\alpha dV_A^\alpha + x_B^\alpha dV_B^\alpha = 0 \tag{13}$$

When applied to the Gibbs energy, this is identical to the Gibbs-Duhem relation. A corresponding relation thus holds for other quantities as well.

7. Internal Stability

A state of equilibrium at given temperature, pressure and composition is characterized by a minimum in Gibbs energy. This criterion will now be discussed in relation to the molar diagram for Gibbs energy.

In order for a single phase to have internal stability, every separation of the two components must result in an increase of G_m for the system. Let us first consider a splitting of the phase in two parts of equal size, one slightly richer in A, the other slightly richer in B. According to the tie-line rule the molar Gibbs energy for such a two-phase state is obtained on the tie-line between the two points representing the two parts and the change in G_m is thus obtained as the length of the arrows for two cases demonstrated in Fig. 9. It is evident that the criterion for stability is

$$d^2G_m/dx_B^2 > 0 \qquad (14)$$

The α phase in Fig. 9 is thus stable but not the β phase.

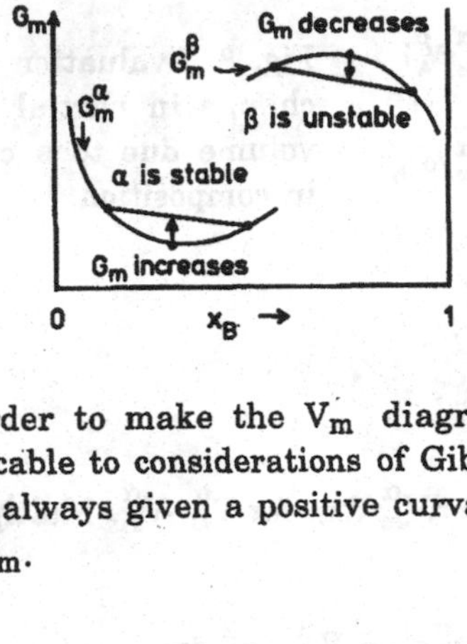

Fig. 9. Test of internal stability of phases.

In order to make the V_m diagram in the preceding paragraph directly applicable to considerations of Gibbs energy in stable systems, the V_m curves were always given a positive curvature although there is no such requirement for V_m.

8. Mathematical Representation of a G_m Curve

In view of the importance of the curvature of the G_m curve, one must keep at least three terms if one wants to represent the G_m curve in the neighborhood of a certain composition ${}^1x_B^\alpha$ by a Taylor series expansion;

$$G_m^\alpha = a + b \cdot \Delta x_B^\alpha + c \cdot (\Delta x_B^\alpha)^2 \qquad (15)$$

where $\Delta x_B^\alpha = x_B^\alpha - {}^1x_B^\alpha$. The coefficient b is the slope of the tangent and c is the curvature of the curve, $c = \frac{1}{2}d^2G_m^\alpha/dx_B^2$. In general, such a parabolic representation is satisfactory only for small changes but it is sometimes illustrative to use the parabolic shape for a phase with a narrow range of existence. For instance, a very narrow parabola can be used to describe a stoichiometric phase with a well defined composition. The G_m diagram in Fig. 10 demonstrates that there is no unique way of placing the tangent to the curve of such a phase. The individual values of G_A^α and G_B^α are not defined for such a phase, but a strict relation exists between them and is defined by the value of G_m^α at the minimum

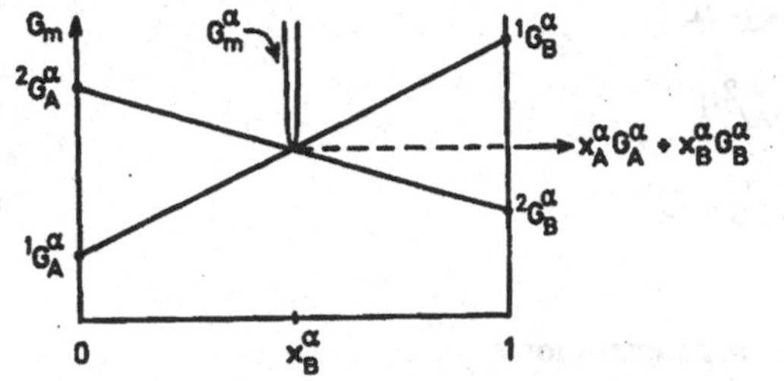

Fig. 10. Relation between the two partial Gibbs energies, G_A and G_B, in a stoichiometric phase.

$$G_m^\alpha = x_A^\alpha\, G_A^\alpha + x_B^\alpha\, G_B^\alpha \tag{16}$$

The fact that one cannot represent the whole of a G_m^α curve by a parabola becomes particularly evident as one approaches the side of the diagram. It may seem self-evident that the chemical potential of B, G_B^α, cannot have a definite value in a phase that contains no B atoms. However, a definite value is predicted by a parabola in the limit $x_B^\alpha \rightarrow 0$, as illustrated in the left-hand part of Fig. 11. Instead, the shape of the G_m^α curve must be such that the tangent becomes vertical at the limit as illustrated by the right-hand part of Fig.11 but it should be realized that this effect has here been greatly exaggerated. Unless drawn with great care, a realistic curve will rather look like the left-hand part. Mathematically this property is adequately described by the term containing the ideal entropy of mixing derived by statistical thermodynamics,

$$-TS^{ideal} = RT[x_A\ln(x_A) + x_B\ln(x_B)] \tag{17}$$

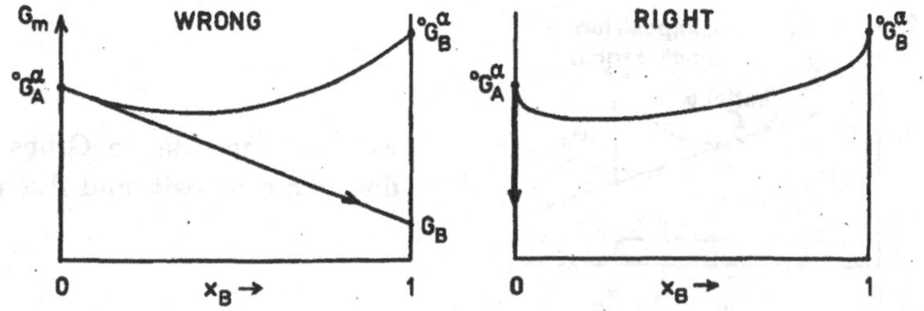

Fig. 11. Shape of molar Gibbs energy curves.

The remaining part of G_m^α can sometimes be adequately described by a power series. It should be noted that, in order to obtain the correct values at the sides of the diagram, one must represent the power series by the following expression, $x_A^\alpha\,{}^oG_A^\alpha + x_B^\alpha\,{}^oG_B^\alpha + x_A^\alpha x_B^\alpha \cdot L^\alpha$. The quantities ${}^oG_A^\alpha$ and ${}^oG_B^\alpha$ are the molar Gibbs energy for pure A and B, respectively, and L^α is a simple power series. In the regular solution approximation it is truncated after the first, constant term. The following expressions are thus obtained,

$$G_m^\alpha = x_A^\alpha\,{}^oG_A^\alpha + x_B^\alpha\,{}^oG_B^\alpha + RT[x_A^\alpha \ln(x_A^\alpha) + x_B^\alpha \ln(x_B^\alpha)] + x_A^\alpha x_B^\alpha \cdot L^\alpha \quad (18)$$

$$G_A^\alpha = {}^oG_A^\alpha + RT\ln(x_A^\alpha) + (x_B^\alpha)^2 \cdot L^\alpha \quad (19)$$

$$G_B^\alpha = {}^oG_B^\alpha + RT\ln(x_B^\alpha) + (x_A^\alpha)^2 \cdot L^\alpha \quad (20)$$

$$c = \frac{1}{2} d^2G_m^\alpha/dx_B^2 = \frac{1}{2} RT/x_A^\alpha x_B^\alpha - L^\alpha \quad (21)$$

9. Fluctuations

Let us now calculate the increase in Gibbs energy in an α phase of composition ${}^1x_B^\alpha$ due to a compositional fluctuation by which a small region has changed its composition to x_B. According to the tie-line rule, there would be no change in Gibbs energy if the point representing that region were situated on the tangent. However, if the region can still be regarded as α phase, it is situated on the G_m^α curve and the distance to the tangent represents the increase in Gibbs energy per mole of atoms in the region, Fig. 12. If the G_m^α curve is represented by Eq. 15, the tangent is given by the equation

$$G_m^\alpha = a + b \cdot \Delta x_B \quad (22)$$

and the distance is obtained from the difference between Eqs. 15 and 22,

$$\Delta G_m = c \cdot (\Delta x_B)^2 = \frac{1}{2} d^2G_m^\alpha/dx_B^2 \cdot (\Delta x_B)^2 \quad (23)$$

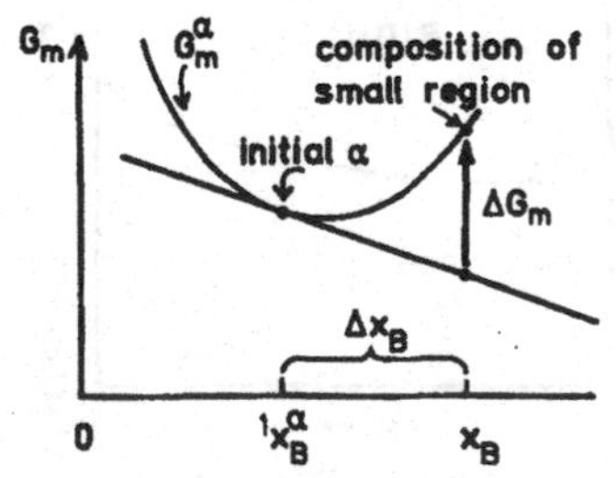

Fig. 12. Increase in Gibbs energy due to a compositional fluctuation.

With the regular solution model one obtains from Eq. 21

$$\Delta G_m = c\cdot(\Delta x_B)^2 = (\tfrac{1}{2}RT/x_A^\alpha x_B^\alpha - L^\alpha)\cdot(\Delta x_B)^2 \tag{24}$$

This approximation requires that the curvature is fairly constant in the Δx_B region and it is thus valid for small Δx_B. It should be noticed that the size of Δx_B should be judged by its effect on $x_A^\alpha x_B^\alpha$ in Eq. 24. The criterion is thus $\Delta x_B \ll x_B^\alpha$ and $\Delta x_B \ll x_A^\alpha$. Actually, Δx_B should be compared to x_B^α and x_A^α before as well as after the fluctuation. This limits the use of the approximation when the fluctuation approaches a pure component, as demonstrated in Fig. 13. The exact value of ΔG_m can, of course, be obtained as follows. The point on the α curve is $x_A\cdot G_A^\alpha(x_B)+x_B\cdot G_B^\alpha(x_B)$ and the point on the tangent is $x_A\cdot G_A^\alpha({}^1x_B^\alpha)+x_B\cdot G_B^\alpha({}^1x_B^\alpha)$. The distance is thus

$$\Delta G_m = x_A[G_A^\alpha(x_B) - G_A^\alpha({}^1x_B^\alpha)] + x_B[G_B^\alpha(x_B) - G_B^\alpha({}^1x_B^\alpha)] \tag{25}$$

and, by application of the regular solution model, one obtains after some manipulation

$$\Delta G_m = RT[x_A\ln(x_A/{}^1x_A^\alpha) + x_B\ln(x_B/{}^1x_B^\alpha)] + L^\alpha\cdot x_A x_B[1-({}^1x_A^\alpha)^2/x_A-({}^1x_B^\alpha)^2/x_B] \tag{26}$$

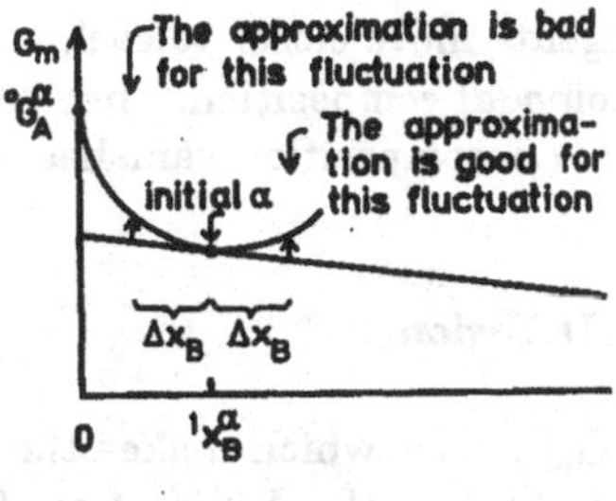

Fig. 13. Validity of the approximate method of evaluating a change in Gibbs energy.

The limit of stability is reached when $d^2G_m^\alpha/dx_B^2 = 0$ and in the regular solution model this occurs when $RT = 2x_A^\alpha x_B^\alpha\cdot L^\alpha$ according to Eq. 21. This defines the so-called spinodal curve in the phase diagram. Inside this curve the $G_m^\alpha(x_B)$ function has a negative curvature. The molar Gibbs energy diagram can be used to demonstrate that, in an alloy outside the spinodal curve, a compositional fluctuation can also become stable and grow spontaneously but only after passing a critical composition, which is situated inside the spinodal curve. The series of vertical arrows in Fig. 14 represents the change in Gibbs energy of the system, ΔG_m, accompanying the formation of fluctuations of different compositions. ΔG_m will be negative if the change in composition is large enough. See the arrow pointing downward on the right-hand side.

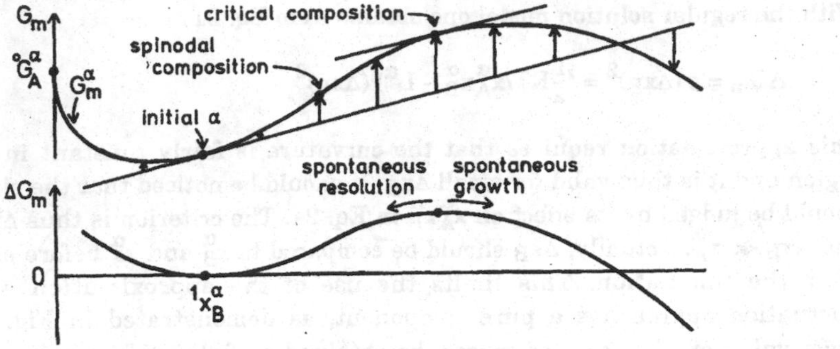

Fig. 14. Criterion for stability of a compositional fluctuation in a miscibility gap.

However, if one considers a fluctuation with a gradual change in composition, one will find that ΔG_m will start to decrease while it is still positive. See the lower part of Fig. 14. The maximum point on that curve represents the critical composition beyond which a fluctuation can grow spontaneously because, after the fluctuation has been activated to that point, the Gibbs energy of the system can start to decrease. The critical composition is easily found for each alloy composition $^1x^{\alpha}_B$ by constructing a second tangent parallel to that at $^1x^{\alpha}_B$. See the upper part of Fig. 14. The vertical distance between the two parallel tangents defines the activation energy necessary in order to form a fluctuation of critical composition. It is easy to see that the two tangents move closer to each other as the alloy composition is moved closer to the spinodal composition. They coincide at the spinodal composition where the barrier to decomposition vanishes.

10. Driving Force for Diffusion

It is natural that the positive curvature of a G_m curve, which makes the phase stable against fluctuations in composition, also provides the driving force for the elimination of differences in composition within the phase if such differences are present, i.e. the driving force for diffusion. We shall now discuss the force which makes the individual B atoms diffuse from a region with a high B content to a region with a low B content. Each region may be regarded as a reservoir of B with its own value of G_B and the difference in G_B is identical to the decrease in Gibbs energy when one mole of B is transferred, Fig. 15.

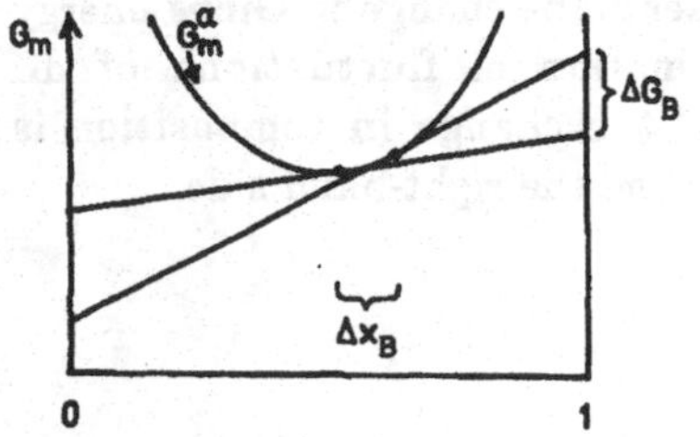

Fig. 15. Evaluation of the driving force for diffusion of B atoms in a binary phase.

Assuming that the rate of transfer is proportional to the decrease in Gibbs energy and to the number of B atoms per volume, $N_B/V = x_B/V_m$, and inversely proportional to the transport distance, Δy, one obtains the following expression for the flux of B atoms

$$J_B = -M_B \cdot x_B/V_m \cdot \Delta G_B/\Delta y = -M_B/V_m \cdot x_B \cdot dG_B/dx_B \cdot \Delta x_B/\Delta y \tag{27}$$

The constant of proportionality, M_B, may be regarded as the mobility of the B atoms. The curvature of the G_m curve can be introduced by evaluating $x_B \cdot dG_B/dx_B$ from an equation equivalent to Eq. 12,

$$J_B = -M_B/V_m \cdot x_A x_B \cdot d^2G_m/dx_B^2 \cdot \Delta x_B/\Delta y \tag{28}$$

This is recognized as Fick's law and the diffusion constant for B is obtained as

$$D_B = M_B \cdot x_A x_B \cdot d^2G_m/dx_B^2 \tag{29}$$

The mobility is thus multiplied by a thermodynamic factor. By introducing the activity or the activity coefficient for B, this factor can be transformed into the form introduced by Darken,

$$x_A x_B \cdot d^2G_m/dx_B^2 = x_B \cdot dG_B/dx_B = dG_B/d\ln(x_B)$$

$$= RT d\ln(a_B)/d\ln(x_B) = RT[1 + d(\ln(f_B)/d\ln(x_B))] \tag{30}$$

where f_B is the activity coefficient of B. It is evident that a similar derivation can be carried out for the other component and that the same factor is obtained,

$$D_A = M_A \cdot x_A x_B \cdot d^2G_m/dx_B^2 \tag{31}$$

11. Calculation of Two-Phase Equilibria

The requirement that the Gibbs energy is a minimum in a state of equilibrium, can be used in calculations of the compositions of two phases in equilibrium with each other. It is becoming increasingly more common to solve such problems by computer-based calculations and it is instructive to use G_m diagrams to illustrate some methods for such calculations. Consider a system containing one mole of matter with the average composition x_B^{av}. Let it first be in a two-phase state $\alpha_1 + \beta_1$. Its average molar Gibbs energy, G_m^{av}, is given at x_B^{av} on the tie-line α_1–β_1 in Fig. 16.

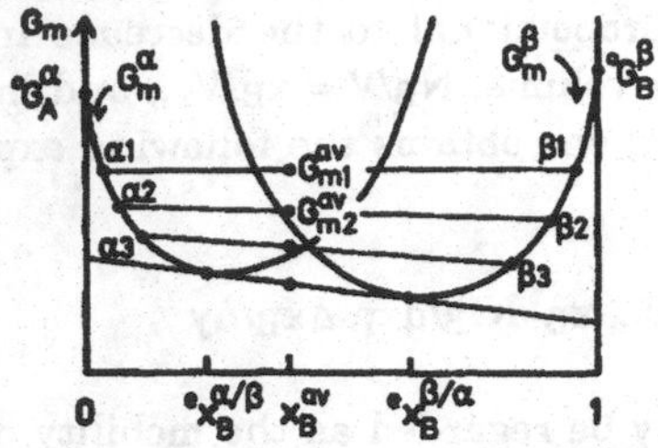

Fig. 16. Calculation of a two-phase equilibrium by finding the minimum in Gibbs energy.

The compositions of α and β are then respectively increased and decreased in B so that G_m^{av} gradually decreases. This diagram demonstrates that the minimum in G_m^{av} will finally be reached when α and β are situated on the points of tangency on the common tangent. By this consideration we have thus shown that the equilibrium compositions, ${}^e x_B^{\alpha/\beta}$ and ${}^e x_B^{\beta/\alpha}$, can be found by constructing the common tangent.

If the average composition is outside the two-phase region, e.g. at α_3 in the diagram, the calculation would stop when α_3 is reached because the amount of β phase would then be zero. That will be the stable state for such an alloy.

An alternative method to proceed from the initial $\alpha_1 + \beta_1$ state to more stable states is to compare the values of G_B and G_A, respectively, in the two phases and to transfer B to the phase with the lower G_B value and A to the phase with the lower G_A value. Fig. 17 shows that $\Delta^1 G_B$ would drive the transfer of B from β_1 to α_1 and $\Delta^1 G_A$ would drive the transfer of A from α_1 to β_1. In this particular case, the two phases would thus move closer to each other and the driving forces would gradually decrease, as illustrated by $\Delta^2 G_B$ and $\Delta^2 G_A$ when the phases are at α_2 and β_2, compared to $\Delta^1 G_B$ and $\Delta^1 G_A$. Both driving forces will finally vanish when the two tangents coincide. We have thus again shown that the equilibrium compositions, ${}^e x_B^{\alpha/\beta}$ and ${}^e x_B^{\beta/\alpha}$, can be found graphically by the common tangent construction. This procedure simulates the physical process of attaining equilibrium and may thus be regarded as a natural method of iteration. It may be a safer way to avoid divergence in the numerical calculations than the method to be described next. Sometimes it may even be advantageous to adjust the rate of transfer of A and B atoms to the size of ΔG_A and ΔG_B.

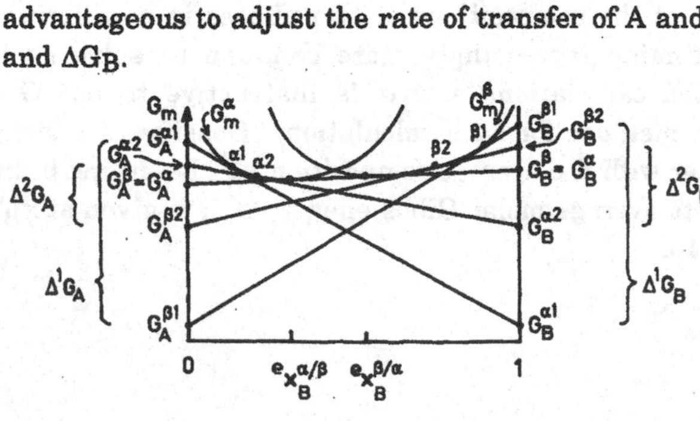

Fig. 17. Calculation of a two-phase equilibrium by moving atoms in the direction of decreasing partial Gibbs energies.

In principle, the two-phase equilibrium can be calculated directly by solving the two equilibrium conditions given by $G_A^\alpha = G_A^\beta$ and $G_B^\alpha = G_B^\beta$, which of course define the common tangent. However, this can seldom be done analytically. If one chooses to carry out the numerical calculations with a Newton-Raphson method, one may actually follow an iteration procedure related to the natural method just described. Alternatively, an iteration procedure illustrated in Fig. 18 can be used. Start by guessing β_1. Calculate $G_B^{\beta 1}$. Calculate α_1 from $G_B^{\alpha 1} = G_B^{\beta 1}$. Calculate $G_A^{\alpha 1}$. Calculate β_2 from $G_A^{\beta 2} = G_A^{\alpha 1}$. Etc.

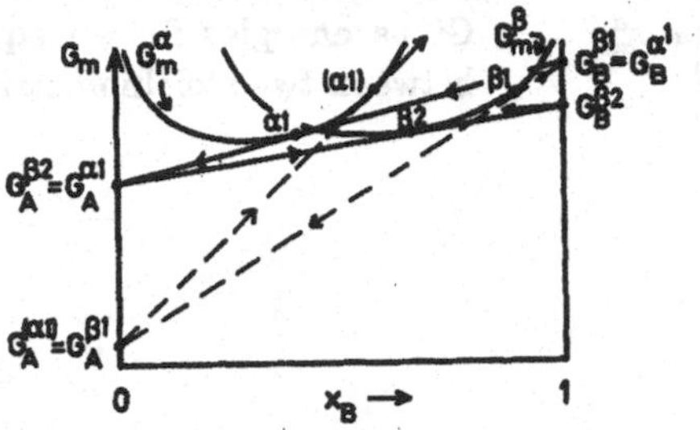

Fig. 18. Calculation of a two-phase equilibrium by an iteration method.

According to the diagram in Fig. 18 this iteration seems to converge to the common tangent, but it would have diverged if α were calculated from $G_A^\alpha = G_A^\beta$ and β from $G_B^\beta = G_B^\alpha$. This is illustrated by the dashed lines. The iteration procedure, like any other numerical method, must be constructed carefully in order to converge.

12. Calculation for Fixed Compositions

The calculation of an equilibrium is greatly simplified if, for some reason, the composition of one of the phases, e.g. β, is known, Fig. 19. One can then calculate the common tangent from the fact that the β phase is situated on the tangent to the α phase. Compare Eq. 9.

$$x_A^\beta \cdot G_A^\alpha(x_B^\alpha) + x_B^\beta \cdot G_B^\alpha(x_B^\alpha) = G_m^\beta(x_B^\beta) \tag{32}$$

The composition of the α phase, x_B^α, is the only unknown quantity and only one equation is thus required. This will be the case when one of the phases has a fixed composition. Such a phase is often called a stoichiometric phase because the coefficients can usually be expresed as small numbers.

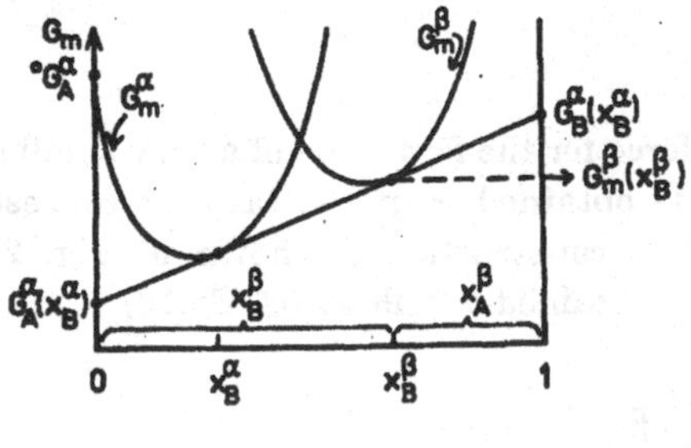

Fig. 19. Calculation of the composition of an α phase in equilibrium with a β phase of known composition.

There will of course be no unknown equilibrium composition to calculate if both phases are stoichiometric. However, it may still be interesting to calculate the values of G_A and G_B established by such an equilibrium. The equilibrium is defined by the common tangent. By comparing triangles in Fig. 20 one obtains

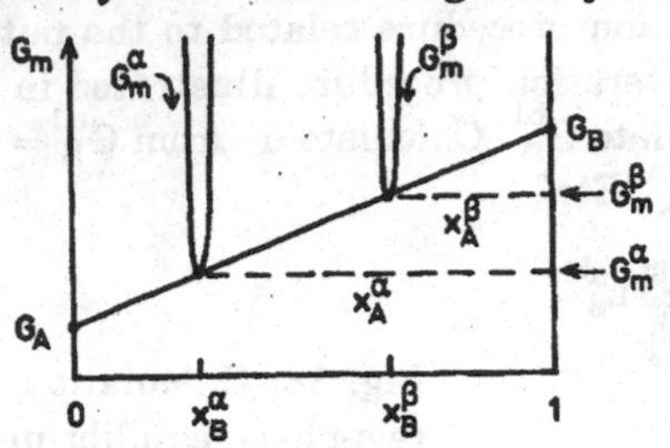

Fig. 20. Evaluation of the partial Gibbs energies for an equilibrium between two stoichiometric phases.

$$\frac{G_B - G_m^\beta}{G_B - G_m^\alpha} = \frac{x_A^\beta}{x_A^\alpha} \tag{33}$$

and by rearranging this equation one finds

$$G_B = \frac{x_A^\alpha G_m^\beta - x_A^\beta G_m^\alpha}{x_A^\alpha - x_A^\beta} \tag{34}$$

In the same way one obtains

$$G_A = \frac{x_B^\alpha G_m^\beta - x_B^\beta G_m^\alpha}{x_B^\alpha - x_B^\beta} \tag{35}$$

13. Driving Force for Precipitation

Consider a one-phase alloy, α_1, which is inside the two-phase region, $\alpha+\beta$. The Gibbs energy would decrease by precipitation of β and finally fall on the common tangent which connects the stable α and β phases. The total driving force for the complete reaction in one mole of an alloy is given by the arrow in Fig. 21. According to Eq. 23 one has approximately

$$\Delta G_m = \tfrac{1}{2} d^2 G_m^\alpha / dx_B^2 \cdot (\Delta x_B^\alpha)^2 \tag{36}$$

On the other hand, the driving force for the formation of a very small quantity of β from a large quantity of α_1 is obtained from the tangent representing the supersaturated α_1 matrix. This construction is shown in Fig. 22 and the magnitude of this driving force is obtained as follows (cf. Eq.10),

$$\Delta G_m = x_A^\beta G_A^{\alpha 1} + x_B^\beta G_B^{\alpha 1} - G_m^\beta \tag{37}$$

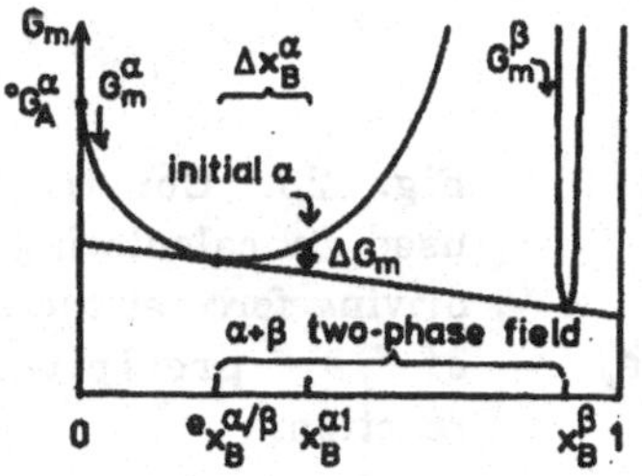

Fig. 21. Demonstration of the driving force for a complete precipitaion reaction.

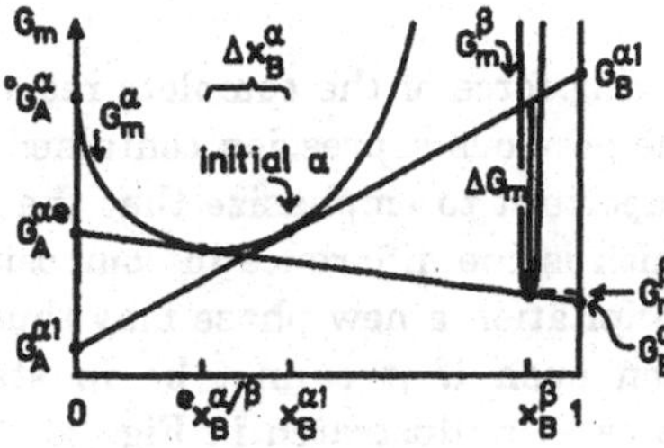

Fig. 22. Demonstration of the driving force at the start of a precipitation reaction.

$$G_m^\beta = x_A^\beta G_A^{\alpha e} + x_B^\beta G_B^{\alpha e} \tag{38}$$

$$\Delta G_m = x_A^\beta \cdot (G_A^{\alpha 1} - G_A^{\alpha e}) + x_B^\beta \cdot (G_B^{\alpha 1} - G_B^{\alpha e}) \tag{39}$$

This is the driving force during the early stages of precipitation, in particular during the nucleation stage. It is expressed per mole of β phase formed.

For low supersaturations, i.e. small Δx_B^α, one can introduce the curvature of the G_m^α curve. By comparing triangles in Fig. 23 one obtains,

$$\frac{\Delta G_m}{x_B^\beta - x_B^\alpha} = \frac{\Delta G_B - \Delta G_A}{1} \tag{40}$$

The exact value of x_B^α should be taken at the intersection between $^e x_B^{\alpha/\beta}$ and $x_B^{\alpha 1}$. However, its exact position is not critical when Δx_B^α is small. Application of an equation equivalent to Eq. 8 yields

$$\Delta G_m = (x_B^\beta - x_B^\alpha) \cdot (\Delta G_B - \Delta G_A) = (x_B^\beta - x_B^\alpha) \cdot \Delta(G_B - G_A)$$

$$= (x_B^\beta - x_B^\alpha) \cdot \Delta(dG_m^\alpha / dx_B) \tag{41}$$

$$\Delta G_m = (x_B^\beta - x_B^\alpha) \cdot d^2 G_m^\alpha / dx_B^2 \cdot \Delta x_B^\alpha \tag{42}$$

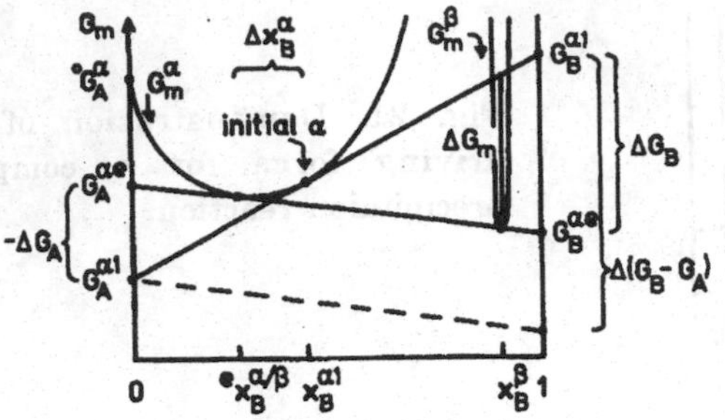

Fig. 23. Construction used in calculating the driving force at the start of a precipitation reaction.

Eq. 42 resembles the expression for the driving force of the complete reaction, Eq. 36, but it is important to notice that the previous expression contained the square of the supersaturation. It is also important to emphasize that the last expression contains the factor $(x_B^\beta - x_B^\alpha)$ which is the difference in composition between the two phases. At the start of precipitation a new phase may thus be favoured if it differs much in composition even if it cannot be in stable equilibrium with the matrix phase. Such a case is illustrated in Fig. 24. It is thus conceivable that the nucleation of a new stable phase can be very difficult because its composition is very close to the parent phase, but a new metastable phase with a very different composition can nucleate and it may later assist in nucleating the stable phase.

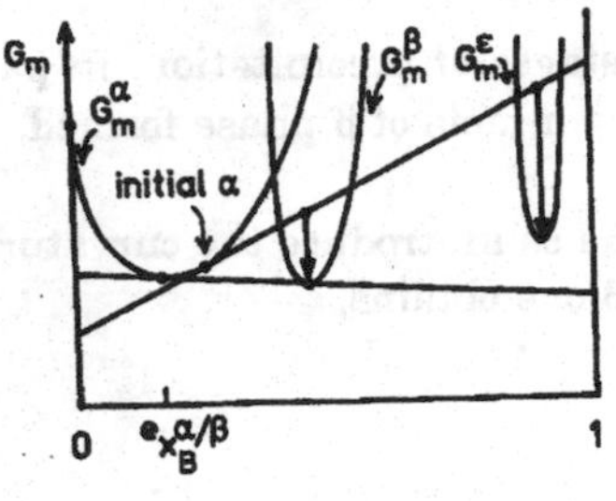

Fig. 24. Demonstration of the effect of phase composition on the driving force at the start of precipitation.

14. Effect of Surface Tension on Two-Phase Equilibria

According to the definition, the Gibbs energy depends upon the pressure, $G_m(P) = G_m(0) + PV_m$ where 0 denotes atmospheric pressure and P represents the "overpressure". For condensed phases the PV_m term can often be neglected if the pressure is not very high. However, it plays an important role for the equilibrium between two phases when they are under different pressures. This occurs when the interface is curved and the pressure difference is caused by the surface tension of the curved interface. For instance, a spherical α/β interface gives $P^\beta - P^\alpha = 2\sigma/r$ where σ is the surface tension and r is the radius of curvature.

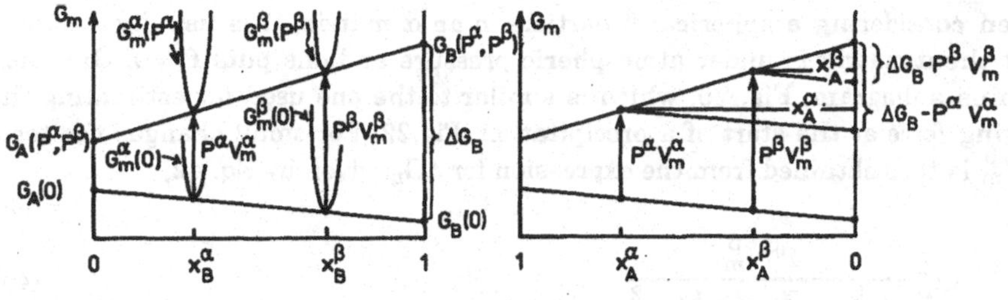

Fig. 25. The effect of pressure on the partial Gibbs energies for an equilibrium between two stoichiometric phases.

Let us consider how the equilibrium between two phases is changed by the introduction of increased pressures. The two G_m curves are displaced upwards by the amounts $P^\alpha V_m^\alpha$ and $P^\beta V_m^\beta$, respectively. The left-hand part of Fig. 25 illustrates that the values of G_A and G_B will change even if the compositions of the phases are so well defined that they do not change. The right-hand part illustrates that the change of G_B can be evaluated by the method used in deriving Eq. 34 from Fig. 20. By comparing triangles one obtains

$$\frac{\Delta G_B - P^\beta V_m^\beta}{\Delta G_B - P^\alpha V_m^\alpha} = \frac{x_A^\beta}{x_A^\alpha} \tag{43}$$

where ΔG_B stands for $G_B(P^\alpha,P^\beta) - G_B(0)$. Eq. 43 can be rearranged to yield

$$G_B(P^\alpha,P^\beta) - G_B(0) = \frac{x_A^\alpha \cdot P^\beta V_m^\beta - x_A^\beta \cdot P^\alpha V_m^\alpha}{x_A^\alpha - x_A^\beta} \tag{44}$$

In the same way one obtains

$$G_A(P^\alpha,P^\beta) - G_A(0) = \frac{x_B^\alpha \cdot P^\beta V_m^\beta - x_B^\beta \cdot P^\alpha V_m^\alpha}{x_B^\alpha - x_B^\beta} \tag{45}$$

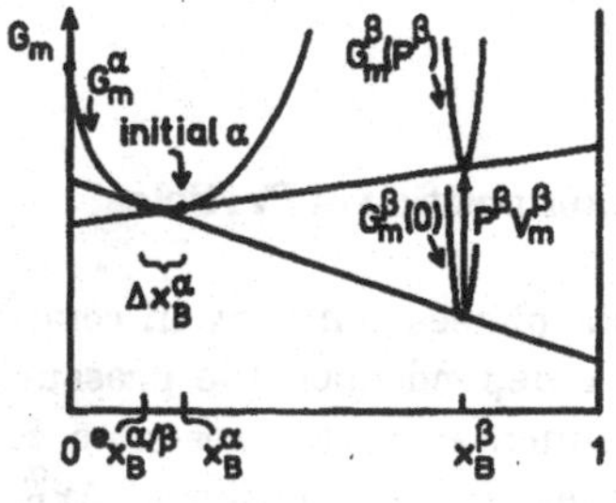

Fig. 26. Change in composition of a phase in equilibrium with a second phase on which a pressure is applied.

When considering a spherical β particle in an α matrix, it is usually assumed that the α matrix is under atmospheric pressure and one puts $P^\alpha=0$. One then obtains a diagram, Fig. 26, which is similar to the one used for estimating the driving force at the start of a precipitation, Fig.22. For small changes the term $P^\beta V_m^\beta$ is thus obtained from the expression for ΔG_m given by Eq. 42,

$$\Delta x_B^\alpha = \frac{P^\beta V_m^\beta}{(x_B^\beta - x_B^\alpha)\cdot d^2G_m^\alpha/dx_B^2} \tag{46}$$

and by inserting $P^\beta = 2\sigma/r$ and applying Eq. 21 according to the regular solution model, one obtains

$$\Delta x_B^\alpha = \frac{2\sigma V_m^\beta x_A^\alpha x_B^\alpha}{r\cdot(x_B^\beta - x_B^\alpha)\cdot(RT - 2L^\alpha x_A^\alpha x_B^\alpha)} \tag{47}$$

This is an extended form of the Gibbs-Thomson equation. For large Δx_B^α the calculation must involve an integration which becomes particularly easy when the matrix α phase is a dilute solution of B in A. In that case x_A^α can be approximated as unity and $2L^\alpha x_A^\alpha x_B^\alpha$ can be neglected in comparison with RT because x_B^α is small:

$$dx_B^\alpha = \frac{dP^\beta V_m^\beta x_B^\alpha}{(x_B^\beta - x_B^\alpha)\cdot RT} \tag{48}$$

The following result is obtained by integrating dx_B^α/x_B^α from atmospheric pressure if the difference in composition between the two phases, $x_B^\beta - x_B^\alpha$, is not affected markedly by pressure.

$$\ln\left[\frac{x_B^{\alpha/\beta}}{{}^e x_B^{\alpha/\beta}}\right] = \frac{P^\beta V_m^\beta}{RT\cdot(x_B^\beta - x_B^\alpha)} = \frac{2\sigma V_m^\beta}{rRT\cdot(x_B^\beta - x_B^\alpha)} \tag{49}$$

${}^e x_B^{\alpha/\beta}$ is the equilibrium composition of the α phase when both phases are under atmospheric pressure.

15. Effect of Surface Tension on Composition of Particle

The calculation becomes more involved if both phases can vary in composition. Then the composition of the β phase will also depend upon the pressure P^β as demonstrated in Fig. 27. It would take too much space to present a complete calculation for this case but for small changes the calculation of Δx_B^α is still

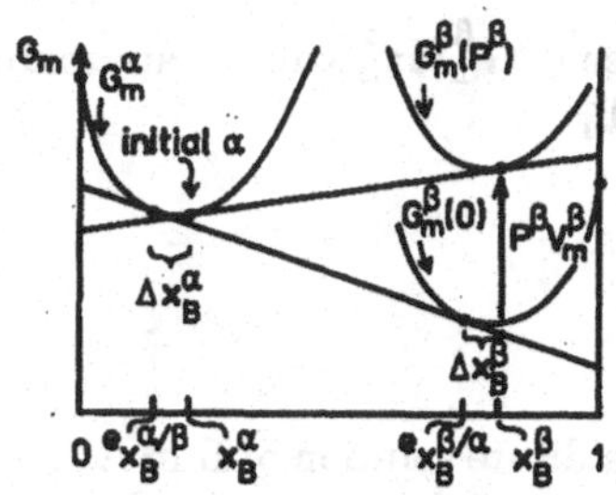

Fig. 27. Change in compositions when a pressure is applied to one of the phases in a two-phase equilibrium.

valid and for the β phase one can find a similar expression. See Fig. 28. Let us first compare the tangent to the G_m^β curve at P^β with the tangent to the ordinary G_m^β curve at the same composition. Their distances on the two component axes are $\Delta G_A = P^\beta V_A^\beta$ and $\Delta G_B = P^\beta V_B^\beta$ and their distance at the composition of the α phase is

$$\Delta G_m = P^\beta (x_A^\alpha V_A^\beta + x_B^\alpha V_B^\beta) \tag{50}$$

This distance is compensated by the change in β composition, which rotates the tangent. At the α composition this yields the following effect on the position of the tangent,

$$\Delta G_m = (x_B^\beta - x_B^\alpha) \cdot d^2 G_m^\beta / dx_B^2 \cdot \Delta x_B^\beta \tag{51}$$

Combination yields

$$\Delta x_B^\beta = \frac{P^\beta \cdot (x_A^\alpha V_A^\beta + x_B^\alpha V_B^\beta)}{(x_B^\beta - x_B^\alpha) \cdot d^2 G_m^\beta / dx_B^2} \tag{52}$$

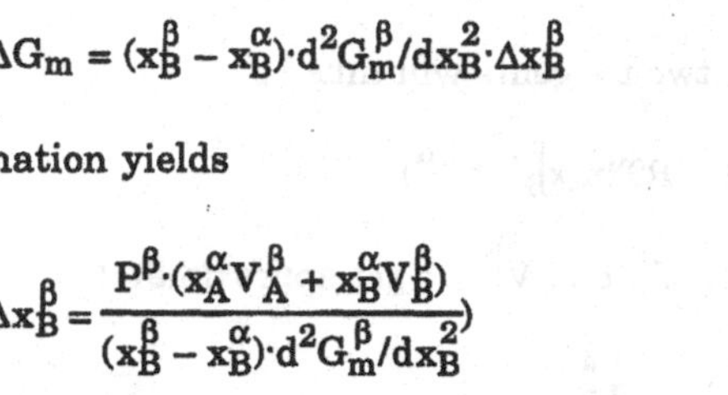

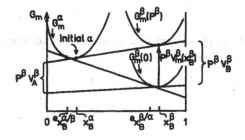

Fig. 28. Construction for finding the new compositions when a pressure is applied to one of the phases in a two-phase equilibrium.

If the β phase is a dilute solution of B in A,

$$x_A^\alpha V_A^\beta + x_B^\alpha V_B^\beta \cong V_A^\beta \cong V_m^\beta \tag{53}$$

and Eq. 52 can be rewritten as follows because $d^2G_m^\beta/dx_B^2$ can be approximated by $RT/x_A^\alpha x_B^\alpha$ and in turn by RT/x_B^α if x_B^α is small,

$$\Delta x_B^\beta = \frac{P^\beta V_m^\beta x_B^\beta}{RT\cdot(x_B^\beta - x_B^\alpha)} \tag{54}$$

This resembles Eq. 48 for the α phase. If α is a dilute solution of B in A,

$$G_m^\alpha = x_A^\alpha\,{}^oG_A^\alpha + x_B^\alpha\,{}^oG_B^\alpha + RT[x_A^\alpha \ln(x_A^\alpha) + x_B^\alpha \ln(x_B^\alpha)] \tag{55}$$

$$dG_m^\alpha/dx_B = {}^oG_B^\alpha - {}^oG_A^\alpha + RT\ln(x_B^\alpha/x_A^\alpha) \cong {}^oG_B^\alpha - {}^oG_A^\alpha + RT\ln(x_B^\alpha) \tag{56}$$

The difference in slope for the two tangents to the α curve in Fig. 28 is

$$\Delta(dG_m^\alpha/dx_B) = RT\ln(x_B^\alpha/\,{}^ex_B^{\alpha/\beta}) \tag{57}$$

If β is also a dilute solution of B in A, we get for the difference in slope between the two tangents to the β curve,

$$\Delta(dG_m^\beta/dx_B) = RT\ln(x_B^\beta/\,{}^ex_B^{\beta/\alpha}) \tag{58}$$

The difference in slope between the two tangents will thus be

$$P^\beta(V_B^\beta - V_A^\beta) = RT\ln(x_B^\alpha/\,{}^ex_B^{\alpha/\beta}) - RT\ln(x_B^\beta/\,{}^ex_B^{\beta/\alpha}) \tag{59}$$

By neglecting the difference between V_B^β and V_A^β we get approximately

$$\frac{x_B^\alpha}{{}^ex_B^{\alpha/\beta}} = \frac{x_B^\beta}{{}^ex_B^{\beta/\alpha}}\cdot \exp\frac{P^\beta(V_B^\beta - V_A^\beta)}{RT} \cong \frac{x_B^\beta}{{}^ex_B^{\beta/\alpha}} \tag{60}$$

Then the factors $x_B^\alpha/(x_B^\beta - x_B^\alpha)$ and $x_B^\beta/(x_B^\beta - x_B^\alpha)$ in Eqs. 48 and 54 are constant and can be replaced by the equilibrium values,

$$\Delta x_B^\alpha = \frac{2\sigma V_m^\beta\cdot {}^ex_B^{\alpha/\beta}}{rRT\cdot({}^ex_B^{\beta/\alpha} - {}^ex_B^{\alpha/\beta})} \tag{61}$$

$$\Delta x_B^\beta = \frac{2\sigma V_m^\beta\cdot {}^ex_B^{\beta/\alpha}}{rRT\cdot({}^ex_B^{\beta/\alpha} - {}^ex_B^{\alpha/\beta})} \tag{62}$$

It should again be emphasized that these equations apply only to cases where both phases are dilute solutions of B in A, whereas Eq. 49 applies when the compositions of the two phases are so well defined that $x_B^\beta - x_B^\alpha$ can be treated as a constant. That case is found when α is a dilute solution of B in A and β is a stoichiometric phase or a dilute solution of A in B.

16. Nucleation in a Binary System

A critical β nucleus may be regarded as being in equilibrium with the α matrix although the α matrix is supersaturated with respect to the β phase. The nucleus would shrink spontaneously if it were a little smaller and grow if it were a little larger. The treatment presented in the preceding paragraph can be applied directly to calculate the size and composition of a critical nucleus. A diagram identical to the one shown in Fig. 27 can be used to illustrate the unstable equilibrium between a critical nucleus and the matrix phase, Fig. 29. For instance, at small supersaturations, Δx_B^α, one obtains by applying Eq. 46 to a critical nucleus,

$$(x_B^\beta - x_B^\alpha)\cdot d^2G_m^\alpha/dx_B^2\cdot\Delta x_B^\alpha = P^*V_m^\beta = \frac{2\sigma V_m^\beta}{r^*} \quad (63)$$

and by inserting this in Eq. 52, applied to a critical nucleus,

$$\Delta x_B^{\beta*} = \frac{P^*(x_A^\alpha V_A^\beta + x_B^\alpha V_B^\beta)}{(x_B^\beta - x_B^\alpha)(d^2G_m^\beta//dx_B^2)} = \frac{x_A^\alpha V_A^\beta + x_B^\alpha V_B^\beta}{V_m^\beta}\cdot\frac{d^2G_m^\alpha/dx_B^2}{d^2G_m^\beta/dx_B^2}\cdot\Delta x_B^\alpha \quad (64)$$

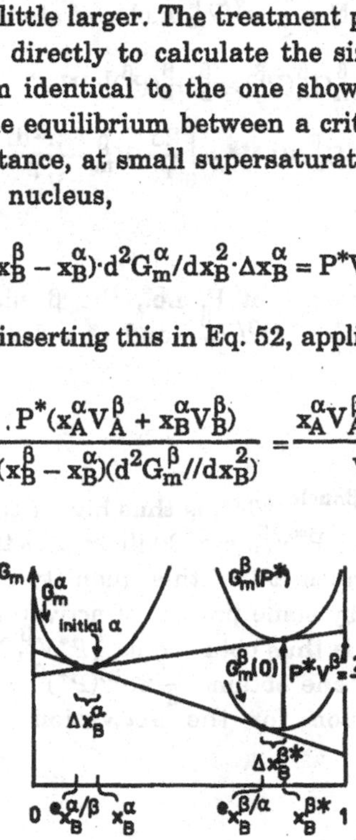

Fig. 29. Evaluation of size and composition of critical nucleus.

It should be emphasized that complications arise if the surface tension varies with the composition of α and β. For instance, the composition of the nucleus would move closer to the composition of the matrix if the surface tension could thus decrease. This might be the case for coherent precipitation in a miscibility gap and it results in a transition from a nucleation-controlled reaction to spinodal decomposition at the spinodal curve in the phase diagram.

The discussion has not yet concerned the activation energy for nucleation. At the first glance it may seem puzzling that the diagram in Fig. 29 shows that the Gibbs energy of a system composed of α phase will not change if one takes a small quantity of A and B atoms with an average composition of $x_B^{\beta*}$ from the α

phase and transforms it into β phase of a higher, constant pressure P^*. The point is that a new, spherical nucleus must form from zero size and grow gradually. Its pressure varies with the size and is thus higher than P^* all the time until it reaches the critical size $r^* = 2\sigma/P^*$. In order to add A and B atoms to the nucleus during its formation, one must thus use higher values for G_A and G_B than are available from the α matrix. If we would evaluate the Gibbs energy of the nucleus by integrating over its growth from zero size and all the time add balanced amounts of A and B atoms to keep the composition constant by taking $dN_A = x_A^\beta \cdot dN$ and $dN_B = x_B^\beta \cdot dN$, we should obtain the following for the Gibbs energy of a spherical nucleus which contains $N = \frac{4}{3}\pi r^3/V_m^\beta$ moles of atoms.

$$G^{\beta nucleus} = \int [G_A^\beta(P^\beta)dN_A + G_B^\beta(P^\beta)dN_B] = \int_0^{N^*} [x_A^\beta G_A^\beta(P^\beta) + x_B^\beta G_B^\beta(P^\beta)]\cdot dN$$

$$= \int_0^{N^*} G_m^\beta(P^\beta)\cdot dN = \int_0^{N^*} [G_m^\beta(0) + P^\beta)V_m^\beta]\cdot dN = G_m^\beta(0)\cdot N^* + \int_0^{r^*} \frac{2\sigma}{r}\cdot V_m^\beta \cdot \frac{4\pi r^2 dr}{V_m^\beta}$$

$$= G_m^\beta(0)\cdot N^* + 4\pi\sigma(r^*)^2 \tag{65}$$

It was here assumed that V_m^β is independent of P, i.e., the β phase is incompressible. By combinations with $N^* = \frac{4}{3}\pi(r^*)^3/V_m^\beta$ and $r^* = 2\sigma/P^*$ one obtains

$$G^{\beta nucleus} = [G_m^\beta(0) + \tfrac{3}{2}P^*V_m^\beta]\cdot N^* \tag{66}$$

The Gibbs energy per mole of the nucleus, $G^{\beta nucleus}/N^*$, is thus higher than the ordinary value for the β phase by the amount $\frac{3}{2}P^*V_m^\beta$. Fig. 30 illustrates that the driving force for precipitation, ΔG_m, provides 2/3 of this quantity and the remaining part, $\frac{1}{2}P^*V_m^\beta$, must be provided by some process of activation. The activation energy of a nucleus of the size N^* is thus obtained as $\frac{1}{2}P^*V_m^\beta\cdot N^*$ and by inserting the expressions for N^* and r^* one obtains $\frac{16}{3}\pi\sigma^3/(P^*)^2$ which is identical to the more common expression for the activation energy, $\frac{16}{3}\pi\sigma^3/(\Delta G_m/V_m^\beta)^2$.

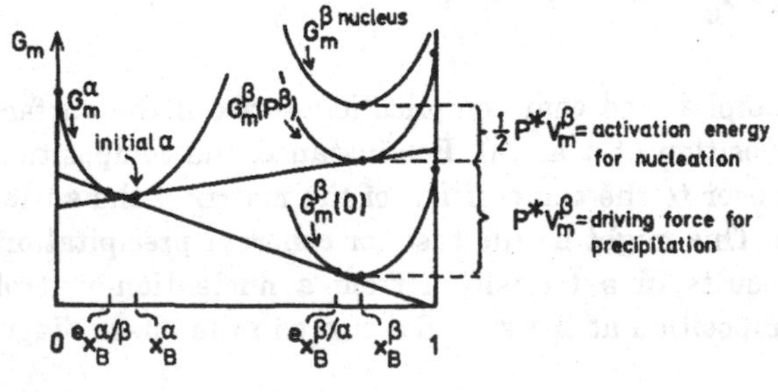

Fig. 30. Demonstration of the activation energy for nucleation.

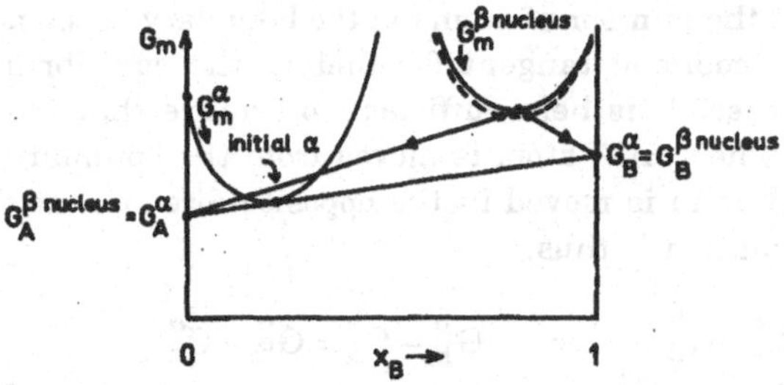

Fig. 31. Partial Gibbs energies of a critical nucleus.

In connection with Fig. 30 it should be pointed out that a tangent to the curve for $G_m^{\beta nucleus}$ does not give the values of G_A and G_B on the component axes. If we add a small quantity of B atoms to the nucleus without changing its A content in view of an equation equivalent to Eq. 5, it will move to the right on the curve but the nucleus will also be larger and the whole curve will move downwards. It is shown in Fig. 31 that the net movement will be in the direction towards the end point of the tangent representing the α matrix, in agreement with our starting point that the critical nucleus is in chemical equilibrium with the matrix.

17. Segregation to Grain Boundaries

Simple physical models of a grain boundary in metals have been used many times in discussions of segregation to grain boundaries and the effect of this segregation on grain boundary energy. The properties of such a model in a binary system will now be discussed with the help of the molar diagram. The model to be discussed simply states that the grain boundary has a constant thickness, that the material in the boundary can be regarded as belonging to a separate phase, which will here be denoted by a superscript "b", and that it has a constant molar volume, V_m^b. A G_m^b function can thus be assigned to the boundary material and by applying the regular solution model one would have:

$$G_m^b = x_A^b\,{}^oG_A^b + x_B^b\,{}^oG_B^b + RT[x_A^b\ln(x_A^b) + x_B^b\ln(x_B^b)] + x_A^b x_B^b\cdot L^b \quad (67)$$

The value for a boundary in pure A, ${}^oG_A^b$, is higher than the value for the matrix in pure A, ${}^oG_A^\alpha$, by an amount which defines the grain boundary energy in pure A, σ_A. By transforming to energy per mole and introducing "t", the thickness of the boundary, we obtain

$$ {}^oG_A^b - {}^oG_A^\alpha = \sigma_A V_m^b/t \quad (68)$$

For an alloy one must consider the equilibrium between the boundary phase and the α matrix. However, by assuming that V_m^b is independent of composition, the model requires that the number of atoms in the boundary is constant. The usual construction of the common tangent for finding the equilibrium is thus too restrictive in this case. It is here sufficient to require that the Gibbs energy should not change when an A atom is moved from the boundary and placed in the matrix and a B atom is moved in the opposite direction at the same time. The equilibrium condition is thus,

$$G_A^\alpha - G_A^b = G_B^\alpha - G_B^b \qquad \text{or} \qquad G_B^b - G_A^b = G_B^\alpha - G_A^\alpha \tag{69}$$

In view of Fig. 4 and Eq. 8 one obtains

$$dG_m^b/dx_B = dG_m^\alpha/dx_B \tag{70}$$

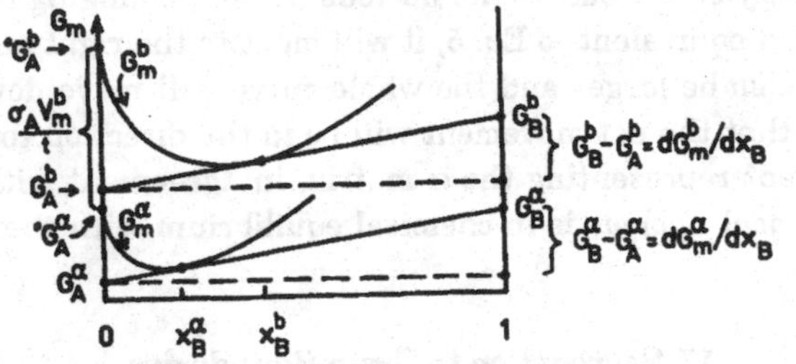

Fig. 32. Evaluation of the segregation to a grain boundary.

The equilibrium composition of the boundary material is thus found by a parallel-tangent construction based on the composition of the matrix, Fig. 32. With the regular solution model, Eq. 18, the equilibrium condition yields

$$RT\ln\frac{x_B^b x_A^\alpha}{x_A^b x_B^\alpha} = {}^\circ G_A^b - {}^\circ G_A^\alpha - {}^\circ G_B^b + {}^\circ G_B^\alpha + L^\alpha(1-2x_B^\alpha) - L^b(1-2x_B^b) \tag{71}$$

The distance between ${}^\circ G_A^b$ and ${}^\circ G_A^\alpha$ on the left-hand axis in Fig. 32 represents the increase in Gibbs energy if one mole of new boundary is created in a system of pure A in the α state, $\sigma_A V_m^b/t$. It is given by Eq. 68 and a corresponding expression holds for pure B. By inserting these expressions in Eq. 71 we obtain for low B contents,

$$\ln(x_B^b/x_B^\alpha) \cong (\sigma_A - \sigma_B)V_m^b/RTt + (L^\alpha - L^b)/RT \tag{72}$$

For ordinary metals σ might be of the order of 1 J/m² and, with $V_m = 7 \cdot 10^{-6}$ m³/mol, $t = 10^{-9}$ m and T=1000 K, the first term on the right-hand side of Eq. 72 is less than 1. It is evident that strong segregation (large x_B^b/x_B^α) must be due to the

L term and in particular to a large negative value of L^b, i.e., a strong tendency for A and B atoms to mix in the boundary.

18. Gibbs' Adsorption Equation

So far, we have not discussed the significance of the distance between the two tangents. It can be obtained from the length of the arrow in Fig. 33. By comparison with Fig. 5 it is evident that the arrow represents the increase in Gibbs energy if one mole of new boundary with a composition x_B^b is created in a large system consisiting of α phase with the composition x_B^α. The distance is thus $\sigma V_m^b/t$ where σ is the surface energy in a material where the matrix has the composition x_B^α.

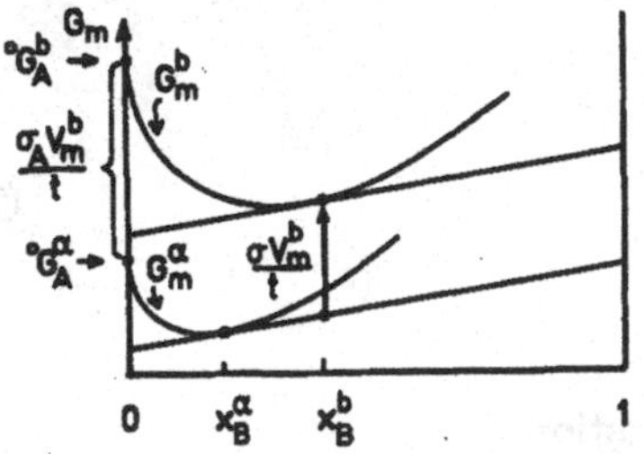

Fig. 33. Evaluation of the surface energy of a boundary in a binary alloy.

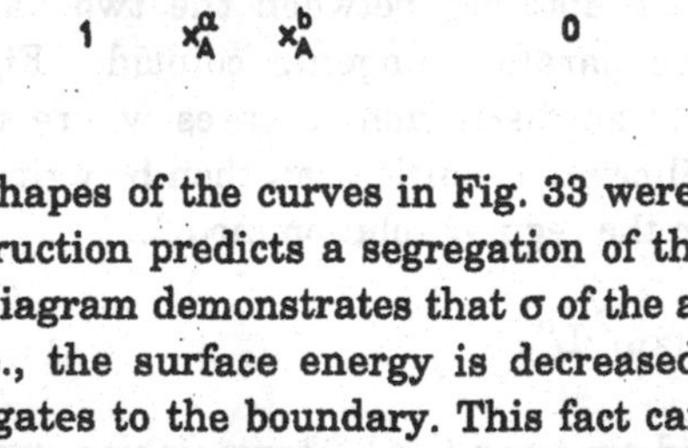

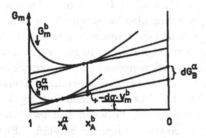

Fig. 34. Relation between the changes in surface energy and partial Gibbs energy.

The shapes of the curves in Fig. 33 were chosen such that the parallel-tangent construction predicts a segregation of the alloying element B to the boundary. The diagram demonstrates that σ of the alloy is lower than the value σ_A for pure A, i.e., the surface energy is decreased by the addition of an element that segregates to the boundary. This fact can be described in mathematical terms. Consider a small increase of the B content in the α matrix and make the parallel-tangent construction in Fig. 34 before and after the change. The change increases the value of G_B^α by dG_B^α and decreases the distance between the parallel tangents by $-d\sigma V_m^b/t$. The relation between these two changes is directly given by comparing triangles,

$$\frac{-d\sigma V_m^b/t}{x_A^\alpha - x_A^b} = \frac{dG_B^\alpha}{x_A^\alpha} \qquad (73)$$

By rearranging and inserting x_B instead of x_A, we find

$$-d\sigma = \frac{x_B^b - x_B^\alpha}{1 - x_B^\alpha} \cdot \frac{t}{V_m^b} \cdot dG_B^\alpha \qquad (74)$$

This is a version of Gibbs' adsorption equation. That equation is usually written $-\sigma = \Gamma_{B(A)} \cdot d\mu_B$ where $d\mu_B$ is equal to dG_B^α and $\Gamma_{B(A)}$ stands for the following combination of parameters, $\Gamma_B - \Gamma_A x_B^\alpha/(1-x_B^\alpha)$. The quantities Γ_B and Γ_A are the excess amounts of B and A per unit area. For the present model, where the total number of atoms in the boundary is constant, one obtains $-\Gamma_A = \Gamma_B = (x_B^b - x_B^\alpha)t/V_m^b$ and thus

$$\Gamma_{B(A)} = \frac{x_B^b - x_B^\alpha}{1 - x_B^\alpha} \cdot \frac{t}{V_m^b} \qquad (75)$$

19. Strong Segregation

The change of σ due to an appreciable addition of B can be calculated by integrating Gibbs' adsorption equation, but can also be obtained directly from the molar diagram. For example, suppose that one is interested to estimate the maximum segregation that can possibly occur to a grain boundary. It may be safe to assume that it is less than the segregation needed in order to make the surface energy vanish, because that case is not normally observed experimentally. An upper limit may thus be obtained by considering such a hypothetical case. It would occur if the distance between the two parallel tangents goes to zero, i.e., if the two parallel tangents coincide. Fig. 35 illustrates that situation. Let us limit the discussion to cases where the B content in the α matrix is low. The equilibrium condition can then be written as follows if one applies Eq. 19 according to the regular solution model.

$${}^oG_A^\alpha \cong G_A^b = {}^oG_A^b + RT\ln(1-x_B^b) + (x_B^b)^2 \cdot L^b \qquad (76)$$

By replacing ${}^oG_A^b - {}^oG_A^\alpha$ with $\sigma_A V_m^b/t$ and omitting the L^b term, which must be negative if B segregates strongly (see Section 17), we obtain

$$x_B^b < -\ln(1-x_B^b) = \sigma_A V_m^b/RTt + (x_B^b)^2 \cdot L^b < \sigma_A V_m^b/RTt \qquad (77)$$

This value of x_B^b in a layer of thickness t is equivalent to a layer of pure B of a

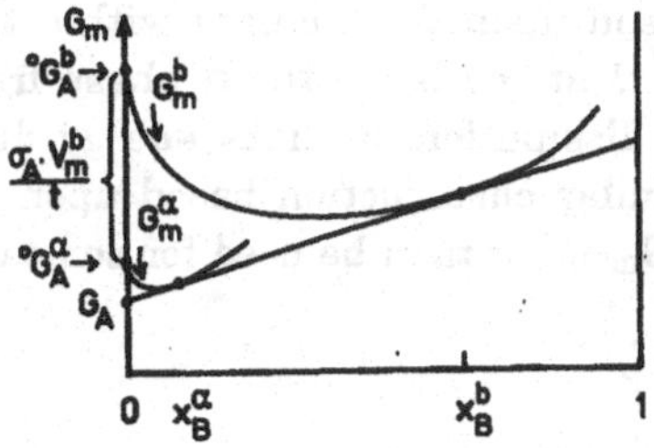

Fig. 35. Hypothetical case of vanishing surface energy.

thickness of $x_B^b \cdot t$. With the reasonable values σ_A=1 J/m^2, V_m^b=7·10^{-6} m^3/mol and T=1000 K we get

$$x_B^b \cdot t < \sigma_A V_m^b / RT = 8 \cdot 10^{-10} \text{ m} = 0.8 \text{ nm} \tag{78}$$

One may thus conclude that, no matter how thick a boundary may be, in order to obtain segregation corresponding to much more than a monolayer, a much larger σ_A or a much lower temperature is required. Of course, this conclusion was reached with a very simple model but it should be realized that the essential part of the derivation was based on the simple fact that the entropic part of G_A^b must be smaller than the distance between the two G_m curves at pure A. The conclusion should thus be qualitatively correct.

20. Dissipation of Gibbs Energy

When discussing equilibrium between two phases we have so far presumed that each one of the phases is homogeneous. In this section we shall consider diffusion inside phases with a variation in composition. A very useful approximation will then be to assume that there is equilibrium between two phases across their interface. This is called the "local equilibrium approximation" and it allows us to use the common tangent construction to find the compositions of the two phases at the interface. We can thus find important boundary conditions in the mathematical treatment of diffusion.

Let us again consider the total driving force for precipitation of β from a supersaturated α phase, ΔG_m in Fig. 21. It would be completely dissipated during the reaction if the composition of the α matrix actually decreased to the equilibrium value ${}^e x_B^{\alpha/\beta}$, i.e., if all the supersaturation vanished during the reaction. Assuming that the composition of the α matrix at the α/β interface is always ${}^e x_B^{\alpha/\beta}$ during the reaction, one can conclude that all the extra B atoms must diffuse down-hill in the α matrix and reach a region of the composition ${}^e x_B^{\alpha/\beta}$ before being incorporated in the growing β phase. All the driving force is then consumed by driving that diffusion. Of course, the dissipation of Gibbs energy due to diffusion in α can only depend upon the changes within the α phase itself. If, by some reason, the extra B atoms leave the α phase from a region of a different composition $x_B^{\alpha/\beta}$, the dissipation of Gibbs energy due to diffusion in α can be evaluated from a similar construction based upon $x_B^{\alpha/\beta}$, Fig. 36. Evidently, the remaining part of ΔG_m must then be used for some other purpose. Such cases will now be discussed.

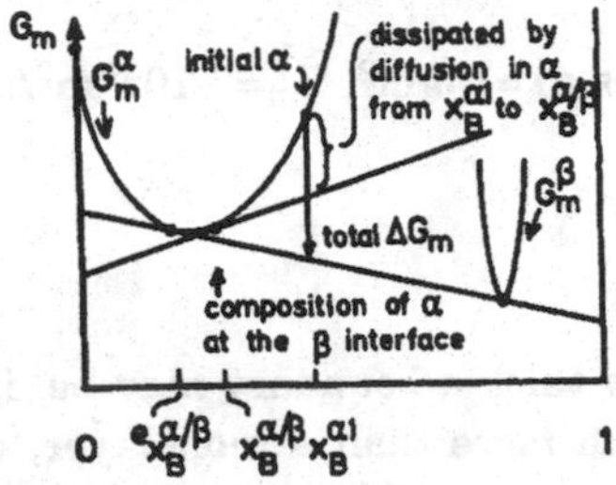

Fig. 36. Evaluation of dissipation of Gibbs energy due to diffusion.

There is a kind of reaction called discontinuous precipitation. It takes place by a steady-state process. Alternating lamellae of the precipitate and of the partially depleted matrix grow side by side. The growth occurs at the edge of the lamellae and the shape stays constant. Knowing the curvature of the edges of the β lamellae, r^β, one knows the pressure under which the β phase is growing, $P^\beta = \sigma/r^\beta$. One can then evaluate the term $P^\beta V_m^\beta$ by which the G_m curve for β is raised, Fig. 37. The common-tangent construction shows that all the supersaturation cannot disappear by such a reaction. The composition of the α matrix in contact with the growing β phase, $x_B^{\alpha/\beta}$, differs from the equilibrium composition, $^e x_B^{\alpha/\beta}$. The part of the total driving force that is dissipated by diffusion is denoted by 1. In the depleted α matrix there still remain some supersaturation, $x_B^{\alpha/\beta} - {}^e x_B^{\alpha/\beta}$, and a driving force for further precipitation of β which is denoted by 4. If evaluated per mole of the original alloy instead of per mole of the depleted α phase, this remaining part of the driving force would correspond to the distance denoted by 3. The part of the total driving force which is denoted by 2 remains to be discussed. It corresponds to the extra work done during the growth of the β phase under the pressure P^β and it is exactly equal to

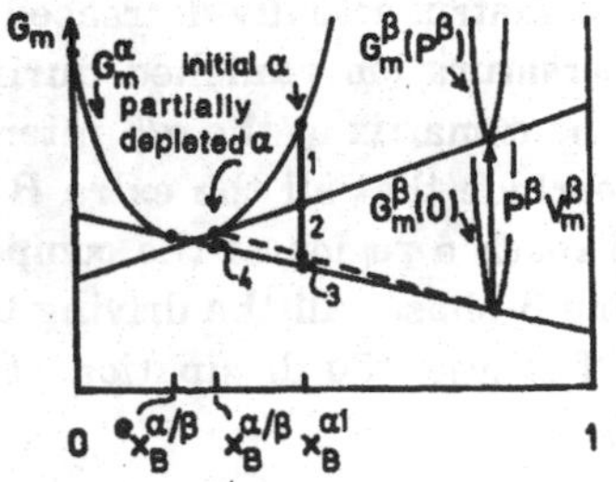

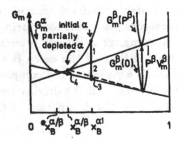

Fig. 37. Identification of different sinks for Gibbs energy when a β phase is growing under pressure.

$P^\beta V_m^\beta$, if evaluated per mole of β that is formed, instead of per mole of the original alloy. That part may all be transformed into interfacial energy of the boundaries between the α and β lamellae or it may be partly dissipated by some irreversible reaction inside the interface.

21. Interface-Controlled Reactions

We shall now assume that the β phase can grow under ordinary pressure and shall discuss what happens if the transfer of atoms across the α/β interface requires a driving force. The local equilibrium approximation and the common-tangent construction can no longer be used. Different cases can be imagined. If the A and B atoms move across the interface individually, it is necessary to have some driving force for the transfer of each element if the growing β phase should contain both. It then seems natural to apply a two-tangent construction with one tangent for each of the phases at the interface. The driving forces ΔG_A and ΔG_B can be evaluated between the two tangents on the sides of the system. When combined with the mobilities of the two elements in the interface, the construction should yield balanced amounts of A and B in the growing β phase. Fig. 38 illustrates such a case and it is evident that the α phase at the interface cannot have the equilibrium composition. Some supersaturation remains and all the driving force available in the initial α phase cannot be used for diffusion. It is again divided into three parts. As in the previous case, distance 1 represents the loss due to diffusion in the α phase and distance 4 represents the driving force remaining in the partially depleted matrix. Distance 3 is the same driving force, counted per mole of the total system. Distance 2 now represents the loss due to the interface reaction, i.e. the driving force for that reaction, whether it consists of the transport of atoms across the interface by diffusion or some other mechanism. The relative sizes of part 1 and part 2 must be such that the diffusional flow of atoms in α down to the interface keeps pace with the interface reaction.

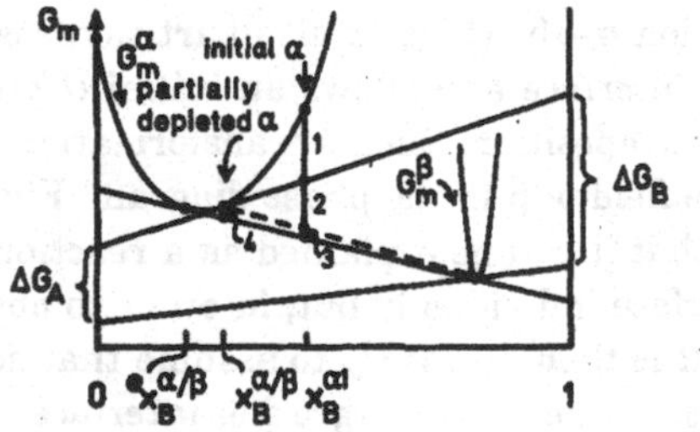

Fig. 38. Identification of different sinks for Gibbs energy during interface-controlled growth of a β phase.

22. Martensitic Transformation

In many systems the α phase may have a higher Gibbs energy than the β phase of the same composition, provided that the supersaturation of α is high enough. This condition occurs above or below the line of equal Gibbs energy in the phase diagram, sometimes called the T_0 line or the allotropic phase boundary. At such a temperature there is a driving force for a partitionless transformation α→β and it is interesting to ask what factor could hold the reaction back. Fig. 39 illustrates such a case and it demonstrates that G_A is lower in the initial α phase than in the product β phase. Under the conditions described by this diagram, the β phase cannot grow if its growth depends upon the individual transfer of A and B atoms across the interface. However, the β phase could grow if the two kinds of atoms cross the interface by some cooperative process. This is probably what happens in a martensitic transformation and it may be due to the special type of interface of which a martensitic transformation makes use. When examining, from the thermodynamic point of view, whether a martensitic transformation could proceed, it is sufficient to evaluate the magnitude of ΔG_m and to examine if it is large enough to overcome various factors such as mechanical stresses.

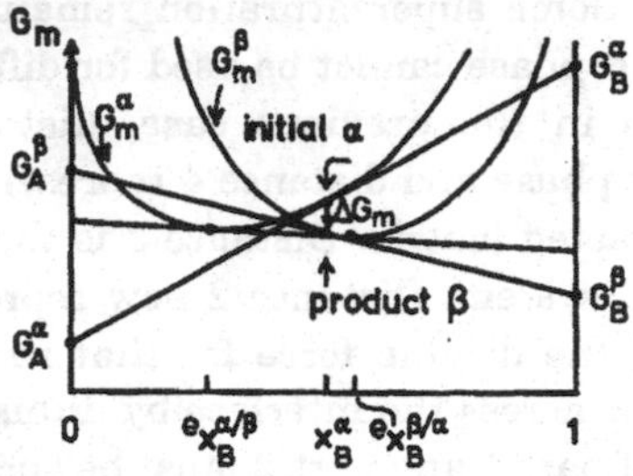

Fig. 39. Evaluation of the driving force for a partitionless transformation of α to β.

23. Massive Transformation

There is a massive type of transformation α→β which is also partitionless. It supposedly makes use of an incoherent interface and allows an initial α phase to transform into a β phase of the same composition. Such a transformation has often been observed inside the one-phase field of β in the phase diagram. Fig. 40 illustrates such a case and it is shown that it can be explained as a reaction by which the A and B atoms cross the interface individually but, in order to obtain a positive value of ΔG_A as well as ΔG_B, it is then necessary to assume that some diffusion takes place in the α matrix just ahead of the migrating interface. The composition of α at the interface is thus changed to a point, $x_B^{\alpha/\beta}$, such that the α tangent is now completely above the β tangent. It is easy to see that this construction is possible only if the β tangent does not intersect the G_m^α curve, i.e., if the alloy composition falls within the one-phase field of β in the phase diagram. Fig. 40 also illustrates what part of ΔG_m is dissipated by diffusion in α, part 1, and how much is left to drive the atoms across the interface and thus make the interface migrate, part 2. The relative sizes of the two parts must be such that the two reactions keep pace with each other.

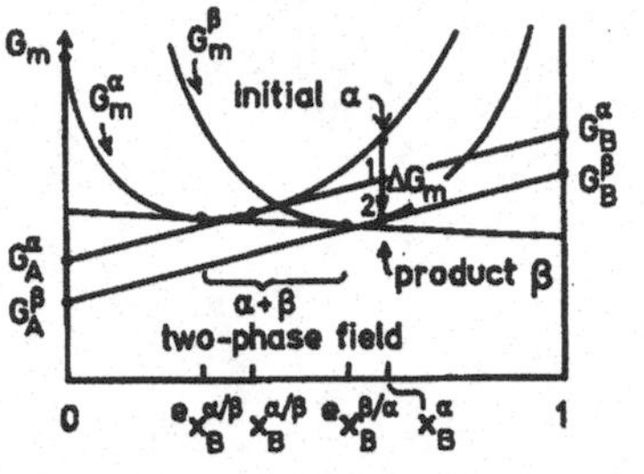

Fig. 40. Demonstration of the conditions for massive transformation of α to β when the two kinds of atoms are transferred individually.

Fig. 40 thus illustrates a possible mechanism for massive growth by individual transfer of atoms across the interface. However, it should be realized that the construction may not apply to a rapidly moving interface. Then it is quite possible that local equilibrium will not be established across the interface between the two phases. Less Gibbs energy will dissipate by diffusion in the α phase and instead there will be dissipation of Gibbs energy inside the interface. The total dissipation may be less than before and the massive transformation may occur inside the two-phase field of the phase diagram. In order to understand such cases it seems necessary to have detailed knowledge about the reactions inside the interface.

The structure of the interface may also be important. The properties of a martensitic interface was discussed in Section 22 and it is well known that a martensitic transformation can occur inside the two-phase field, the thermodynamic limit being the T_0 line. With a less coherent interface the limit may fall closer to the β phase boundary. With a fully incoherent interface it may fall on the β phase boundary, as described above, but only if the rate of reaction is low enough to allow local equilibrium across the interface.

24. Summary

In this chapter it has been demonstrated that thermodynamic considerations of phenomena in alloys can be illustrated with molar Gibbs energy diagrams and a more physical insight may thus be obtained. Such diagrams may be helpful in penetrating the mechanism of a process. In many cases it is even possible to derive thermodynamic relationships from the properties of molar Gibbs energy diagrams.

diagram. Fig. 40 also illustrates what part of ΔG_m is dissipated by diffusion in α phase, and how much is left to drive the atoms across the interface and thus make the interface a mobile part. The relative sizes of the two parts must be such that the two reactions keep pace with each other.

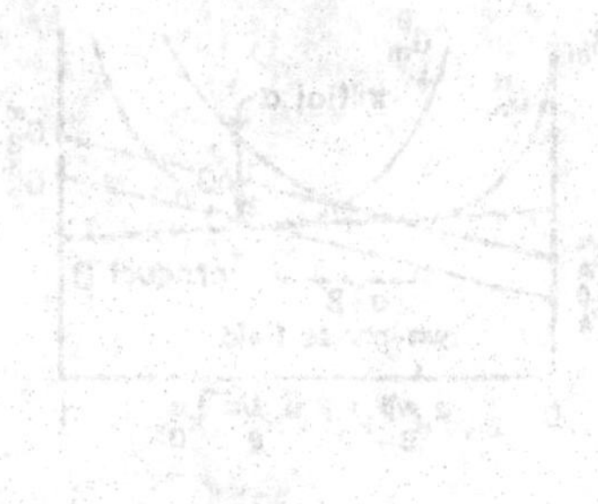

Fig. [illegible]. Demonstration of the conditions for massive transformation [illegible] to which the two kinds of atoms are transferred individually.

Fig. 40 thus illustrates a possible mechanism for massive growth by individual transfer of atoms across the interface. However, it should be realized that the construction may not apply to a rapidly moving interface. Then it is quite possible that local equilibrium will not be established across the interface between the two phases. Less Gibbs energy will dissipate by diffusion in the α phase and there will be dissipation of Gibbs energy inside the interface. The local situation may be less [illegible] and the massive transformation may occur inside the two-phase field of the phase diagram. In order to understand such cases it seems necessary to have detailed knowledge about the reactions inside the interface.

The structure of the interface may also be important. The properties of a martensitic interface was discussed in section [illegible] and it is well known that a martensitic transformation can occur inside the two-phase field, the thermodynamic limit being the T_0 line. With a less coherent interface the limit may fall closer to the phase boundary. With a fully incoherent interface it may fall in the two-phase region as described above but only if the rate of reaction is [illegible] local equilibrium across the interface.

2[illegible]. Summary

In this chapter it has been demonstrated that thermodynamic considerations of [illegible] in alloys can be illustrated with molar Gibbs energy diagrams and [illegible] may [illegible]. Such diagrams may be helpful in [illegible] the mechanism of a process. In many cases it has been possible to [illegible] thermodynamic conditions from the properties of molar Gibbs energy diagrams.

Influence of Elastic Stress on Phase Transformations

William C. Johnson

Department of Materials Science and Engineering
University of Virginia
Charlottesville, VA 22903-2442 USA

September 20, 1999

1 Introduction

Elastic stresses arise naturally during solid-state phase transformations. Compositional heterogeneity developing during diffusional transformations engenders stress when the partial molar volumes of the diffusing species are different. Stresses also develop during displacive transformations, when the lattices of the parent and product phase possess different dimensions or symmetries, and during various second-order magnetic, ferroelectric and order-disorder transitions.

Elastic stresses, whether arising from an externally applied load, compositional heterogeneity or a transformation strain, manifest themselves in various ways during a phase transformation. The elastic deformation affects the free energy and, hence, the thermodynamic description of the system [1-10]. The strain field arising from a spatially localized transformation is long range; it results in both local and nonlocal distortions of the matrix that influence the kinetics of the phase transformation, the domain and/or precipitate morphology, and the spatial distribution of precipitates.

There is significant experimental evidence that elastic stress affects microstructural evolution in two-phase systems. Figure 1 illustrates the effect of precipitate misfit strain on equilibrium precipitate shape with increasing particle radius for a Ni-Al binary alloy. The interfacial energy density is essentially isotropic in this alloy and the equilibrium precipitate shape in the absence of stress is expected to be a sphere [11]. The elastic energy of this elastically anisotropic system is minimized, however, when the precipitate assumes a plate-like shape [12, 13, 14], suggesting that the equilibrium shape will be determined through competition between the interfacial and elastic energies. Equiaxed particles are observed at small particle sizes in Fig. 1a. With increasing particle radius, the precipitate develops facets along the {100} directions and the precipitates begin to resemble cubes as seen in Fig. 1b. At yet larger sizes, Fig. 1c, the four-fold symmetry of the precipitate in the (001) projection is broken and plate-shaped cuboids develop; the aspect ratio increasing with increasing particle size, Fig. 1d. In addition, the precipitates begin to align along the elastically soft $< 100 >$ directions of the matrix phase. Other, more complicated evolution paths are also observed. For example, the splitting of a single particle into two or more particles has been reported in this alloy [15, 16, 17].

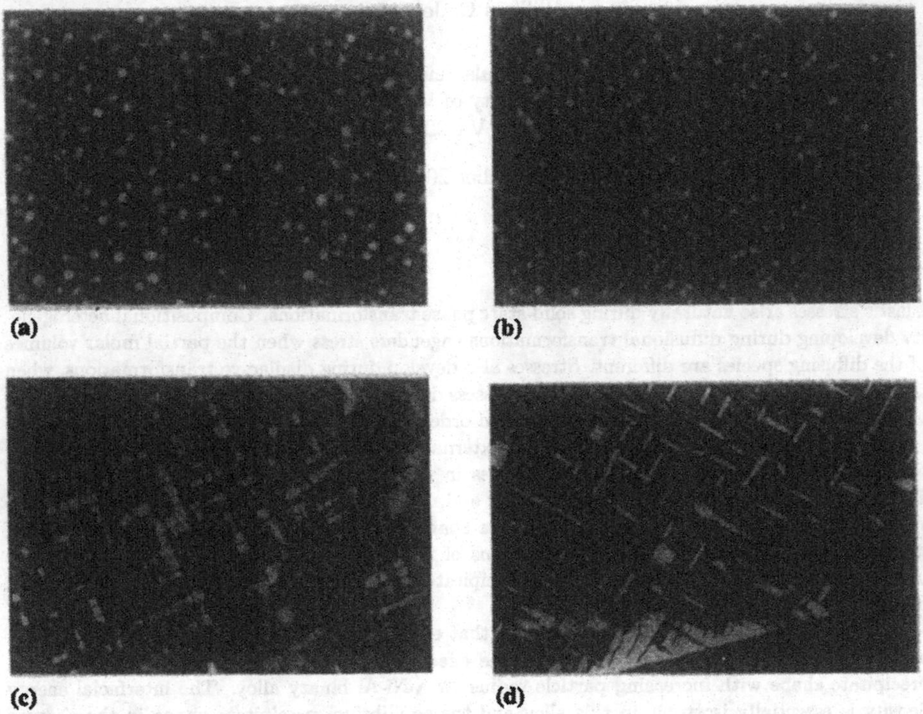

Figure 1: Centered dark-field images show precipitate shape evolution in a Ni-6.71w%Al alloy aged for various times at 750°C: (a) 15 min, equiaxed γ' (Ni_3Al) precipitates; (b) 4h, cub-shaped precipitates; (c) 72h, cubical precipitates become plate-shaped and aligned along $< 100 >$ elastically soft directions; and (d) 450h. Precipitate shape transitions occur with increasing precipitate size in coherent systems owing to the different scaling of the elastic strain and interfacial energies with the precipitate radius: The interfacial energy dominates at small precipitate radii giving equiaxed precipitate shapes in this example while the elastic energy dominates at large precipitate radii. Micrographs courtesy of A. J. Ardell.

Table 1: Materials parameters for experimental Ni-Al-Mo alloys: Figs. 2 and 3

Alloy	Al $[at\%]$	Mo $[at\%]$	$f_{\gamma'}[\%]$	γ' solvus ^oC	ϵ^T (misfit)
A1	12.5	2.0	15 ± 5	≈ 1000	+0.65%
A2	9.9	5.0	19 ± 6	≈ 1000	+0.40%
A3	7.7	7.9	15 ± 5	≈ 950	< 0.10%
A4	5.7	10.9	7 ± 2	?	−0.15%
A5	5.3	13.0	8 ± 2	≈ 900	−0.30%

Elastic stresses also influence the spatial correlation between particles as illustrated in the set of centered dark-field micrographs appearing in Figs. 2 and 3. In each figure, the microstructure of five different $(\gamma + \gamma')$ Ni-Al-Mo ternary alloys is depicted after identical aging treatments. The volume fraction of the γ' precipitates in each alloy is roughly 10% but the misfit strain, ϵ^T, becomes progressively more negative with increasing Mo content. The alloy composition (atomic percent) and corresponding misfit strain of the alloys are given in Table I.

After 67h of aging at 775^oC, the alloys of Fig. 2 exhibit distinctive microstructures. The alloys with the largest magnitude of misfit, A1 (a) and A2 (b), show microstructures in which the precipitates are aligned along the elastically soft $< 100 >$ directions of the matrix phase. The lowest misfit alloys A3 (c) and A4 (d), still show a random dispersion of spherical precipitates while the precipitates of alloy A5 (e) have begun to assume a cube-like shape. After 430h of aging at 775^oC, the precipitates of alloys A1 (a) and A2 (b) have lost their four-fold symmetric shapes and those of alloy A4 (d) have begun to assume a more facetted shape. The precipitates of alloy A3 (c), which have a vanishingly small misfit strain with respect to their matrix are still spherical, indicating that minimization of the interfacial energy is controlling their equilibrium shape [18].

The influence of elastic stress on solid-state phase transformations is sometimes complex and is far from being completely understood. It is also difficult to model, as stress effects are usually nonlocal in nature. In the following, several simple models will be developed for examining the influence of elastic deformation on solid-solid phase transformations. First, the formalism for treating elastic deformation is developed and then incorporated into the thermodynamic treatment of crystals. This formalism is used to derive the conditions for thermodynamic equilibrium which are then applied to phase equilibria in stressed systems. Finally, precipitate shape transitions and microstructural evolution in stressed systems similar to those exhibited in Figs. (1)-(3) are examined using simple theory and various computer simulation techniques.

2 Basics

In this section, some of the tools necessary for examining the effect of elastic stress on phase transformations are developed. Particular attention is devoted to the understanding and connection between elastic stress and strain. Since the stress and strain fields acting at a material point are best described in terms of second-rank tensors, a brief introduction to tensors is provided in subsection 2.1 for those unfamiliar with tensor manipulations. Others can proceed directly to subsections 2.2 and 2.3 where a description of elastic deformation, including the deformation gradient tensor and its relationship to the strain tensor, is given. The connection between the mechanical

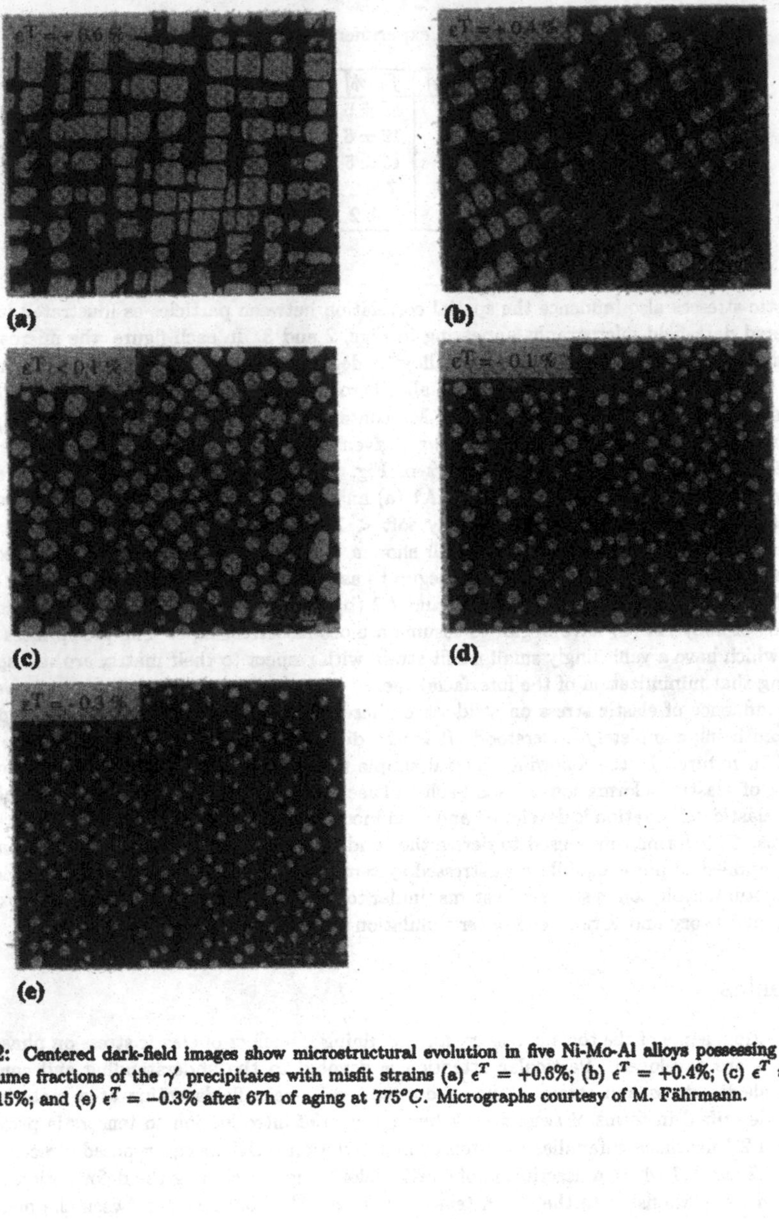

Figure 2: Centered dark-field images show microstructural evolution in five Ni-Mo-Al alloys possessing roughly equal volume fractions of the γ' precipitates with misfit strains (a) $\epsilon^T = +0.6\%$; (b) $\epsilon^T = +0.4\%$; (c) $\epsilon^T \approx 0$; (d) $\epsilon^T = -0.15\%$; and (e) $\epsilon^T = -0.3\%$ after 67h of aging at 775^oC. Micrographs courtesy of M. Fährmann.

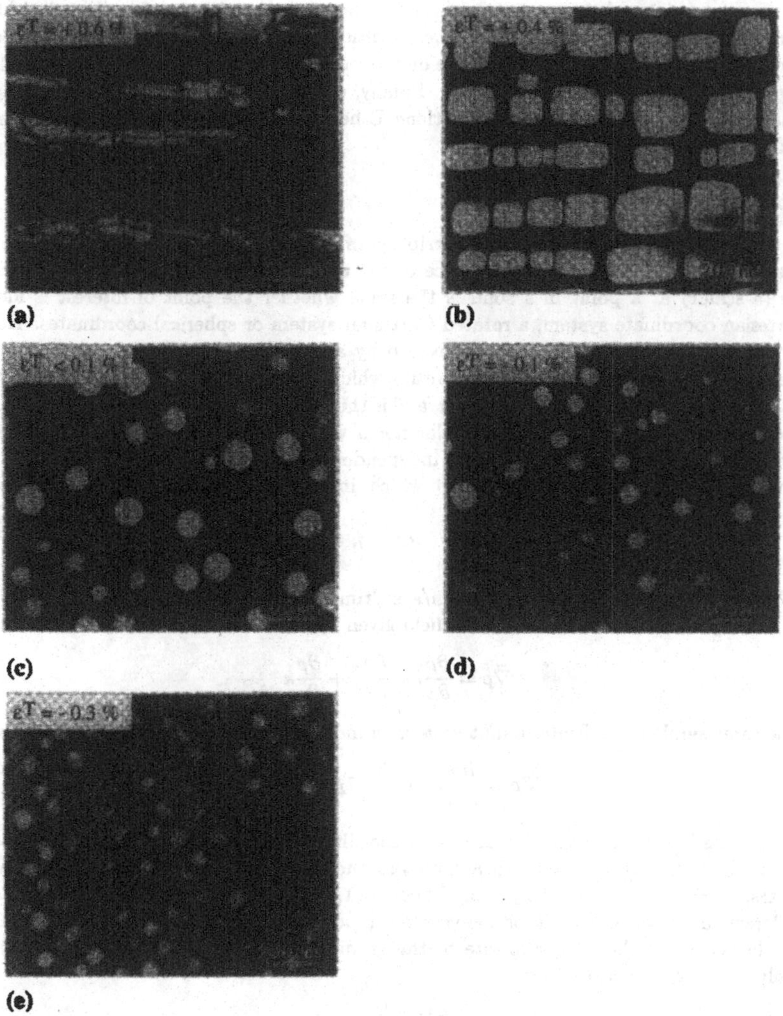

Figure 3: Centered dark-field images show microstructural evolution in five Ni-Mo-Al alloys of roughly equal volume fraction with misfit strains (a) $\epsilon^T = +0.6\%$; (b) $\epsilon^T = +0.4\%$; (c) $\epsilon^T \approx 0$; (d) $\epsilon^T = -0.15\%$; and (e) $\epsilon^T = -0.3\%$ after 430h of aging at 775°C. Micrographs courtesy of M. Fährmann.

forces acting at a point and the stress tensor is developed in subsection 2.4. In subsection 2.5, general expressions for the elastic energy and the conditions for mechanical equilibrium are derived. In subsection 2.6, simple constitutive laws are developed which connect the elastic stress and strain to changes in temperature and composition. Finally, owing to its frequent use in studying both diffusional and diffusionless phase transformations, Eshelby's misfitting sphere problem is presented in subsection 2.7.

2.1 Introduction to Tensors

Tensors occur frequently in the physical description of crystals [19]. Like a scalar, they represent quantities that are independent of the choice of the coordinate system. For example, the temperature (a scalar) at a point in a solid is the same whether the point of interest is identified by a Cartesian coordinate system, a rotated Cartesian system or spherical coordinates. However, necessary information cannot always be conveyed by a single number as in the case of a scalar. The velocity of an element of material, a quantity which is also independent of coordinate system, depends on both a magnitude and a direction, and is thus best conveyed as a vector. Other physical quantities can be conveyed as neither a scalar nor a vector, but rather as a tensor. In order to extend the idea of physical quantitites being independent of coordinate system to tensors, consider Fick's first law of diffusion for isotropic solids which, in vector and indicial form appears as:

$$\vec{j} = -D\vec{\nabla}\rho \quad \text{or} \quad j_i = -D\frac{\partial \rho}{\partial x_i} \tag{1}$$

where D is the diffusivity, $\vec{j}$ is the flux (atoms/area/time), ρ is the concentration in atoms/volume, and $\vec{\nabla}\rho$ is the gradient of the concentration field given by

$$\vec{\nabla}\rho = \frac{\partial \rho}{\partial x}\hat{i} + \frac{\partial \rho}{\partial y}\hat{j} + \frac{\partial \rho}{\partial z}\hat{k} \tag{2}$$

where the carat symbol, $\hat{}$, indicates a unit vector. In more general notation, the gradient is written

$$\vec{\nabla}\rho = \frac{\partial \rho}{\partial x_1}\hat{e}_1 + \frac{\partial \rho}{\partial x_2}\hat{e}_2 + \frac{\partial \rho}{\partial x_3}\hat{e}_3 \tag{3}$$

where $\hat{e}_i$ are the basis vectors of the Cartesian coordinate system (sometimes written as $\hat{e}_1 = \hat{i}$, $\hat{e}_2 = \hat{j}$ and $\hat{e}_3 = \hat{k}$.) Both $\vec{j}$ and $\vec{\nabla}\rho$ are vectors and D is a scalar. As is usual for Cartesian coordinates, $x_1 = x$, $x_2 = y$, and $x_3 = z$. Since $i = 1, 2, 3$ in Eq. (1), Eq. (1) actually stands for three different functions, one each for the vector components j_1, j_2 and j_3. Equation (1) indicates that the flux is in the direction opposite to the gradient in concentration. If the concentration varies only in the x_3 direction, then

$$j_3 = -D\frac{\partial \rho}{\partial x_3} \tag{4}$$

and the components of the flux in the x_1 and x_2 directions, j_1 and j_2, vanish.

In many anisotropic crystals, however, the flux in the x_3 direction can also be affected by concentration gradients in directions other than x_3. Assuming a linear relationship, the flux in the x_3 direction in such a case can be written

$$j_3 = -D_{31}\frac{\partial \rho}{\partial x_1} - D_{32}\frac{\partial \rho}{\partial x_2} - D_{33}\frac{\partial \rho}{\partial x_3} \tag{5}$$

where one diffusion coefficient is needed for each component of the spatial derivative of the concentration. In general, we can express the relationship between the flux and concentration gradient as

$$\begin{pmatrix} j_1 \\ j_2 \\ j_3 \end{pmatrix} = \begin{pmatrix} D_{11} & D_{12} & D_{13} \\ D_{21} & D_{22} & D_{23} \\ D_{31} & D_{32} & D_{33} \end{pmatrix} \begin{pmatrix} \frac{\partial \rho}{\partial x_1} \\ \frac{\partial \rho}{\partial x_2} \\ \frac{\partial \rho}{\partial x_3} \end{pmatrix} \tag{6}$$

There are nine components of the diffusivity when all three vector components of the flux are considered. If we now use D_{ik} to represent the element of the diffusivity tensor (or matrix) in the *ith* row and *kth* columen, then the relationship can be written more compactly as:

$$j_i = -\sum_{k=1}^{3} D_{ik} \frac{\partial \rho}{\partial x_k}. \tag{7}$$

Here, i is a *free indice* and can assume any value from 1 to 3, depending on the component of the flux vector of interest. The vector $\vec{j}$ is given in Cartesian coordinates by

$$\vec{j} = j_1 \hat{e}_1 + j_2 \hat{e}_2 + j_3 \hat{e}_3 = \sum_{k=1}^{3} j_k \hat{e}_k. \tag{8}$$

Since the flux is a physical quantity that must remain unchanged when expressed in a different coordinate system, the flux could also be expressed in a rotated (primed) Cartesian system as

$$\vec{j} = j'_1 \hat{e}'_1 + j'_2 \hat{e}'_2 + j'_3 \hat{e}'_3 = \sum_{k=1}^{3} j'_k \hat{e}'_k \tag{9}$$

where $\hat{e}'_i$ are the basis vectors in the rotated Cartesian coordinate system.

As the example above illustrates, a second-rank tensor can be viewed as connecting one vector with a second vector in such a way that the result is independent of coordinate system. The diffusivity, which is a scalar in Fick's first law but is expressed as a matrix of nine components in the more general formulation of Eq. (7), is a second-rank tensor as it connects the concentration gradient (a vector) with the flux (a vector). Since both of these vectors are independent of coordinate system, the diffusivity must also be independent of coordinate system and is, therefore, a second-rank tensor. An equivalent definition is that the diffusivity matrix is defined to be a tensor if the general relationship, Eq. (7), holds for all possible orientations of the Cartesian system; i.e., the diffusivity is independent of the coordinate system.

The fact that a tensor must be independent of the choice of coordinate system does not imply that the individual tensor components are the same in each of the component systems. In order to clarify this point, consider first the two-dimensional position vector $\vec{p}$ in Fig. 4. Two coordinate systems are shown, the Cartesian system with basis vectors $\hat{e}_1$ and $\hat{e}_2$, and a rotated (primed) Cartesian system with basis vectors $\hat{e}'_1$ and $\hat{e}'_2$. The vector $\vec{p}$ has a magnitude and direction independent of the coordinate system in which it is expressed. In the original system:

$$\vec{p} = p_1 \hat{e}_1 + p_2 \hat{e}_2, \tag{10}$$

while in the rotated coordinate system the position vector is given by:

$$\vec{p} = p'_1 \hat{e}'_1 + p'_2 \hat{e}'_2. \tag{11}$$

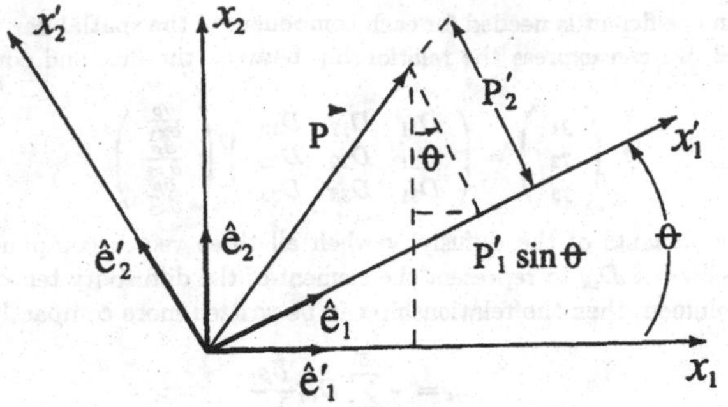

Figure 4: The position vector $\vec{p}$ can be represented in either the Cartesian system ($\hat{e}_1$ and $\hat{e}_2$) or the rotated (primed) Cartesian system ($\hat{e}_1'$ and $\hat{e}_2'$). θ is the rotation angle between the two coordinate systems.

If the primed coordinate system is rotated counter-clockwise by an angle θ with respect to the unprimed system, the relationship between the primed and unprimed components of $\vec{p}$ can be obtained:

$$p_1' = p_1\cos\theta + p_2\sin\theta \tag{12}$$

and

$$p_2' = -p_1\sin\theta + p_2\cos\theta. \tag{13}$$

In matrix form, Eqs. (12) and (13) become

$$\begin{pmatrix} p_1' \\ p_2' \end{pmatrix} = \begin{pmatrix} \cos\theta & \sin\theta \\ -\sin\theta & \cos\theta \end{pmatrix} \begin{pmatrix} p_1 \\ p_2 \end{pmatrix}. \tag{14}$$

If the rotation matrix, a_{ij}, (i denotes row, j denotes column) is defined as

$$a_{ij} = \begin{pmatrix} \cos\theta & \sin\theta \\ -\sin\theta & \cos\theta \end{pmatrix}, \tag{15}$$

then the components of $\vec{p}$ expressed in the two coordinate systems are related by:

$$p_i' = \sum_{k=1}^{2} a_{ik}p_k. \tag{16}$$

For example, for $i = 1$ in Eq. (16), the resulting sum yields Eq. (12). Likewise, setting $i = 2$ yields Eq. (13). The components of $\vec{p}$ in the primed coordinate system are connected to those of the Cartesian (unprimed) system by the rotation matrix a_{ij}. Examination of Fig. 4 shows that the rotation matrix represents the direction cosines between the two coordinate axes. The elements of a_{ij} are given by the dot product of the basis vectors (of unit magnitude):

$$a_{ij} = \hat{e}_i' \cdot \hat{e}_j. \tag{17}$$

Equation (17) is equally valid in three dimensions.

It can be proven that a two-by-two matrix, such as that given by the diffusivity matrix, is a tensor and independent of coordinate system, if the elements of the matrix satisfy the following relationship [20]

$$D_{ij}' = \sum_{m=1}^{3}\sum_{n=1}^{3} a_{im}a_{jn}D_{mn} \tag{18}$$

where the D'_{ij} are the elements of the diffusivity matrix when referred to the primed coordinate system. The diffusivity is a tensor of rank two, since two sums are necessary to connenct the components of the diffusivity in the primed and unprimed systems. Equivalently, a vector can be considered a tensor of rank one, since a vector is independent of coordinate system and its components in the primed and unprimed systems are connected by one rotation matrix as in Eq. (16). A matrix does not necessarily represent the components of a tensor, but all second-rank tensors can be expressed in matrix form. A scalar is a tensor of rank zero.

Second-rank tensors, such as the diffusivity, can be considered a type of vector function. The tensor associates with each argument vector another vector which is the value of the function. For the general form of Fick's first law, Eq. (7), the diffusivity tensor associates a flux with the concentration gradient (argument vector). A third-rank tensor is also a function that could associate each argument vector with a second-rank tensor. Likewise, a fourth-rank tensor could associate a second-rank tensor with another second-rank tensor (or a vector with a third-rank tensor, etc.). The components of a fourth-rank tensor, such as the elastic constants, C_{ijkl}, transform in the same way as a second-rank tensor, except four summations involving the rotation matrix are required:

$$C'_{ijkl} = \sum_{m=1}^{3}\sum_{n=1}^{3}\sum_{p=1}^{3}\sum_{q=1}^{3} a_{im}a_{jn}a_{kp}a_{lq}C_{mnpq} \tag{19}$$

where the C'_{ijkl} are the components of the elastic constants tensor expressed in the primed coordinate system.

Since tensors can be represented in matrix form, many of the definitions and operations which apply to matrices are also applicable to tensors. For example, a *symmetric* tensor is one for which the tensor components obey

$$D_{ij} = D_{ji}; \tag{20}$$

i.e., the components of the tensor are symmetric about the diagonal elements D_{ii}. The components of the *transpose* of the tensor, D^T_{ij}, are obtained by interchanging the rows and columns of the tensor such that

$$D^T_{ij} = D_{ji}. \tag{21}$$

Tensors of like rank can be added and subtracted, just like matrices. For example, the addition of two second-rank tensors to give a third requires that each of the respective components be added together:

$$T_{ij} = R_{ij} + S_{ij}, \quad \text{e.g.,} \quad T_{13} = R_{13} + S_{13}. \tag{22}$$

If a tensor is multiplied by a scalar, then each component of the tensor must be multiplied by the scalar. Likewise, when the derivative of a tensor is evaluated, one must take the derivative of each component of the tensor. It is also possible to combine tensors in various ways to form other tensors. For example, two second-rank tensors (**A** and **B**) can be *contracted* to yield a scalar f as follows

$$f = \sum_{m=1}^{3}\sum_{n=1}^{3} A_{mn}B_{mn} \tag{23}$$

where we again observe that the repeated indices, m and n, are summed over. The two second-rank tensors could also be multiplied together to form a new second-rank tensor:

$$H_{ij} = \sum_{k=1}^{3} A_{ik}B_{kj} \tag{24}$$

where the sum is over the one repeated indice k. Note that the tensor $\mathbf{H}$ is simply the product of the matrices of components of tensors $\mathbf{A}$ and $\mathbf{B}$:

$$\begin{pmatrix} H_{11} & H_{12} & H_{13} \\ H_{21} & H_{22} & H_{23} \\ H_{31} & H_{32} & H_{33} \end{pmatrix} = \begin{pmatrix} A_{11} & A_{12} & A_{13} \\ A_{21} & A_{22} & A_{23} \\ A_{31} & A_{32} & A_{33} \end{pmatrix} \begin{pmatrix} B_{11} & B_{12} & B_{13} \\ B_{21} & B_{22} & B_{23} \\ B_{31} & B_{32} & B_{33} \end{pmatrix}. \tag{25}$$

Another useful tensor is the *unit tensor* of second rank. In matrix form, the diagonal components of the unit tensor are one and the off-diagonal components vanish. In three dimensions, the unit tensor is often designated as $\mathbf{E}$ or $\mathbf{1}$ and is written as

$$\mathbf{1} = \begin{pmatrix} 1 & 0 & 0 \\ 0 & 1 & 0 \\ 0 & 0 & 1 \end{pmatrix}. \tag{26}$$

A particularly common notation for the unit tensor is the *Kronecker delta*, δ_{ij}. The Kronecker delta is defined such that $\delta_{ij} = 1$, if $i = j$, and $\delta_{ij} = 0$, if $i \neq j$. For example, component $\delta_{12} = 0$ and $\delta_{22} = 1$. This definition reproduces the unit matrix of Eq. (26), since $\delta_{11} = \delta_{22} = \delta_{33} = 1$.

When a tensor of second rank is multiplied by the unit tensor in accordance with Eq. (24), the original tensor is recovered. A unit tensor of second rank also transforms an argument vector into itself. For example, using matrix multiplication,

$$\begin{pmatrix} x_1 \\ x_2 \\ x_3 \end{pmatrix} = \begin{pmatrix} 1 & 0 & 0 \\ 0 & 1 & 0 \\ 0 & 0 & 1 \end{pmatrix} \begin{pmatrix} x_1 \\ x_2 \\ x_3 \end{pmatrix}. \tag{27}$$

Using indicial notation, Eq. (27) is written

$$x_i = \sum_{k=1}^{3} \delta_{ik} x_k. \tag{28}$$

The *inverse* of a second-rank tensor F_{ij} is denoted F_{ij}^{-1}, and is defined such that

$$\delta_{ij} = \sum_{k=1}^{3} F_{ik} F_{kj}^{-1} = \sum_{k=1}^{3} F_{ik}^{-1} F_{kj}. \tag{29}$$

The components of the inverse tensor are determined by first expressing the tensor in matrix form and then determining the inverse matrix [20].

Tensor notation is very compact and it is possible to express many equations in a very concise form. The above equations can be even further simplified by introducing the concept of *free* and *dummy* indices and the Einstein summation convention. Free indices appear only once in each term and, in three dimensions, can assume any value from 1 to 3 that is assigned to them. In Eq. (29) above, i and j are free indices. There are nine independent equations represented by Eq. (29), one equation for each combination of i and j. The indice k in Eq. (29) is known as a dummy indice and assumes the value assigned to it by the summation. Since dummy indices are always summed and always appear in pairs, the summation sign is omitted and indices that are repeated in a term are understood to be summed. For example, the equation

$$f_i = \sum_{j=1}^{3} L_{ij} g_j + \sum_{k=1}^{3} H_{ki} p_k \tag{30}$$

would be expressed using the summation convention as:

$$f_i = L_{ij}g_j + H_{ki}p_k. \tag{31}$$

i is the free indice since it appears only once in each term. j and k are dummy indices since they are repeated within a term and, therefore, must be summed. Equation (31) can also be written as:

$$f_i = L_{i1}g_1 + L_{i2}g_2 + L_{i3}g_3 + H_{1i}p_1 + H_{2i}p_2 + H_{3i}p_3. \tag{32}$$

When $i = 3$, one has

$$f_3 = L_{3j}g_j + H_{k3}p_k = L_{31}g_1 + L_{32}g_2 + L_{33}g_3 + H_{13}p_1 + H_{23}p_2 + H_{33}p_3. \tag{33}$$

The choice of letter to use as a dummy indice is arbitrary, so long as it is not the same as a free indice. For example, $L_{ij}g_j = L_{ik}g_k = L_{i1}g_1 + L_{i2}g_2 + L_{i3}g_3$. Furthermore, the summation does not depend on the order in which the terms appear; $L_{ij}g_j = g_jL_{ij}$. Note, however, that $g_jL_{ij} \neq g_iL_{ij}$, unless L_{ij} is a symmetric tensor. Some other relationships that we will encounter include

$$\delta_{ik}A_{kj} = A_{ij}, \tag{34}$$

$$\delta_{ij}x_j = x_i, \tag{35}$$

$$\delta_{jj} = \delta_{11} + \delta_{22} + \delta_{33} = 3, \tag{36}$$

$$\frac{\partial x_i}{\partial x_j} = \delta_{ij}, \tag{37}$$

and

$$\delta_{ij}A_{ij} = A_{jj} = A_{11} + A_{22} + A_{33}. \tag{38}$$

The trace of a matrix of $2-d$ tensor is defined as the sum of its diagonal elements. The symbol A_{jj}, within the Einstein summation convention, represents the summation over elements with identical indices and is, therefore, identical to the trace of the tensor **A**. The trace of the tensor **A** and is sometimes written as Tr **A**. The vector product can be expressed in terms of the Kronecker delta as

$$\vec{x} \cdot \vec{x} = x_ix_i = x_i(\delta_{ij}x_j). \tag{39}$$

2.2 Elastic Deformation

In this subsection, a means for characterizing the elastic deformation of a material when acted upon by external forces is developed. No assumptions are made concerning the origin of the forces or, at least initially, about the deformation being "linear" or "small." The elastic deformation will be represented in terms of different second-rank tensors including the deformation gradient tensor and the strain tensor.

When forces are applied to a material, the material responds by deforming. The extent to which a material deforms depends upon the material's stiffness and the magnitude of the forces applied. If the material quickly reverts to its original state when the applied forces are removed, the deformation is termed elastic. If the material does not return to its original state, then some amount of plastic deformation has occurred.

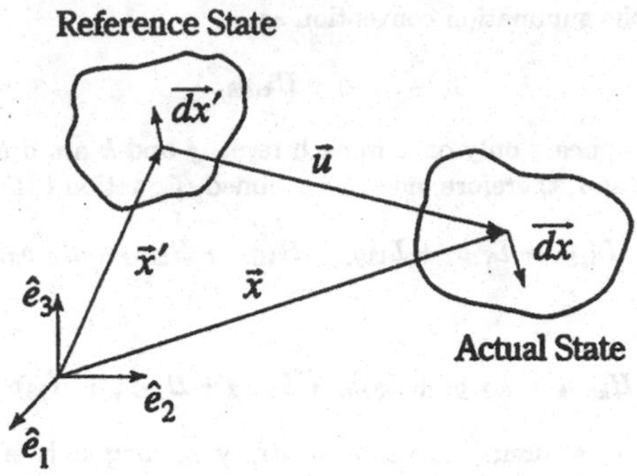

Figure 5: **The vectors $\vec{x}'$ and $\vec{x}$ locate the same material element in the reference and deformed states, respectively, and are connected by the displacement vector $\vec{u}$. A line element $d\vec{x}'$ is rotated and stretched into the line element $d\vec{x}$ by the deformation.**

When quantifying deformation, it is necessary to define a state, termed the reference state, from which the deformation can be measured. The reference state is defined to be free of deformation, but it is not necessarily free from stress or external forces acting upon it. While there is no unique choice for the reference state, a judicious choice of reference state can simplify significantly the solution of a given problem.

Consider the material body shown in Fig. 5. The material is assumed initially to be free of all external forces. The vector $\vec{x}'$ gives the position of a point or volume element in the stress-free material, which is denoted the reference state in Fig. 5. $d\vec{x}'$ is an infinitesimal vector located at the position $\vec{x}'$ and whose direction is arbitrary. If the material is a crystal, then $\vec{x}'$ could define a lattice site and $d\vec{x}'$ an infinitesimal vector pointing along a certain crystallographic direction. When a distribution of forces is applied to the surface of the material, the material can undergo both a translation and elastic deformation. During this process, the element of material originally at point $\vec{x}'$ is displaced to the point $\vec{x}$. The vector $d\vec{x}'$ is also displaced, stretched and rotated and appears in the deformed state as the vector $d\vec{x}$. The deformation of the material can be quantified, if a relationship between $d\vec{x}'$ and $d\vec{x}$ can be found.

For most of the problems to be discussed here, the stress-free state of a particular phase will be chosen as the reference state for the measurement of deformation. The position of a material element in the reference state configuration and its position in the deformed state are related by the displacement vector, $\vec{u}$, as

$$\vec{x} = \vec{x}' + \vec{u}. \tag{40}$$

The position of a material element in the deformed state is considered to be a function of its position in the reference state. This relation can be expressed mathematically as:

$$\vec{x} = \vec{x}(\vec{x}') \quad \text{or} \quad x_i = x_i(x_1', x_2', x_3') \tag{41}$$

where, in the usual Cartesian coordinates, $x_1 = x$, $x_2 = y$ and $x_3 = z$. So long as the deformation does not open up gaps in the material or result in material being folded over upon itself, the inverse relation is also valid:

$$\vec{x}' = \vec{x}'(\vec{x}) \quad \text{or} \quad x_i' = x_i'(x_1, x_2, x_3). \tag{42}$$

Equations (41) and (42) are used to obtain a relationship between the infinitesimal vectors $d\vec{x}'$ and $d\vec{x}$. Using the chain rule and Eq. (41), the dx_1 component of the vector $d\vec{x}$ can be written

$$dx_1 = \left(\frac{\partial x_1}{\partial x_1'}\right) dx_1' + \left(\frac{\partial x_1}{\partial x_2'}\right) dx_2' + \left(\frac{\partial x_1}{\partial x_3'}\right) dx_3'. \tag{43}$$

Similar expressions are obtained for dx_2 and dx_3. Using indicial notation[1], the components of vectors $d\vec{x}$ and $d\vec{x}'$ are related by:

$$dx_i = \left(\frac{\partial x_i}{\partial x'_j}\right) dx'_j \tag{44}$$

or:

$$dx_i = F_{ij} dx'_j \tag{45}$$

where F_{ij} is the deformation gradient tensor and is defined by [21]:

$$F_{ij} = \frac{\partial x_i}{\partial x'_j}. \tag{46}$$

The deformation gradient tensor can be considered as an operator that takes the infinitesimal vector $d\vec{x}'$ at the point $\vec{x}'$ and transforms it into the deformed vector $d\vec{x}$ located at the point $\vec{x}$. It contains all information on the deformation (stretch and rotation) imparted to $d\vec{x}'$ as a result of applied forces. It does not contain information on rigid-body displacements[2]. The deformation gradient can only be used to transform infinitesimal vectors; "longer" vectors could also be "bent" in the deformation process.

2.3 Strain

Consider a point in the reference state located at $\vec{x}'$. When forces are applied, the material in the vicinity of $\vec{x}'$ is stretched. The extent of the stretching depends upon the displacement direction in the material; two small vectors, $d\vec{x}'_1$ and $d\vec{x}'_2$, initially of equal length and located at $\vec{x}'$, can have different lengths after the deformation, if their initial orientations are different. The strain tensor is constructed from the deformation gradient tensor and is defined so as to provide information on how an infinitesimal vector $d\vec{x}'$ located at $\vec{x}'$ in the reference state is stretched during the deformation. In defining the strain, only the actual stretching of the material, and not rigid-body displacements, are of concern. In what follows, it will be simpler initially to consider the square of the length of the line elements rather than the length itself.

2.3.1 Lagrangian strain tensor

Let dS be the magnitude or length of the material vector $d\vec{x}'$ in the undeformed state. Then:

$$dS^2 = d\vec{x}' \cdot d\vec{x}' = dx'_j dx'_j = dx'_j \delta_{jk} dx'_k \tag{47}$$

where δ_{ij} is the Kronecker delta[3] and Eq. (35) was used in obtaining the last term. If ds is the length of line element $d\vec{x}$ corresponding to line element $d\vec{x}'$ in the deformed state, then:

$$ds^2 = d\vec{x} \cdot d\vec{x} = dx_i dx_i. \tag{48}$$

Using Eq. (45) to express $d\vec{x}$ in terms of $d\vec{x}'$, allows Eq. (48) to be written as:

$$ds^2 = F_{ij} dx'_j F_{ik} dx'_k. \tag{49}$$

[1]Repeated indices are summed from 1 to 3, as discussed in sec. 2.1.

[2]Rigid-body displacements or translations are spatial translations of the material body accomplished without imparting any deformation or rotation to the material.

[3]The Kronecker delta is defined such that $\delta_{ij} = 1$ if $i = j$, and $\delta_{ij} = 0$ if $i \neq j$. See Eq. (26).

The "stretch" that $d\vec{x}'$ undergoes is related to the difference between Eqs. (47) and (49):

$$ds^2 - dS^2 = dx'_j F_{ij} F_{ik} dx'_k - dx'_j \delta_{jk} dx'_k \tag{50}$$

or:

$$ds^2 - dS^2 = dx'_j \left(F_{ij} F_{ik} - \delta_{jk}\right) dx'_k. \tag{51}$$

Equation (51) can be rewritten as

$$ds^2 - dS^2 = 2dx'_j E_{jk} dx'_k \tag{52}$$

where the strain tensor, E_{jk}, has been defined as

$$E_{jk} = \frac{1}{2}\left(F_{ij} F_{ik} - \delta_{jk}\right). \tag{53}$$

When E_{ij} is formulated using the undeformed line elements ($d\vec{x}'$) in Eq. (52), it is defined as the Lagrangian strain tensor. Other formulations of the strain tensor are also possible. For example, if the term $ds^2 - dS^2$ were written in terms of the deformed configuration ($d\vec{x}$), the Eulerian strain tensor is obtained [21].

Using the definition for the deformation gradient tensor, Eq. (46), the strain tensor of Eq. (53) can be expressed as:

$$E_{jk} = \frac{1}{2}\left[\left(\frac{\partial x_i}{\partial x'_j}\right)\left(\frac{\partial x_i}{\partial x'_k}\right) - \delta_{jk}\right]. \tag{54}$$

Replacing the free indices j and k of Eq. (54) by the free indices i and j, and the dummy indice i by k yields the equivalent equation:

$$E_{ij} = \frac{1}{2}\left[\left(\frac{\partial x_k}{\partial x'_i}\right)\left(\frac{\partial x_k}{\partial x'_j}\right) - \delta_{ij}\right]. \tag{55}$$

The strain tensor can also be expressed in terms of the displacement vector. Substituting Eq. (40) into Eq. (46) and using Eq. (38) allows the deformation gradient tensor to be expressed as:

$$F_{ij} = \left(\frac{\partial x_i}{\partial x'_j}\right) = \left(\frac{\partial(x'_i + u_i)}{\partial x'_j}\right) = \left(\frac{\partial x'_i}{\partial x'_j}\right) + \left(\frac{\partial u_i}{\partial x'_j}\right) = \delta_{ij} + \left(\frac{\partial u_i}{\partial x'_j}\right). \tag{56}$$

Combining Eq. (56) with Eq. (55) after making the appropriate changes in the free indices gives

$$E_{ij} = \frac{1}{2}\left\{\left[\delta_{ki} + \left(\frac{\partial u_k}{\partial x'_i}\right)\right]\left[\delta_{kj} + \left(\frac{\partial u_k}{\partial x'_j}\right)\right] - \delta_{ij}\right\}. \tag{57}$$

Multiplying term by term and noting that $\delta_{ki}\delta_{kj} = \delta_{ij}$, gives for the strain

$$E_{ij} = \frac{1}{2}\left[\left(\frac{\partial u_i}{\partial x'_j}\right) + \left(\frac{\partial u_j}{\partial x'_i}\right) + \left(\frac{\partial u_k}{\partial x'_i}\right)\left(\frac{\partial u_k}{\partial x'_j}\right)\right]. \tag{58}$$

It is important to remember that Eq. (58), giving the Lagrangian formulation of the strain, is an exact expression and not a second-order approximation to the strain tensor.

2.3.2 Small strain tensor

For many problems in solid-state phase transformations, the derivatives of the displacement with respect to the material coordinates $\vec{x}'$ are small, and the term containing the product of the displacement gradients can be dropped from Eq. (58) since, in such cases, the product is very small with respect to the linear terms. In this limiting case of the small-strain approximation, the difference between the derivatives with respect to x_i' and x_i is negligible. The small strain tensor, ϵ_{ij}, is thus defined as:

$$\epsilon_{ij} = \frac{1}{2}\left[\left(\frac{\partial u_i}{\partial x_j}\right) + \left(\frac{\partial u_j}{\partial x_i}\right)\right] \approx \frac{1}{2}\left[\left(\frac{\partial u_i}{\partial x_j'}\right) + \left(\frac{\partial u_j}{\partial x_i'}\right)\right]. \tag{59}$$

Equation (59) indicates that the strain field can be uniquely determined from a known displacement field through differentiation. However, the inverse problem is not so simple: there may be many displacement fields that give the same strain field.

Physical meaning can be imparted to the small strain components by considering the actual change in length of the vector $d\vec{x}'$ during deformation. Using Eq. (49):

$$ds - dS = \left[dx_j' F_{ij} F_{ik} dx_k'\right]^{1/2} - dS. \tag{60}$$

If a unit vector, $\hat{n}'$, is defined such that it lies in the direction in which $d\vec{x}'$ points, then $d\vec{x}' = dS\hat{n}'$. Using Eq. (35), one has

$$\hat{n}' \cdot \hat{n}' = n_j' n_j' = n_j' \delta_{jk} n_k' = 1. \tag{61}$$

Equation (60) can be written, after adding and subtracting 1 to the term inside the square brackets and using Eq. (61):

$$ds - dS = dS\left[n_j' F_{ij} F_{ik} n_k' - n_j' \delta_{jk} n_k' + 1\right]^{1/2} - dS. \tag{62}$$

Introducing the strain tensor from Eq. (53) into Eq. (62) allows the change in length per-unit-length of line element, or unit extension, to be expressed as:

$$\frac{(ds - dS)}{dS} = \left[1 + 2n_j' E_{jk} n_k'\right]^{1/2} - 1. \tag{63}$$

Suppose that the unit extension of a line element, initially parallel to the x_1 axis, is to be determined. In this case, $\hat{n}' = \hat{e}_1$, where $\hat{e}_1$ is the unit basis vector in the x_1 direction. The components of the unit vector $\hat{n}'$ in this case are thus $n_1' = 1$ and $n_2' = n_3' = 0$. Matrix multiplication of $E_{ij} n_j'$ yields the column vector:

$$\begin{pmatrix} E_{11} & E_{12} & E_{31} \\ E_{21} & E_{22} & E_{23} \\ E_{31} & E_{32} & E_{33} \end{pmatrix} \begin{pmatrix} 1 \\ 0 \\ 0 \end{pmatrix} = \begin{pmatrix} E_{11} \\ E_{21} \\ E_{31} \end{pmatrix}. \tag{64}$$

Contracting the resulting vector of Eq. (64) with the normal n_i' (taking the dot product of the two vectors); $(E_{11}\hat{e}_1 + E_{21}\hat{e}_2 + E_{31}\hat{e}_3) \cdot (\hat{e}_1) = E_{11}$ and Eq. (63) becomes:

$$\frac{(ds - dS)}{dS} = [1 + 2E_{11}]^{1/2} - 1. \tag{65}$$

If the magnitude of E_{11} is sufficiently small, the unit extension, Eq. (65), can be expanded in a Taylor series about $E_{11} = 0$ to give:

$$\frac{(ds - dS)}{dS} \approx 1 + \frac{1}{2}(2E_{11}) - 1 = E_{11} = \epsilon_{11} \tag{66}$$

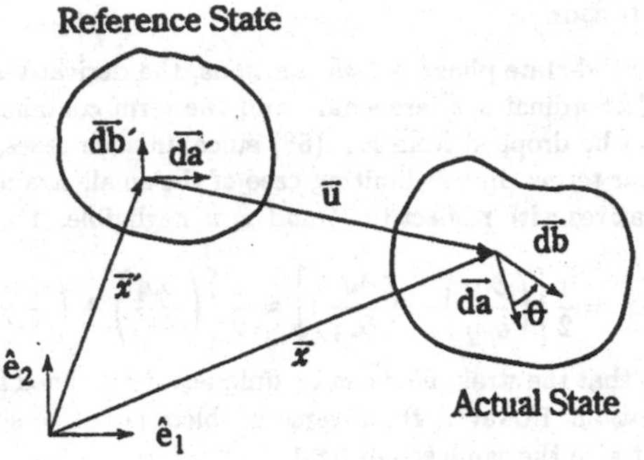

Figure 6: By examining the relative distortion of two line elements in the $x_1 - x_2$ plane, a physical interpretation of the shear strain can be obtained. Line elements $d\vec{a}'$ and $d\vec{b}'$ are initially perpendicular. After deformation, the angle between the two line elements is θ.

where the small-strain component ϵ_{11} has been substituted for its corresponding term in the Lagrangian formulation E_{11}. Since strains have been assumed to be small, the small-strain component ϵ_{11} can, therefore, be interpreted as the approximate extension, or change in length per-unit-length of a line element initially parallel to the x_1 axis. The same physical interpretation can be given to the other diagonal strain components; ϵ_{22} and ϵ_{33} give the change in length per unit-length of a line element initially parallel to the x_2 and x_3 axes, respectively.

In order to understand the physical meaning of the off-diagonal components of the strain tensor, the relative deformation of two line elements is considered. Figure 6 shows the $x_1 - x_2$ cross section of a material body in its reference (undeformed) and deformed states. The deformation is assumed to be confined to the $x_1 - x_2$ plane for illustration. Two line elements, $d\vec{a}'$ and $d\vec{b}'$, are each initially orthogonal and located at $\vec{x}'$. $d\vec{a}'$ and $d\vec{b}'$ are directed along the principal directions $\hat{e}_1$ and $\hat{e}_2$, respectively, such that

$$d\vec{a}' = dS_a\hat{e}_1 \quad \text{and} \quad d\vec{b}' = dS_b\hat{e}_2 \tag{67}$$

where dS_a and dS_b are the magnitudes of the undeformed line elements $d\vec{a}'$ and $d\vec{b}'$. $d\vec{a}$ and $d\vec{b}$ are the corresponding line elements in the deformed state. If the magnitudes of $d\vec{a}$ and $d\vec{b}$ are ds_a and ds_b, respectively, the angle between $d\vec{a}$ and $d\vec{b}$ in the deformed state, θ, is obtained from the dot product as:

$$d\vec{a} \cdot d\vec{b} = ds_a ds_b \cos\theta = da_k db_k. \tag{68}$$

Expressing the line elements in the deformed state with respect to the undeformed state using the deformation gradient tensor, Eq. (45), gives:

$$ds_a ds_b \cos\theta = F_{ki} da'_i F_{kj} db'_j. \tag{69}$$

Using Eq. (67) and noting that $F_{ki}da'_i = F_{k1}da'_1 = F_{k1}dS_a$, gives:

$$ds_a ds_b \cos\theta = dS_a F_{k1} F_{k2} dS_b. \tag{70}$$

The magnitudes ds_a and ds_b can be expressed in terms of dS_a and dS_b, respectively, using Eq. (63) (or Eq. (65)) as

$$ds_a = dS_a(1 + E_{11}) \quad \text{and} \quad ds_b = dS_b(1 + E_{22}). \tag{71}$$

Substituting for ds_a and ds_b and solving Eq. (70) for $\cos\theta$ gives

$$\cos\theta = \frac{2E_{12}}{[1 + 2E_{11}]^{1/2}\,[1 + 2E_{22}]^{1/2}} \tag{72}$$

where we have used the definition of the Kronecker delta funcition ($\delta_{12} = 0$) and $F_{k1}F_{k2} = 2E_{12}$ ($= 2E_{21}$) from the definition of strain, Eq. (53). Since the line elements $d\vec{a}'$ and $d\vec{b}'$ were initially orthogonal, the change in angle (ϕ) between $d\vec{a}'$ and $d\vec{b}'$ caused by the deformation is $\phi = \pi/2 - \theta$. Using the trigonometric identity, $\cos\theta = \cos(\pi/2 - \phi) = \sin\phi$. If the displacement gradient is assumed to be small, the angle ϕ is small and the small-strain approximation can be employed. Expanding $\sin\phi$ in a Taylor series about $\phi = 0$ and the right-hand-side of Eq. (72) in a Taylor series in the strain components about $E_{ij} = 0$, one obtains, to first-order:

$$\cos\theta = \sin\phi \approx \phi \approx 2E_{12} = 2\epsilon_{12} = 2\epsilon_{21}. \tag{73}$$

Thus the off-diagonal terms of the small strain tensor are symmetric and equal to one-half the change in angle owing to the deformation; i.e., $\epsilon_{12} = \epsilon_{21} = \phi/2$.

Either the deformation tensor, F_{ij}, or the strain tensor, E_{ij}, can be used to describe the deformation of the material. The small strain tensor, ϵ_{ij}, is a valid description when the displacement gradients are small. This is a reasonable assumption for many problems in solid-state phase transformations but is certainly not always valid, as will be seen.

2.3.3 Volume change - dilatation

The ratio of the volume of an element in the deformed state to the volume of the corresponding element in the reference state can be deduced by considering a simple deformation. Assume a rectangular volume element in the reference state with sides parallel to the reference axes and of length dS_1, dS_2 and dS_3, respectively. The unit extension of the side originally parallel to the x_1 axis is given by Eq. (65), $ds_1 = (1+2E_{11})^{1/2}dS_1$; similar expressions obtain for the other two sides. If a simple deformation is imagined in which the volume element remains a cuboid, then the volume of the element in the deformed state, dV, is:

$$dV = ds_1 ds_2 ds_3 = [(1+2E_{11})(1+2E_{22})(1+2E_{33})]^{1/2}\, dS_1 dS_2 dS_3. \tag{74}$$

Defining J as the ratio of the volumes of the element in the deformed and reference states gives:

$$\frac{dV}{dV'} = \frac{ds_1 ds_2 ds_3}{dS_1 dS_2 dS_3} = J = [(1+2E_{11})(1+2E_{22})(1+2E_{33})]^{1/2}. \tag{75}$$

If Eq. (75) is expanded in a Taylor series in the strain components about $E_{ij} = 0$, and only the leading term in the strain is retained, the small-strain approximation is obtained:

$$\frac{dV}{dV'} = J \approx 1 + E_{11} + E_{22} + E_{33} = 1 + \epsilon_{kk} \tag{76}$$

where $\epsilon_{kk} = \epsilon_{11} + \epsilon_{22} + \epsilon_{33}$ is the trace of the strain tensor. In the small-strain approximation, the trace of the strain tensor gives the change in volume per-unit-volume owing to the deformation:

$$\frac{dV - dV'}{dV'} \approx \epsilon_{kk}. \tag{77}$$

A more formal procedure considers the unit extension in a coordinate system taken parallel to the principal axes of the strain tensor, E_{ij}. The resulting ratio of the volume elements is shown to be [21]

$$\frac{dV}{dV'} = J = \det F_{ij} \tag{78}$$

where J is the determinant of the deformation gradient tensor.

As shown by Euler in 1762, if $f(\vec{x})$ is a continuous scalar function of position, then the integral of f over the deformed state can be expressed as an integral over the reference state ($\vec{x}'$) coordinates by

$$\int_V f(\vec{x})dx_1dx_2dx_3 = \int_{V'} f(\vec{x}(\vec{x}'))Jdx_1'dx_2'dx_3' = \int_{V'} f(\vec{x}')dx_1'dx_2'dx_3'. \tag{79}$$

For example, the mass of a region V of material, $\mathcal{M}$, can be calculated from volume integrals of the density ρ in either the deformed or reference states:

$$\mathcal{M} = \int_V \rho(\vec{x})dx_1dx_2dx_3 = \int_{V'} \rho(\vec{x}(\vec{x}'))Jdx_1'dx_2'dx_3' = \int_{V'} \rho(\vec{x}')dx_1'dx_2'dx_3'. \tag{80}$$

A relationship between the elements of area in the reference and deformed states can also be obtained. If dS_o is the area of a surface element in the reference state with outward pointing unit normal $\hat{n}'$, its corresponding area in the deformed state dS with outward pointing unit normal $\hat{n}$ is [21]:

$$\hat{n}_j dS = J\hat{n}_i' F_{ij}^{-1} dS_o \tag{81}$$

where F_{ij}^{-1} is the inverse of the deformation gradient tensor. Equation (81) is known as Nanson's formula [21].

2.4 Stress

In this subsection, the second-rank stress tensor, which describes the forces acting on an element of material, is developed.

Consider a small, cubical volume element whose center is situated at the point $\vec{x}$ in the deformed state of the material body. The faces of the cube are taken to be parallel to the coordinate axes as shown in Fig. 7. In general, two types of forces can be imagined to act on the volume element [22]. The first type arises from an interaction of the volume element with a force at a distance. Examples of this type of interaction include forces resulting from a gravitational or electromagnetic field. The second type of force is the mechanical force exerted by an adjacent block of material on the original cube element. This force acts across the cube surface. It does not necessarily act perpendicular to the surface, but has components lying in each of the three principal directions. If the area of the cube face is sufficiently small, then the traction or stress vector can be defined as the force acting on the surface per unit area of surface. The traction is a vector and has dimensions of force per unit area or energy per unit volume. The tractions acting on those cube surfaces with outward pointing normals lying along the positive axes are designated in Fig. 7 as $\vec{t}_1$, $\vec{t}_2$ and $\vec{t}_3$. The components of the tractions projected in the directions of the principal axes are designated by σ_{ij}, where i denotes the face on which the traction acts and j denotes the direction in which the force component is acting. The force components σ_{ij} form the elements of the stress tensor.

By considering an element of the material body in static equilibrium, a general relationship between the components of the stress tensor and the traction acting on a surface with outward pointing unit normal $\hat{n}$ different from one of the principal directions can be obtained. A small material element in the shape of a tetrahedron is shown in Fig. 8. The traction $\vec{t}$ acts on the surface with normal $\hat{n}$ and area ds. If the volume element is in static equilibrium, the sum of the forces acting on the element must sum to zero. Using the forces acting in the x_2 direction as an

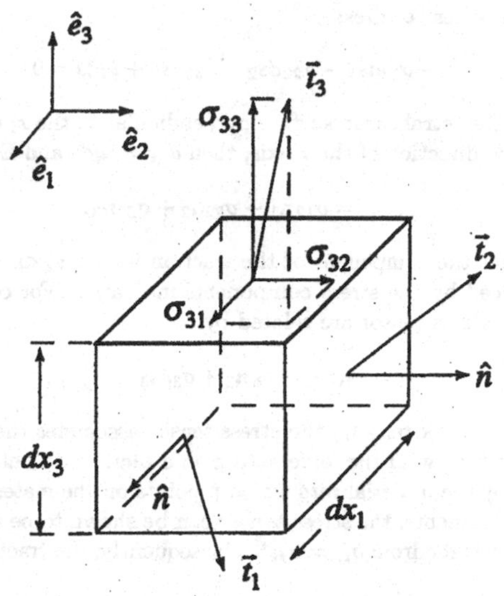

Figure 7: The volume element located at $\vec{x}$ and of dimensions dx_1, dx_2 and dx_3, is subjected to two types of forces: one arising from a force at-a-distance, $\vec{b}$, and the other a mechanical force exerted on the volume element by adjacent material. This latter force is designated by the traction vector $\vec{t}$. The traction vector projected in each of the principal directions gives the components of the stress tensor.

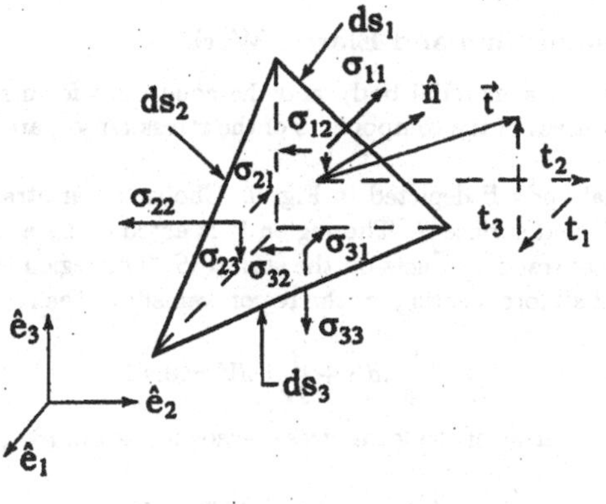

Figure 8: The relationship between the stress tensor at a point (σ_{ij}) and the traction ($\vec{t}$) acting on a surface element with normal $\hat{n}$. The vector components of $\vec{t}$ are t_1, t_2 and t_3.

example, the balance of forces requires:

$$-\sigma_{12}ds_1 - \sigma_{22}ds_2 - \sigma_{32}ds_3 + t_2 ds = 0 \quad (82)$$

where ds_i is the area of the tetrahedron surface perpendicular to the x_i axis. If n_i is the component of the unit normal in the direction of the i axis, then $ds_i = n_i ds$ and Eq. (82) becomes:

$$t_2 = \sigma_{12}n_1 + \sigma_{22}n_2 + \sigma_{32}n_3. \quad (83)$$

Equation (83) states that the component of the traction in the x_2 direction acting on the surface with normal $\hat{n}$ is balanced by the stress components indicated. For component i of the traction vector, the traction and stress tensor are related by:

$$t_i = \sigma_{1i}n_1 + \sigma_{2i}n_2 + \sigma_{3i}n_3 = n_j\sigma_{ji}. \quad (84)$$

In the language of second-rank tensors, the stress tensor associates the normal vector (argument) with the traction vector; i.e., with the force acting on a plane with outward pointing normal $\hat{n}$. It gives the force exerted by the material *into* which $\hat{n}$ points on the material *from* which $\hat{n}$ points. In the absence of external moments, the stress tensor can be shown to be symmetric by balancing the moments acting on the tetrahedron; $\sigma_{ij} = \sigma_{ji}$[4]. Consequently, the traction could also be computed from $t_i = \sigma_{ij}n_j$.

The above analysis was performed by considering the material in its deformed state and the stress thereby defined is termed the Cauchy stress tensor. The Cauchy stress tensor gives the actual force per unit area of deformed material. Other definitions of the stress are possible. For example, the actual force could be referred to the corresponding unit area in the reference state (the first Piola-Kirchhoff stress tensor) or the actual force could be rotated along with the material back to the reference state and expressed with respect to a unit area of the reference state (second Piola-Kirchhoff stress tensor [21]). We will use the first Piola-Kirchhoff stress tensor when treating the thermodynamics of stressed crystals.

2.5 Mechanical Equilibrium and Elastic Work

The elastic energy stored in a material body and the conditions for mechanical equilibrium, expressed in terms of derivatives of the components of the stress tensor, are developed in this subsection.

Consider the material body $\mathcal{B}$ depicted in Fig. 9. Choose an arbitrary region of the material defined by the volume V and surface S. The region V is acted on by a body force $\vec{b}(\vec{x})$ (force per unit volume) while a set of tractions $\vec{t}$ acts on the surface S. The region is said to be in mechanical equilibrium if the sum of all forces acting on the region vanishes. The mathematical expression for the balance of forces is:

$$\int_S t_i dS + \int_V b_i dV = 0. \quad (85)$$

Expressing the traction in terms of the local stress tensor using Eq. (84), Eq. (85) becomes:

$$\int_S \sigma_{ji}n_j dS + \int_V b_i dV = 0 \quad (86)$$

[4]If, for a given volume element, $\sigma_{12} \neq \sigma_{21}$, the volume element would tend to spin about the x_3 axis.

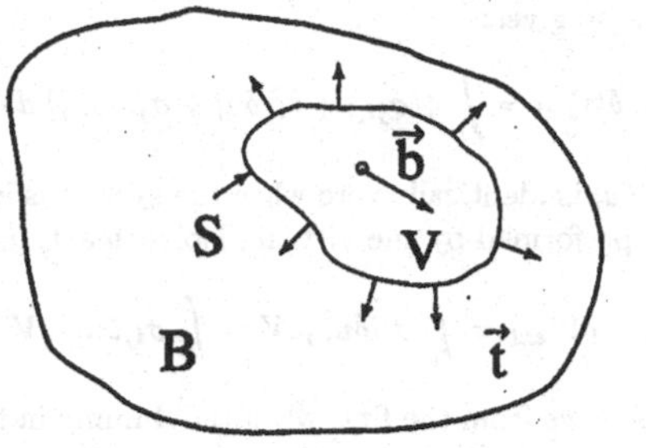

Figure 9: An arbitrary volume V with surface S is selected within the material $\mathcal{B}$ in order to derive the conditions for mechanical equilibrium. $\vec{b}$ is a body force and $\vec{t}$ is the traction acting on surface S.

where n_j is the outward-pointing unit normal to the region V. Applying the divergence theorem[5]:

$$\int_S \sigma_{ji} n_j dS = \int_V \frac{\partial \sigma_{ji}}{\partial x_j} dV, \tag{87}$$

to the surface integral of Eq. (86) yields:

$$\int_V \left\{ \frac{\partial \sigma_{ji}}{\partial x_j} + b_i \right\} dV = 0. \tag{88}$$

Since the volume V is arbitrary, Eq. (88) is identically satisfied only when the integrand vanishes. Mechanical equilibrium is therefore obtained when

$$\frac{\partial \sigma_{ji}}{\partial x_j} + b_i = \sigma_{ji,j} + b_i = 0. \tag{89}$$

where a comma appearing in the subscript implies differentiation[6]. Equation (89) is a partial differential equation which must be satisfied at all points in the material if the material is in mechanical equilibrium.

Assume the material body shown in Fig. 9 to be in mechanical equilibrium. If each volume and surface element is imagined to undergo a small, or virtual, displacement[7], $\delta\vec{u}(\vec{x})$, then the corresponding virtual work (force multiplied by displacement), δW_{ext}, performed by the surface traction and body forces is:

$$\delta W_{ext} = \int_S t_i \delta u_i dS + \int_V b_i \delta u_i dV. \tag{90}$$

The virtual displacement field is arbitrary, as long as the structural integrity of the material is maintained (i.e., no gaps or material overlap occur.) Expressing once again the traction vector in terms of the local stress tensor (using Eq. (84)) and invoking the divergence theorem, Eq. (87), gives:

$$\delta W_{ext} = \int_V \left\{ \frac{\partial (\sigma_{ji} \delta u_i)}{\partial x_j} + b_i \delta u_i \right\} dV. \tag{91}$$

[5]This equation is an extension of the more familiar form of the divergence theorem applicable to vectors: $\int_V \vec{\nabla} \cdot \vec{u} dV = \int_S \vec{u} \cdot \hat{n} dS$ which, in indicial notation, becomes $\int_V \frac{\partial u_i}{\partial x_i} dV = \int_S u_i n_i dS$.

[6]The summation convention still applies: $\sigma_{ji,j} = \sigma_{1i,1} + \sigma_{2i,2} + \sigma_{3i,3}$.

[7]The use of $\delta\vec{u}$ represents an imaginary displacement performed in such a way that the forces acting on the body are not changed.

Differentiating and rearranging gives:

$$\delta W_{ext} = \int_V \{(\sigma_{ji,j} + b_i)\,\delta u_i + \sigma_{ji}\delta u_{i,j}\}\,dV. \tag{92}$$

Since the term multiplying δu_i is identically zero when the system is in mechanical equilibrium (see Eq. (89)), the virtual work performed by the virtual displacement, δu_i, is:

$$\delta W_{ext} = \int_V \sigma_{ji}\delta u_{i,j} dV = \int_V \sigma_{ij}\delta u_{j,i} dV \tag{93}$$

where the second equality follows from the first when the dummy indices are exchanged. Since the stress tensor σ_{ij} is symmetric in i and j, $\sigma_{ij} = \sigma_{ji}$, and Eq. (93) yields:

$$\delta W_{ext} = \int_V \frac{1}{2}\sigma_{ij}\left(\delta u_{i,j} + \delta u_{j,i}\right) dV \tag{94}$$

or, from Eq. (59),

$$\delta W_{ext} = \int_V \sigma_{ij}\delta\epsilon_{ij} dV. \tag{95}$$

where $\delta\epsilon_{ij}$ is the virtual strain induced by the virtual displacement.

Equation (95) indicates that there is a change in the elastic strain energy of each volume element that is produced by the small displacement δu_i. The change in strain energy of a volume element, δe_v, is given by:

$$\delta e_v = \sigma_{ij}\delta\epsilon_{ij} \tag{96}$$

If the virtual strain is considered infinitesimal, Eq. (96) can be written in terms of derivatives as:

$$de_v = \sigma_{ij} d\epsilon_{ij}. \tag{97}$$

Equation (97) connects a small change in the state of strain with a change in the elastic energy of the volume element. The total elastic energy of the volume element can thus be obtained by integrating Eq. (97) from the unstrained state (the state of zero elastic energy) to the given state of strain. Consequently:

$$\int_0^{e_v} de_v = e_v = \int_0^{\epsilon_{ij}} \sigma_{ij} d\epsilon_{ij}. \tag{98}$$

e_v is the elastic energy associated with an infinitesimal volume element and is, therefore, called the elastic strain energy density. (e_v has units of energy per volume.) Integrating the elastic strain energy density over the entire volume of material (i.e., over all volume elements) gives the elastic strain energy of the material.

Integration of Eq. (98) in order to obtain the elastic energy stored in a unit volume of crystal requires knowledge of how the stress depends on the state of strain. Such constitutive laws are examined in the following section.

2.6 Constitutive Equations

The displacement gradient tensor and strain tensor are used to describe the material deformation. The stress tensor is used to describe the forces acting at a point in the material. In order to relate the material deformation to the applied force, however, a constitutive equation for the material must be postulated. In this subsection, several constitutive laws are formulated which relate deformation arising from temperature variations, compositional inhomogeneity and phase transformations to an elastic stress.

We first consider a solid at fixed composition and temperature. If an appropriate distribution of forces is applied to the material, it deforms. If the forces are removed and the material returns promptly to its original state, the material is called ideally elastic. If the deformation is sufficiently small and the reference state for the measurement of strain is taken as the stress-free state of the material, the relationship between the stress and strain components can be treated as linear. In principle, the application of any one stress component could give rise to any one of the nine strain components. This would imply that there could be up to 81 (9 x 9) elastic constants connecting all possible stress and strain components. The simplest way of expressing this relationship using tensor notation (instead of writing nine separate equations) is:

$$\sigma_{ij} = C_{ijkl}\epsilon_{kl} \tag{99}$$

where C_{ijkl} is the elastic stiffness tensor (elastic constants). C_{ijkl} is a fourth-rank tensor that connects the second-rank strain tensor with the second-rank stress tensor. Each component of the stiffness tensor can be viewed as an elastic constant connecting a component of the strain tensor with a component of the stress tensor. The strain tensor can be expressed in terms of the stress components in like manner:

$$\epsilon_{mn} = S_{mnij}\sigma_{ij} \tag{100}$$

where S_{mnij} is the elastic compliance tensor and is the inverse of the elastic constants tensor. The relationship between the tensors S_{ijkl} and C_{ijkl} is obtained by first contracting Eq. (99) with S_{mnij} to give:

$$S_{mnij}\sigma_{ij} = S_{mnij}C_{ijkl}\epsilon_{kl}. \tag{101}$$

Recognizing that Eq. (100) can be rewritten as:

$$\epsilon_{mn} = \delta_{km}\delta_{ln}\epsilon_{kl} = S_{mnij}\sigma_{ij}, \tag{102}$$

comparison of Eq. (101) and Eq. (102) requires that:

$$S_{mnij}C_{ijkl} = \delta_{km}\delta_{ln}. \tag{103}$$

This is the relationship that shows that S_{ijkl} and C_{ijkl} are inverse tensors. Note that Eq. (103) represents 81 independent equations. It is sufficient that, given the 81 elastic constants, the 81 elastic compliances could be determined.

In a crystal, the 81 individual elastic constants are neither unique nor independent. The symmetry of the elastic strain ($\epsilon_{ij} = \epsilon_{ji}$) and stress ($\sigma_{ij} = \sigma_{ji}$) tensors requires that $C_{ijkl} = C_{ijlk}$ and $C_{ijkl} = C_{jikl}$ so that there are, at most, 36 independent elastic constants. The rotational and translational symmetries of the crystal impose additional constraints on the number of independent elastic constants [23] requiring some of the elastic constants to be identically zero while requiring other components to be equal. The precise form of the elastic constants depends on the crystal symmetry. The elastic constants are a function of the temperature and composition.

Since the stress and strain tensors are symmetric, Eq. (99) represents six independent equations connecting 12 quantities; six stress components and six strain components. Consequently, if the state of strain is given, the six stress components can be determined by solving the six independent

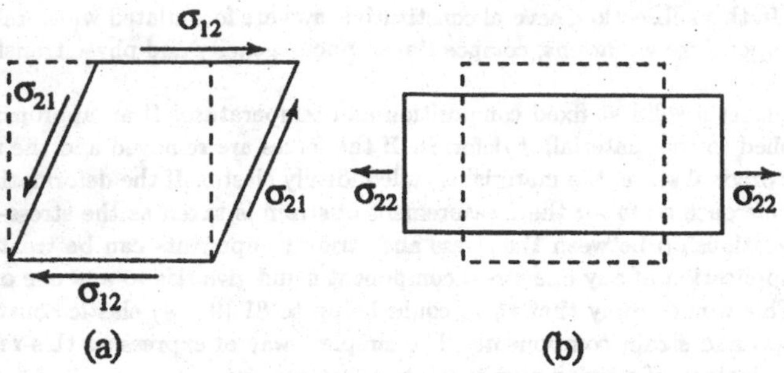

Figure 10: Application of a shear stress $\sigma_{12} = \sigma_{21}$ to an isotropic volume element in (a) induces only a shear strain $\epsilon_{12} = \epsilon_{21}$. In (b) a uniaxial stress, σ_{22}, is applied which does not couple to the shear strains but does induce strains in all three prinicpal directions. Dashed lines depict the original shape of the volume element.

equations of Eq. (99) simultaneously. Likewise, should three strain components and three stress components be given, the remaining stress and strain components could be determined. Of course, Eq. (99) pertains to each volume element or to a material with a homogeneous state of strain.

For an isotropic material, there are only two independent elastic constants: one corresponding to the material's resistance to the application of a shear stress (the shear modulus, μ) and the other to the material's resistance to the application of a uniaxial stress (the elastic modulus or Young's modulus, E). Using these elastic constants, the elastic stiffness tensor for an isotropic material can be written using tensor notation and the Kronecker delta as [23]:

$$C_{ijkl} = \frac{\mu(E - 2\mu)}{(3\mu - E)}\delta_{ij}\delta_{kl} + \mu\left(\delta_{ik}\delta_{jl} + \delta_{il}\delta_{jk}\right). \tag{104}$$

This relationship is often expressed in terms of Lame's constants as:

$$C_{ijkl} = \lambda\delta_{ij}\delta_{kl} + \mu\left(\delta_{ik}\delta_{jl} + \delta_{il}\delta_{jk}\right) \tag{105}$$

where $\lambda = \mu(E - 2\mu)/(3\mu - E)$. This relationship between λ and E will be derived below.

The physical meaning of the shear modulus becomes clear by considering a shear stress imparted to a cubical volume element as shown in Fig. 10. The five other stress components; the two shear stresses ($\sigma_{13} = \sigma_{31}$ and $\sigma_{23} = \sigma_{32}$) and the axial stresses (σ_{11}, σ_{22}, and σ_{33}) are taken to be zero. The strain resulting from application of the shear stress $\sigma_{12} = \sigma_{21}$ is obtained by substituting Eq. (105) into Eq. (99) and setting the free indices $i = 1$ and $j = 2$:

$$\sigma_{12} = C_{12kl}\epsilon_{kl} = \left[\lambda\delta_{12}\delta_{kl} + \mu\left(\delta_{1k}\delta_{2l} + \delta_{1l}\delta_{2k}\right)\right]\epsilon_{kl}. \tag{106}$$

Since $\delta_{12} = 0$ and $\delta_{1k}\delta_{2l}\epsilon_{kl} = \epsilon_{12}$, one obtains:

$$\sigma_{12} = \mu(\epsilon_{12} + \epsilon_{21}) = 2\mu\epsilon_{12} \tag{107}$$

and the resulting shear strain is:

$$\epsilon_{12} = \sigma_{12}/2\mu. \tag{108}$$

The shear stress σ_{12} only induces a shear strain ϵ_{12} in an isotropic material and does not couple to the axial strains or the other shear strains.

In order to understand the connection between Lame's constant (λ) and the elastic modulus (E), consider a uniaxial stress applied in the x_2 direction as shown in Fig. 10b. All shear stresses and the two other axial stresses, σ_{11} and σ_{33}, are taken to be zero. Using Eq. (105), σ_{22} is obtained from Eq. (99) by setting $i = j = 2$:

$$\sigma_{22} = C_{22kl}\epsilon_{kl} = [\lambda\delta_{22}\delta_{kl} + \mu(\delta_{2k}\delta_{2l} + \delta_{2l}\delta_{2k})]\epsilon_{kl}. \tag{109}$$

Simplification of Eq. (109) yields:

$$\sigma_{22} = \lambda(\epsilon_{11} + \epsilon_{22} + \epsilon_{33}) + 2\mu\epsilon_{22}. \tag{110}$$

In order to obtain the relationship between ϵ_{22} and σ_{22}, expressions for ϵ_{11} and ϵ_{33} are needed. These are obtained by first setting $i = j = 1$ in Eq. (99) and using the condition $\sigma_{11} = 0$ to give:

$$\sigma_{11} = 0 = C_{11kl}\epsilon_{kl} = [\lambda\delta_{11}\delta_{kl} + \mu(\delta_{1k}\delta_{1l} + \delta_{1l}\delta_{1k})]\epsilon_{kl} \tag{111}$$

which simplifies to:

$$0 = \lambda(\epsilon_{11} + \epsilon_{22} + \epsilon_{33}) + 2\mu\epsilon_{11}. \tag{112}$$

Likewise, setting $i = j = 3$ in Eq. (99) and using the condition $\sigma_{33} = 0$ leads to:

$$\sigma_{33} = 0 = C_{33kl}\epsilon_{kl} = [\lambda\delta_{11}\delta_{kl} + \mu(\delta_{3k}\delta_{3l} + \delta_{3l}\delta_{3k})]\epsilon_{kl} \tag{113}$$

or

$$0 = \lambda(\epsilon_{11} + \epsilon_{22} + \epsilon_{33}) + 2\mu\epsilon_{33}. \tag{114}$$

Solving Eqs. (110), (112) and (114) simultaneously for the strains gives:

$$\epsilon_{22} = \frac{(\lambda + \mu)\sigma_{22}}{\mu(3\lambda + 2\mu)} \tag{115}$$

and

$$\epsilon_{11} = \epsilon_{33} = -\frac{\lambda\sigma_{22}}{2\mu(3\lambda + 2\mu)}. \tag{116}$$

Equation (115) provides a direct relationship between the axial stress and strain components when a uniaxial stress is applied to the material. Since the definition of the elastic modulus was given as $\sigma_{22} = E\epsilon_{22}$, we see that $E = \mu(3\lambda + 2\mu)/(\lambda + \mu)$ which, when solved for λ gives the expression below Eq. (105).

Poisson's ratio, ν, is defined as the ratio of the contraction in a direction perpendicular to the extension induced by a uniaxial stress; i.e., $\nu = -\epsilon_{11}/\epsilon_{22}$. From Eq. (115) and Eq. (116), one obtains $\nu = \lambda/2(\lambda + \mu)$. Any combination of two elastic constants can be used to describe the elastic state of the material. For the isotropic system in general:

$$\sigma_{ij} = \lambda\epsilon_{kk}\delta_{ij} + 2\mu\epsilon_{ij}. \tag{117}$$

The elastic compliance tensor is written:

$$S_{ijkl} = \frac{-\nu}{E}\delta_{ij}\delta_{kl} + \frac{(1+\nu)}{2E}(\delta_{ik}\delta_{jl} + \delta_{il}\delta_{jk}) \tag{118}$$

which, when combined with Eq. (102), yields:

$$\epsilon_{ij} = \frac{(1+\nu)\sigma_{ij}}{E} - \frac{\nu}{E}\epsilon_{kk}\delta_{ij}. \tag{119}$$

Using the constitutive law connecting the stress and strain fields, it is now possible to calculate the elastic energy stored in each volume element by integrating the strain energy density of Eq. (98). Consider once again the case of the applied shear stress, σ_{12} in Fig. 10a. Since all other stress components vanish, the differential of the strain energy density, Eq. (97), becomes:

$$de_v = \sigma_{ij} d\epsilon_{ij} = \sigma_{12} d\epsilon_{12} + \sigma_{21} d\epsilon_{21} = 2\sigma_{12} d\epsilon_{12}. \tag{120}$$

Using Eq. (108), the strain energy density can be integrated to yield:

$$e_v = \int_0^{\epsilon_{12}} 2\sigma_{12} d\epsilon_{12} = \int_0^{\epsilon_{12}} 4\mu\epsilon_{12} d\epsilon_{12} = 2\mu\epsilon_{12}^2. \tag{121}$$

Similarly, the strain energy density for the case of uniaxial loading in Fig. 10b is obtained as:

$$de_v = \sigma_{ij} d\epsilon_{ij} = \sigma_{22} d\epsilon_{22} = E\epsilon_{22} d\epsilon_{22} \tag{122}$$

where the relationship $\sigma_{22} = E\epsilon_{22}$ was used. Integrating Eq. (122) gives:

$$e_v = \int_0^{\epsilon_{22}} E\epsilon_{22} d\epsilon_{22} = \frac{1}{2} E\epsilon_{22}^2. \tag{123}$$

In each case, the strain energy density depends upon the square of the strain. An expression for the strain energy density of an isotropic material under a general state of strain is obtained by substituting Eq. (117) into Eq. (98) and integrating:

$$e_v = \int_0^{\epsilon_{ij}} \sigma_{ij} d\epsilon_{ij} = \int_0^{\epsilon_{ij}} (\lambda\epsilon_{kk}\delta_{ij} + 2\mu\epsilon_{ij})\, d\epsilon_{ij}. \tag{124}$$

Breaking the integral into two parts and contracting gives:

$$e_v = \int_0^{\epsilon_{kk}} \lambda\epsilon_{kk} d\epsilon_{kk} + \int_0^{\epsilon_{ij}} 2\mu\epsilon_{ij} d\epsilon_{ij}. \tag{125}$$

Integrating each term yields:

$$e_v = \frac{1}{2}\lambda\epsilon_{kk}^2 + \mu\epsilon_{ij}\epsilon_{ij}. \tag{126}$$

The strain energy density can also be obtained for the general case of an anisotropic crystal. Using

$$de_v = C_{ijkl}\epsilon_{kl} d\epsilon_{ij} \tag{127}$$

and recognizing that the components of the elastic constants tensor are constants, one has

$$\int_0^{e_v} de_v = e_v = \int_0^{\epsilon_{ij}} C_{ijkl}\epsilon_{kl} d\epsilon_{ij} = \frac{1}{2} C_{ijkl}\epsilon_{ij}\epsilon_{kl} = \frac{1}{2}\sigma_{ij}\epsilon_{ij} \tag{128}$$

where the strain is measured with respect to the unstressed state of the phase.

For a cubic material, there are three independent elastic constants and

$$C_{ijkl} = C_{12}\delta_{ij}\delta_{kl} + C_{44}\left(\delta_{ik}\delta_{jl} + \delta_{il}\delta_{jk}\right) + (C_{11} - C_{12} - 2C_{44})\delta_{ijkl} \tag{129}$$

where the Voigt notation [19] has been used and $\delta_{ijkl} = 1$, if $i = j = k = l$, and is zero otherwise. When $C_{11} = C_{12} + 2C_{44}$, an isotropic system is recovered with the recognition that $C_{12} = \lambda$ and $C_{44} = \mu$.

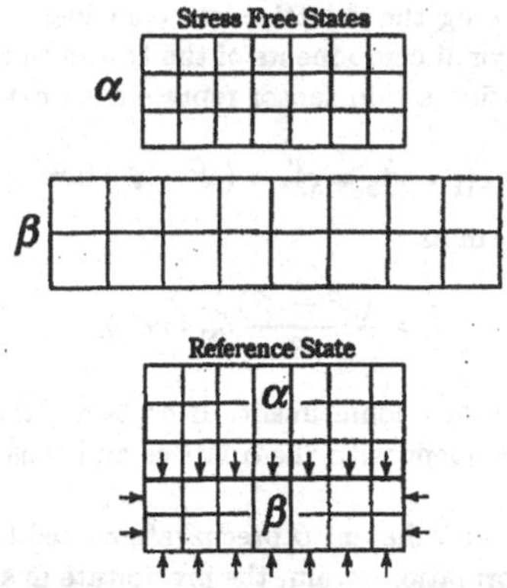

Figure 11: In their unstressed states, the α and β phases, here assumed to be rectangular plates, possess different lattice parameters. If unstressed α is chosen as the reference state for deformation, forces must be applied to the β phase to bring the two lattices into coincidence, as shown in the coherent structure of (b). (Arrows represent constraining forces acting on β.)

2.6.1 Misfit strain

It is sometimes inconvenient to define the reference state for the measurement of strain as the stress-free state of the phase, especially when more than one phase is present in the system. Such cases include the precipitation of a coherent second-phase and phase transformations in epitaxial crystalline films where the unstressed state of one of the phases would be chosen as the reference state. Figure 11 shows the respective stress-free states of two planar phases possessing different lattice constants. If the stress-free state of the α phase is chosen as the reference state for the measurement of deformation, then the stress-free β phase must be deformed to bring it into coincidence with the reference state. If the phases are coherent, with a continuity of lattice planes between the phases, the transformation of the β phase into the reference state can be imagined as the deformation required to make the lattice of the β phase commensurate with the lattice of the α phase. If this transformation strain between the α and β phases is designated ϵ_{ij}^T, as measured with respect to the stress-free α phase, then the stress which the β phase must experience when its lattice is coincident with that of the α phase would be [24, 25, 26]:

$$\sigma_{ij}^{\beta} = -C_{ijkl}^{\beta}\epsilon_{kl}^{T} \tag{130}$$

where C_{ijkl}^{β} are the elastic constants of the β phase. By definition, the strain in the reference state of the β phase is zero. However, since the reference state of β is not the stress-free state of β, stresses are present in the reference state of β. If the transformation strain ϵ_{ij}^T is small, the stress due to the transformation strain and any stress resulting from the deformation of the material from its reference state configuration (quantified by ϵ_{ij}) can be added so that the stress-strain constitutive equation for the β phase becomes:

$$\sigma_{ij}^{\beta} = C_{ijkl}^{\beta}\left(\epsilon_{kl} - \epsilon_{kl}^{T}\right). \tag{131}$$

Once again, the minus sign is necessary because the strain state of the β phase is referred to the stress-free α phase.

Equation (131) can be understood physically from the following simple example. Assume that both the α and β phases are face-centered cubic in structure with lattice parameters a^{α} and a^{β}, respectively, and that stress-free α is to be used as the reference state for the measurement of strain.

The transformation required to bring the β lattice into coincidence with the α lattice requires no shear deformation, so the off-diagonal components of the transformation-strain tensor vanish. The diagonal terms of the transformation-strain tensor represent an extension of the lattice and, from Eq. (66), we have:

$$\epsilon_{11}^T = \epsilon_{22}^T = \epsilon_{33}^T = (a^\beta - a^\alpha)/a^\alpha \tag{132}$$

which can be written in tensor form as

$$\epsilon_{ij}^T = \frac{(a^\beta - a^\alpha)}{a^\alpha}\delta_{ij} = \epsilon^T \delta_{ij} \tag{133}$$

where $\epsilon^T = (a^\beta - a^\alpha)/a^\alpha$ is the dilatational misfit. If $a^\beta > a^\alpha$, then compressive forces must be applied to make the β lattice coincident with the α lattice and thus the origin of the minus sign in Eq. (130).

Equation (131) is applicable to a coherent β precipitate embedded in an α matrix. In this case, ϵ_{ij}^T is called the stress-free transformation strain, the precipitate misfit, misfit strain or eigenstrain. The concept of transformation strain is not restricted to two phases possessing a common crystal structure. For example, the stress field of a precipitate with a tetragonal or monoclinic crystal structure embedded in a cubic matrix can be calculated in this manner.

2.6.2 Thermal strain

The above thought process can be used to establish a stress-strain constitutive equation for a phase in which the temperature or composition are allowed to vary as a function of position in the crystal. The reference state for measurement of deformation assumes a certain temperature and composition. If an element of the material possesses a temperature different from that of the reference state, thermal stresses arise owing to the coefficient of thermal expansion. If a cubic material is assumed, an increase in temperature induces an expansion of the lattice parameter, a. Expanding the lattice parameter in a Taylor series in the temperature about the lattice parameter found in the reference state, the lattice parameter at the absolute temperature θ is approximated as:

$$a(\theta) = a(\theta_r) + \left(\frac{\partial a}{\partial \theta}\right)_{\theta=\theta_r}(\theta - \theta_r) + \frac{1}{2}\left(\frac{\partial^2 a}{\partial \theta^2}\right)_{\theta=\theta_r}(\theta - \theta_r)^2 + \cdots \tag{134}$$

where θ_r is the temperature of the reference state. The coefficient of thermal expansion α_t is defined:

$$\alpha_t = \frac{1}{a(\theta_r)}\left(\frac{\partial a}{\partial \theta}\right). \tag{135}$$

The relative change in lattice parameter induced by the temperature difference is found to first-order (using Eq. (66)) as:

$$\frac{[a(\theta) - a(\theta_r)]}{a(\theta_r)} = \alpha_t(\theta - \theta_r). \tag{136}$$

For a cubic material, no shear strains are introduced by the temperature change so the thermal strain can be written in tensor form as:

$$\frac{[a(\theta) - a(\theta_r)]}{a(\theta_r)}\delta_{ij} = \alpha_t(\theta - \theta_r)\delta_{ij}. \tag{137}$$

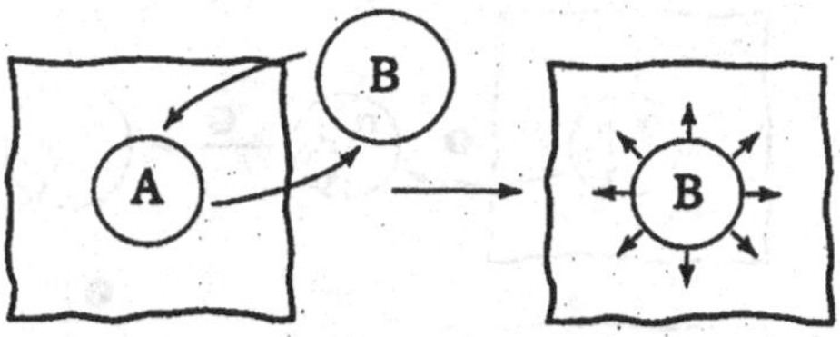

Figure 12: **An exchange of atoms possessing different partial molar volumes will induce a deformation in the surrounding crystal. This deformation is depicted schematically when a B atom replaces a smaller A atom distorting the surrounding lattice planes. The arrows depict the displacement of the surrounding lattice.**

Since the stress at a point results from the superposition of the thermal and mechanical strains:

$$\sigma_{ij} = C_{ijkl}\left[\epsilon_{kl} - \alpha_t(\theta - \theta_r)\delta_{kl}\right]. \tag{138}$$

When the temperature field is non-uniform, the stress at a point depends on both the mechanical deformation, measured through ϵ_{ij}, and the local temperature, θ.

2.6.3 Compositional strain

If the lattice parameter is a function of the local composition, then a change in composition will engender a stress. Consider a binary substitutional alloy of components A and B in which the partial molar volumes of A and B are different. Replacing an A atom with a B atom induces a strain and corresponding stress because the local volume changes. If the composition of the reference state is c_r, where c is the mole fraction of component B, the lattice parameter at composition c, $a(c)$, can be estimated by expanding the lattice parameter in a Taylor series about the reference state composition:

$$a(c) = a(c_r) + \left(\frac{\partial a}{\partial c}\right)_{c_r}(c - c_r) + \frac{1}{2}\left(\frac{\partial^2 a}{\partial c^2}\right)_{c_r}(c - c_r)^2 + \cdots \tag{139}$$

which, to first-order in the composition change, becomes:

$$a(c) = a(c_r)\left[1 + \epsilon^c(c - c_r)\right] \tag{140}$$

where:

$$\epsilon^c = \frac{1}{a(c_r)}\left(\frac{\partial a}{\partial c}\right)_{c=c_r} = \left(\frac{\partial \ln a}{\partial c}\right)_{c=c_r} \tag{141}$$

The compositional strain, ϵ^c, is a measure of the change in the local lattice parameter induced by the change in composition since, from Eq. (140):

$$\frac{a(c) - a(c_r)}{a(c_r)} = \epsilon^c(c - c_r). \tag{142}$$

Analogous to the case of thermally induced strain, Eq. (138), the stress induced at a point by a composition change at the same point can be written:

$$\sigma_{ij} = C_{ijkl}\left[\epsilon_{kl} - \epsilon^c(c - c_r)\delta_{kl}\right]. \tag{143}$$

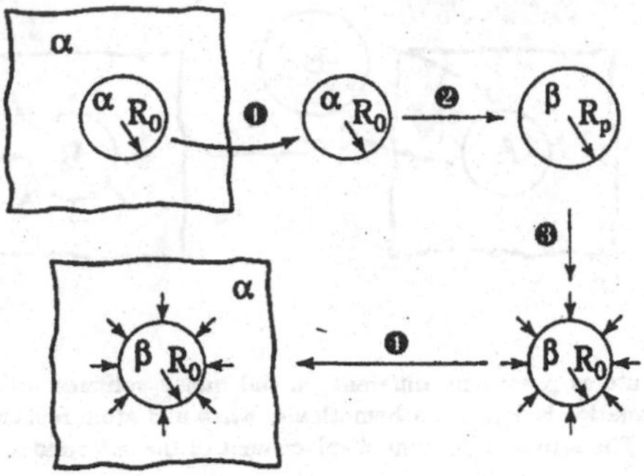

Figure 13: Eshelby's thought experiment imagines a region of the α matrix, here represented as a sphere of radius R_o, to be cut and removed from the matrix (step 1). The removed region undergoes a phase transformation from α to β in step 2. The unstressed β phase has a radius of R_p. After surface forces are applied to the β particle in order to return its radius to R_o in step 3, the particle is reinserted into the original hole of radius R_o in the matrix phase in step 4. Relaxing the surface forces will result in the particle and matrix reaching the equilibrium state of deformation.

Of course, if a composition change does not distort the lattice uniformly in all directions, as would occur in a noncubic material or if shear stresses were induced, then the compositional strain ϵ^c would be expressed as a tensor, rather than a scalar. In such a case, one could write:

$$\sigma_{ij} = C_{ijkl}\left[\epsilon_{kl} - \epsilon^c_{kl}\right]. \tag{144}$$

where the compositional dependence of the compositional strain could be nonlinear. For an isotropic or cubic material in which the lattice parameter depends linearly on composition, Eq. (143) is recovered from Eq. (144) with:

$$\epsilon^c_{kl} = \epsilon^c(c - c_r)\delta_{kl}. \tag{145}$$

Stress induced through a change in composition is often referred to as coherency stress.

2.7 Eshelby's Misfitting Sphere

In this section, the elastic stress and strain fields and the elastic energy associated with the placement of a spherical precipitate of β phase in an infinite, isotropic matrix of α phase are determined. The assumption of a spherical precipitate and the resulting spherical symmetry allows the elastic fields to be calculated directly by expressing the conditions for mechanical equilibrium derived earlier in spherical coordinates. This classical approach has been used extensively to examine the effect of misfit strain on nucleation and growth of precipitates [27]. Calculation of the elastic field associated with nonspherical precipitates or precipitates in an anisotropic system requires use of a Green's function approach [24] which will not be discussed here. The elasticity problem is formulated as follows.

Eshelby's thought experiment [24] is a good starting point for understanding the origin of elastic stress in misfitting precipitates and provides a step-by-step procedure for the calculation of the stress and strain fields. To begin, consider an α phase extending to infinity and which is initially free from all stress, as shown in Fig. 13. A sphere of radius R_o situated at the origin is imagined to be removed from the α matrix and allowed to transform to the β (precipitate) phase without change in the number of lattice sites. Since the β phase has a different crystal structure or molar volume, the radius of the β precipitate in a stress-free environment, R_p, will be different from the radius of the α phase originally removed from the material and allowed to transform, R_o. In order to insert the sphere of β back into the α matrix, forces must be applied to the surface of the precipitate. For example, if $R_p > R_o$, then compressive forces must be applied to the β phase in order to squeeze it back into the original hole. After the β phase has been reinserted and the constraining forces are removed, both the β precipitate and α matrix phases will relax to a state of mechanical equilibrium, as depicted in Fig. 13. Mechanical equilibrium is assured when the stress field satisfies Eq. (89). In the following example, it is assumed that the phases are isotropic and possess different elastic constants.

The unstressed α phase is chosen as the reference state for the measurement of strain. If the spherical coordinates, r, θ, and ϕ are employed, the only non-zero component of the displacement tensor is u_r, the displacement in the radial direction. $u_r(r)$ is a function of the radial position only and all off-diagonal terms of the stress and strain tensors vanish by symmetry. For a spherically symmetric system, the radial strain, ϵ_{rr}, is given by [22]:

$$\epsilon_{rr} = \frac{\partial u_r}{\partial r}. \tag{146}$$

The strain in the θ and ϕ directions, $\epsilon_{\theta\theta}$ and $\epsilon_{\phi\phi}$, are also a measure of the change in length per unit length and are obtained from the radial displacement by:

$$\epsilon_{\theta\theta} = \epsilon_{\phi\phi} = \frac{u_r}{r}. \tag{147}$$

The spherical coordinate system is an orthogonal coordinate system just like the Cartesian coordinate system. This means the tensor relationships we examined in the previous sections are also applicable to the spherical coordinate system. The spherical coordinate system forms a right-hand coordinate system in the same way as the Cartesian coordinate system. Associating the spherical tensor components with those of the Cartesian system, $\sigma_{11} = \sigma_{rr}$, $\sigma_{22} = \sigma_{\theta\theta}$ and $\sigma_{33} = \sigma_{\phi\phi}$, the stress-strain constitutive equation for the α matrix is simply:

$$\sigma^{\alpha}_{ij} = C^{\alpha}_{ijkl}\epsilon_{kl} \tag{148}$$

where the subscripts i and j now stand for the spherical coordinates. For an isotropic system, Eq. (148) can be written as Eq. (117). The non-vanishing stress components of Eq. (148) are obtained using Eq. (117) by setting the free indices $i = j = r$ and $i = j = \theta$ to give:

$$\sigma^{\alpha}_{rr} = \lambda^{\alpha}\epsilon^{\alpha}_{kk} + 2\mu^{\alpha}\epsilon^{\alpha}_{rr} \tag{149}$$

and

$$\sigma^{\alpha}_{\theta\theta} = \sigma^{\alpha}_{\phi\phi} = \lambda^{\alpha}\epsilon^{\alpha}_{kk} + 2\mu^{\alpha}\epsilon^{\alpha}_{\theta\theta} \tag{150}$$

where the trace of the strain tensor, $\epsilon^{\alpha}_{kk} = \epsilon^{\alpha}_{rr} + \epsilon^{\alpha}_{\theta\theta} + \epsilon_{\phi\phi}$, and λ^{α} and μ^{α} are the isotropic elastic constants of the α phase. The stresses can be expressed in terms of the radial displacement using Eqs. (146) and (147):

$$\sigma^{\alpha}_{rr} = \lambda^{\alpha}\left(\frac{\partial u^{\alpha}_r}{\partial r} + \frac{2u^{\alpha}_r}{r}\right) + 2\mu^{\alpha}\left(\frac{\partial u^{\alpha}_r}{\partial r}\right) \tag{151}$$

and

$$\sigma^{\alpha}_{\theta\theta} = \sigma^{\alpha}_{\phi\phi} = \lambda^{\alpha}\left(\frac{\partial u^{\alpha}_{r}}{\partial r} + \frac{2u^{\alpha}_{r}}{r}\right) + 2\mu^{\alpha}\left(\frac{u^{\alpha}_{r}}{r}\right) \tag{152}$$

The stress-strain constitutive law for the β phase must account for the non-zero transformation strain of the β phase as well as any subsequent deformation from the reference state. Using the ideas presented in Eshelby's thought experiment, the misfit strain of the β phase can be defined in precisely the same manner as in Eq. (133):

$$\epsilon^{T}_{ij} = \frac{(R_p - R_o)}{R_o}\delta_{ij} = \epsilon^{T}\delta_{ij}. \tag{153}$$

If a^{α} and a^{β} are the unstressed lattice parameters of α and β, then:

$$\epsilon^{T} = \frac{(a^{\beta} - a^{\alpha})}{a^{\alpha}} \tag{154}$$

since $R_p = na^{\beta}$ and $R_o = na^{\alpha}$ where n is the number of unit cells along the radial direction. The stress-strain constitutive equation for the β phase is, from Eq. (131), thus:

$$\sigma^{\beta}_{ij} = C^{\beta}_{ijkl}\left(\epsilon^{\beta}_{kl} - \epsilon^{T}_{kl}\right) = C^{\beta}_{ijkl}\left(\epsilon^{\beta}_{kl} - \epsilon^{T}\delta_{kl}\right). \tag{155}$$

Using Eq. (117), Eq. (155) can be written:

$$\sigma^{\beta}_{ij} = \lambda^{\beta}(\epsilon^{\beta}_{kk} - \epsilon^{T}_{kk})\delta_{ij} + 2\mu^{\beta}(\epsilon^{\beta}_{ij} - \epsilon^{T}_{ij}). \tag{156}$$

Using Eq. (153), Eq. (156) simplifies to:

$$\sigma^{\beta}_{ij} = \lambda^{\beta}(\epsilon^{\beta}_{kk} - 3\epsilon^{T})\delta_{ij} + 2\mu^{\beta}(\epsilon^{\beta}_{ij} - \epsilon^{T}\delta_{ij}). \tag{157}$$

The non-vanishing stress components of Eq. (157) are:

$$\sigma^{\beta}_{rr} = \lambda^{\beta}\left(\epsilon^{\beta}_{kk} - 3\epsilon^{T}\right) + 2\mu^{\beta}\left(\epsilon^{\beta}_{rr} - \epsilon^{T}\right) \tag{158}$$

and

$$\sigma^{\beta}_{\theta\theta} = \sigma^{\beta}_{\phi\phi} = \lambda^{\beta}\left(\epsilon^{\beta}_{kk} - 3\epsilon^{T}\right) + 2\mu^{\beta}\left(\epsilon^{\beta}_{\theta\theta} - \epsilon^{T}\right). \tag{159}$$

The stresses in the β phase can be expressed in terms of the radial displacement using Eqs. (146) and (147) as:

$$\sigma^{\beta}_{rr} = \lambda^{\beta}\left(\frac{\partial u^{\beta}_{r}}{\partial r} + \frac{2u^{\beta}_{r}}{r} - 3\epsilon^{T}\right) + 2\mu^{\beta}\left(\frac{\partial u^{\beta}_{r}}{\partial r} - \epsilon^{T}\right) \tag{160}$$

and

$$\sigma^{\beta}_{\theta\theta} = \sigma^{\beta}_{\phi\phi} = \lambda^{\beta}\left(\frac{\partial u^{\beta}_{r}}{\partial r} + \frac{2u^{\beta}_{r}}{r} - 3\epsilon^{T}\right) + 2\mu^{\beta}\left(\frac{u^{\beta}_{r}}{r} - \epsilon^{T}\right). \tag{161}$$

Mechanical equilibrium is assured when Eq. (89) is satisfied in each phase. Expressed in spherical coordinates, there is only one mechanical equilibrium condition for each phase given by [21]:

$$\frac{\partial \sigma_{rr}}{\partial r} + \frac{1}{r}\left[2\sigma_{rr} - \sigma_{\theta\theta} - \sigma_{\phi\phi}\right] = 0. \tag{162}$$

Substituting Eqs. (151) and (152) into Eq. (162), the equilibrium condition for the α phase can be expressed in terms of the radial displacement in the α phase as:

$$(\lambda^\alpha + 2\mu^\alpha)\left\{\frac{\partial^2 u_r^\alpha}{\partial r^2} + \frac{2}{r}\frac{\partial u_r^\alpha}{\partial r} - \frac{2u_r^\alpha}{r^2}\right\} = 0. \tag{163}$$

Likewise, substituting Eqs. (160) and (161) into Eq. (162), the equilibrium condition for the β phase expressed in terms of the radial displacement for the β phase is:

$$\left(\lambda^\beta + 2\mu^\beta\right)\left\{\frac{\partial^2 u_r^\beta}{\partial r^2} + \frac{2}{r}\frac{\partial u_r^\beta}{\partial r} - \frac{2u_r^\beta}{r^2}\right\} = 0. \tag{164}$$

Equations (163) and (164) are ordinary differential equations for the radial displacements u_r^α and u_r^β. As can be shown by direct substitution, the general solutions to Eqs. (163) and (164) are, respectively:

$$u_r^\alpha(r) = A_1^\alpha r + \frac{A_2^\alpha}{r^2} \quad \text{and} \quad u_r^\beta(r) = A_1^\beta r + \frac{A_2^\beta}{r^2} \tag{165}$$

where A_1 and A_2 are constants, different for the α and β phases. In the α (matrix) phase, the displacement must remain bounded as the radial position becomes large. This necessitates that $A_1^\alpha = 0$. Likewise, the displacement in the β phase must remain bounded as the origin is approached requiring that $A_2^\beta = 0$. The displacements in each phase are:

$$u_r^\beta(r) = A_1^\beta r \quad \text{for } r \leq R_o \tag{166}$$

and

$$u_r^\alpha(r) = \frac{A_2^\alpha}{r^2} \quad \text{for } r \geq R_o. \tag{167}$$

An immediate consequence of these solutions is that, from Eqs. (146) and (147):

$$\epsilon_{rr}^\beta = \epsilon_{\theta\theta}^\beta = \epsilon_{\phi\phi}^\beta = A_1^\beta \tag{168}$$

and

$$\epsilon_{\theta\theta}^\alpha = \epsilon_{\phi\phi}^\alpha = -\epsilon_{rr}^\alpha/2 = \frac{A_2^\alpha}{r^3} \quad \text{with} \quad \epsilon_{kk}^\alpha = 0. \tag{169}$$

In order to determine the two unknown constants A_1^β and A_2^α, two additional equations are required. These equations are obtained from the continuity of displacement and force (traction) across the interface. The displacement is continuous because no gaps or material overlap at the interface are allowed. The traction across the interface must be continuous to ensure mechanical equilibrium. If tractions are not equal, the interface would have unequal mechanical forces acting on it and equilibrium would not obtain. Continuity of displacement means that, at the interface, the displacement vectors in each phase are equal, $\vec{u}^\alpha(R_o) = \vec{u}^\beta(R_o)$. For the spherically symmetric system considered, only the radial component of the displacement is nonvanishing and the continuity of displacement condition is written:

$$u_r^\alpha(R_o) = u_r^\beta(R_o). \tag{170}$$

The continuity of traction condition requires the force exerted by the α phase on the β phase to equal the force exerted on α by β. Thus:

$$\sigma_{rr}^{\alpha}(R_o) = \sigma_{rr}^{\beta}(R_o). \tag{171}$$

Substituting Eqs. (166) and (167) into Eq. (170) and evaluating at $r = R_o$ gives:

$$A_2^{\alpha} = A_1^{\beta} R_o^3. \tag{172}$$

The second equation is obtained by first substituting Eq. (167) into Eq. (151) and then evaluating at $r = R_o$. The radial stress in the α phase evaluated at the interface is:

$$\sigma_{rr}^{\alpha} = \frac{-4\mu^{\beta} A_2^{\alpha}}{r^3}. \tag{173}$$

Likewise, the radial stress at the interface in the β phase is obtained by substituting Eq. (166) into Eq. (160) and evaluating at $r = R_o$:

$$\sigma_{rr}^{\beta} = (3\lambda^{\beta} + 2\mu^{\beta})\left(A_1^{\beta} - \epsilon^T\right) \tag{174}$$

Finally, substituting Eqs. (173) and (174) into Eq. (171) gives:

$$3K^{\beta}(A_1^{\beta} - \epsilon^T) = -4\mu^{\alpha} A_2^{\alpha}/R_o^3 \tag{175}$$

where $K^{\beta} = \lambda^{\beta} + 2\mu^{\beta}/3$ is the bulk modulus. Solving Eqs. (172) and (175) simultaneously for A_1^{β} and A_2^{α} yields:

$$A_1^{\beta} = \frac{3K^{\beta}\epsilon^T}{(3K^{\beta} + 4\mu^{\alpha})} \tag{176}$$

and

$$A_2^{\alpha} = \frac{3K^{\beta}\epsilon^T R_o^3}{(3K^{\beta} + 4\mu^{\alpha})}. \tag{177}$$

The resulting stress fields are:

$$\sigma_{rr}^{\beta} = \sigma_{\theta\theta}^{\beta} = \sigma_{\phi\phi}^{\beta} = \frac{-12\mu^{\alpha} K^{\beta}\epsilon^T}{(3K^{\beta} + 4\mu^{\alpha})} \tag{178}$$

and

$$\sigma_{rr}^{\alpha} = -2\sigma_{\theta\theta}^{\alpha} = -2\sigma_{\phi\phi}^{\alpha} = \frac{-12\mu^{\alpha} K^{\beta}\epsilon^T R_o^3}{(3K^{\beta} + 4\mu^{\alpha})r^3}. \tag{179}$$

The precipitate is hydrostatically stressed as the three diagonal stress components inside the precipitate are constant and equal. The precipitate is in a state of compression when $\epsilon^T > 0$; the surrounding matrix prevents the precipitate from completely relaxing. In the matrix phase, the elastic field decays as $1/r^3$ and the trace of the stress vanishes, $\sigma_{kk}^{\alpha} = 0$.

In order to calculate the elastic energy associated with introducing the misfitting precipitate into the matrix, the strain energy density must be integrated over the entirety of both phases $(V_{\alpha} + V_{\beta})$. For the linear elastic system, the strain energy density is given by Eq. (128) and the elastic strain energy, E_{strain}, is:

$$E_{strain} = \frac{1}{2}\int_{V_{\alpha}} C_{ijkl}^{\alpha}\epsilon_{ij}^{\alpha}\epsilon_{kl}^{\alpha} dV + \frac{1}{2}\int_{V_{\beta}} C_{ijkl}^{\beta}\epsilon_{ij}^{\beta}\epsilon_{kl}^{\beta} dV \tag{180}$$

where, in keeping with Eq. (128), the superscript indicates that the strain is measured with respect to the unstressed state of the indicated phase. It is customary for the unstressed matrix phase to be chosen as the reference state for the measurement of strain when the precipitate phase is coherent with the matrix. Designating the strain measured with respect to the unstressed matrix[8] as ϵ_{ij}, then $\epsilon^{\alpha}_{ij} = \epsilon_{ij}$ and $\epsilon^{\beta}_{ij} = \epsilon_{ij} - \epsilon^{T}_{ij}$ and the strain energy is written:

$$E_{strain} = \frac{1}{2}\int_{V_\beta} C^{\beta}_{ijkl}(\epsilon_{ij} - \epsilon^{T}_{ij})(\epsilon_{kl} - \epsilon^{T}_{kl})dV + \frac{1}{2}\int_{V_\alpha} C^{\alpha}_{ijkl}\epsilon_{ij}\epsilon_{kl}dV \tag{181}$$

or,

$$E_{strain} = \frac{1}{2}\int_{V_\beta} \sigma^{\beta}_{ij}(\epsilon_{ij} - \epsilon^{T}_{ij})dV + \frac{1}{2}\int_{V_\alpha} \sigma^{\alpha}_{ij}\epsilon_{ij}dV. \tag{182}$$

For a linear elastic solid the stress tensor is symmetric and $\sigma^{\alpha}_{ij}\epsilon_{ij} = \sigma^{\alpha}_{ij}u_{i,j}$. Using the identity:

$$\frac{\partial}{\partial x_j}\left[\sigma^{\alpha}_{ij}u_i\right] = \sigma^{\alpha}_{ij,j}u_i + \sigma^{\alpha}_{ij}u_{i,j} = \sigma^{\alpha}_{ij,j}u_i + \sigma^{\alpha}_{ij}\epsilon_{ij} \tag{183}$$

the integral over the matrix appearing in Eq. (182) becomes:

$$\frac{1}{2}\int_{V_\alpha} \sigma^{\alpha}_{ij}\epsilon_{ij}dV = \frac{1}{2}\int_{V_\alpha}\left\{\frac{\partial}{\partial x_j}\left[\sigma^{\alpha}_{ij}u_i\right] - \sigma^{\alpha}_{ij,j}u_i\right\}dV. \tag{184}$$

For a system in mechanical equilibrium, $\sigma^{\alpha}_{ij,j} = 0$ (Eq. (89)), and the last term of the volume integral of Eq. (184) vanishes. Applying the divergence theorem, Eq. (87), to the remaining integral gives:

$$\frac{1}{2}\int_{V_\alpha} \sigma^{\alpha}_{ij}\epsilon_{ij}dV = \frac{1}{2}\int_{S_\alpha} \sigma^{\alpha}_{ij}u^{\alpha}_i n^{\alpha}_j dS + \frac{1}{2}\int_{S_\infty} \sigma^{\alpha}_{ij}u_i n_j dS \tag{185}$$

where S_∞ is the surface infinitely far from the precipitate, S_α is the surface of the α phase at the $\alpha - \beta$ interface and u^{α}_i is the displacement on the α side of the $\alpha - \beta$ interface[9].

The displacement field at infinity owing to the precipitate misfit is zero, so the second surface integral of Eq. (185) vanishes. Continuity of traction across the precipitate-matrix interface requires $\sigma^{\alpha}_{ij}n^{\alpha}_j = -\sigma^{\beta}_{ij}n^{\beta}_j$ (see Eq. (300)). Likewise, continuity of displacement at the $\alpha - \beta$ interface requires $u^{\alpha}_i = u^{\beta}_i = u_i$. Therefore:

$$\frac{1}{2}\int_{V_\alpha} \sigma^{\alpha}_{ij}\epsilon_{ij}dV = -\frac{1}{2}\int_{S_\beta} \sigma^{\beta}_{ij}u_i n^{\beta}_j dS. \tag{186}$$

The right-hand-side of Eq. (186) is expressed solely in terms of quantities defined for the precipitate phase. The divergence theorem would, therefore, now relate the surface integral to an integral over the precipitate volume:

$$\frac{1}{2}\int_{V_\alpha} \sigma^{\alpha}_{ij}\epsilon_{ij}dV = -\frac{1}{2}\int_{S_\beta} \sigma^{\beta}_{ij}u_i n^{\beta}_j dS = -\frac{1}{2}\int_{V_\beta} \frac{\partial}{\partial x_j}\left[\sigma^{\beta}_{ij}u_i\right]dV \tag{187}$$

or

$$\frac{1}{2}\int_{V_\alpha} \sigma^{\alpha}_{ij}\epsilon_{ij}dV = -\frac{1}{2}\int_{V_\beta}\left\{\sigma^{\beta}_{ij}u_{i,j} + \sigma^{\beta}_{ij,j}u_i\right\}dV = -\frac{1}{2}\int_{V_\beta} \sigma^{\beta}_{ij}\epsilon_{ij}dV \tag{188}$$

---

[8]This strain is Eshelby's constrained strain [24].

[9]The divergence theorem connects the integral over a volume of material to the surface enclosing that volume. For the matrix, this surface is the precipitate-matrix interface and the surface of the matrix located at infinity.

since $\sigma^{\beta}_{ij,j} = 0$ when the system is in mechanical equilibrium. Combining Eq. (188) with Eq. (180) gives for the elastic strain energy associated with a misfitting precipitate:

$$E_{strain} = -\frac{1}{2}\int_{V_\beta} \sigma^{\beta}_{ij}\epsilon^{T}_{ij} dV. \tag{189}$$

Equation (189) is valid for arbitrarily shaped precipitates and anisotropic materials. Using our example of a spherical precipitate in an isotropic matrix, in which the strain is constant within the precipitate, the integrand is a constant and the strain energy is:

$$E_{strain} = -\frac{1}{2}\sigma^{\beta}_{ij}\epsilon^{T}\delta_{ij}V_\beta = -\frac{3}{2}\sigma^{\beta}_{rr}\epsilon^{T}V_\beta \tag{190}$$

where V_β is the volume of the precipitate and Eq. (178) was used. Using the stress field for the spherical precipitate, Eq. (178), Eq. (190) yields:

$$E_{strain} = \frac{18\mu^{\alpha}K^{\beta}(\epsilon^{T})^{2}V_\beta}{(3K^{\beta} + 4\mu^{\alpha})}. \tag{191}$$

The strain energy of a misfitting precipitate increases proportionally with the volume of the precipitate.

3 Thermodynamics of Stressed Systems

In this section, a thermodynamic treatment of elastically stressed crystals is presented and the conditions for thermodynamic equilibrium are derived. These equations provide the basis for analyzing a number of problems concerned with both displacive and diffusional phase transformations. In the first subsection, the thermodynamic framework for treating stressed crystals is presented. This approach is then applied to a multicomponent, single-phase crystal in order to derive the conditions for thermodynamic equilibrium. In the third subsection, an expression for the stress dependence of the diffusion potential is obtained and applied to two simple examples. Finally, the equilibrium conditions for a two-phase, multicomponent crystals are obtained.

3.1 Thermodynamic Framework

The equilibrium conditions are obtained using Gibbs' variational approach [1]. This is a powerful and broadly applicable tool which allows formulation of very general relationships among the thermodynamic variables to be derived. Its starting point is the combined form of the first two laws of thermodynamics which Gibbs stated as [1]: *"For the equilibrium of any isolated system, it is necessary and sufficient that in all possible variations in the state of the system which do not alter its entropy, the variation of its (internal) energy shall either vanish or be positive."* Gibbs' variational principle thus states that an isolated system is in equilibrium if the internal energy of the system is a minimum. It also provides a methodology for finding the equilibrium conditions of a system regardless of the imposed experimental conditions. An arbitrary region of the system is imagined to be isolated from its surroundings. If this region is in thermodynamic equilibrium, then arbitrary perturbations in the thermodynamic state of the system (such as changes in the local composition, entropy or strain) that maintain a constant entropy in the region result in a change in internal energy for the region that is either positive or zero. It is the internal energy which is defined by the first two laws of thermodynamics. Other free energies, such as the Helmholtz and Gibbs free energies, are derived quantities.

Before the equilibrium conditions can be determined, it is necessary to define the type of material being considered as this will affect the thermodynamic work terms appearing in the first law. The present analysis will concern simple crystals which deform elastically and for which there is only one type of substitutional site and one type of interstitial site. In addition, atoms of a given species are found on only one type of site. The only defect considered will be vacancies, which can exist on both types of sites. These restrictions can, of course, be relaxed in order to treat more complex crystals, but the resulting algebra becomes more cumbersome [8, 10].

In general, the thermodynamic state of a crystal is heterogeneous; the composition, strain and entropy are functions of position. As a consequence, the system entropy, internal energy and other extensive properties must be expressed as a volume integral over the system. For example, if e_v is the internal energy density, then the internal energy $\mathcal{E}$ is:

$$\mathcal{E} = \int_V e_v dV. \tag{192}$$

Since the volume of an element of solid is affected by its state of strain, expressing a density as per-unit-volume of the actual or deformed state is usually not very convenient; if the state of strain is varied, the density of the mass components will change as well. Instead, densities expressed per-unit-volume of the reference state are often used. Densities expressed per-unit-volume of the reference state are denoted with a prime superscript. Thus:

$$\mathcal{E} = \int_{V'} e_{v'} dV \tag{193}$$

where, in accord with Eq. (80), $e_{v'} = Je_v$ and $J = dV/dV'$. The internal energy contained in the volume V of the actual state is identical to the internal energy contained in the corresponding volume V' of the reference state.

It is reasonable to assume that the internal energy of any volume element of the crystal will depend on its entropy, composition of each component, and state of deformation[10]. The deformation can be represented by one of the strain tensors but, if the internal energy is referred to the reference state, it is more convenient to use the deformation gradient tensor, at least initially. Therefore, we assume the internal energy density is a function of the following variables:

$$e_{v'} = e_{v'}(s_{v'}, F_{ij}, \rho_i^{I'}, \rho_i^{S'}) \tag{194}$$

where $s_{v'}$ is the entropy density, F_{ij} is the deformation gradient tensor, $\rho_i^{I'}$ is the the number density of interstitial component i, and $\rho_i^{S'}$ is the number density of substitutional component i (atoms of component i per unit reference volume.) Since $\rho_o^{S'}$ and $\rho_o^{I'}$ are the number density of substitutional and interstitial sites in the reference state, respectively, then the following conditions must hold:

$$\rho_o^{S'} = \sum_{i=1}^{N_S} \rho_i^{S'} \tag{195}$$

[10]Of course, the internal energy density can depend on other variables as well. For example, in an ionic crystal, the internal energy is also a function of the electric displacement and the various atomic species can have different charge states.

and

$$\rho_o^{I'} = \sum_{i=1}^{N_I} \rho_i^{I'} \tag{196}$$

where N_I and N_S are the numbers of different species occupying the interstitial and substitutional sublattices (including vacancies), respectively. Equations (195) and (196) are local conditions satisfied at every point (or within every volume element) in the crystal. In addition, the quantities $\rho_o^{S'}$ and $\rho_o^{I'}$ are constants that depend only on the choice of the reference state. These conditions are referred to as the "network constraint" [5].

From Eq. (194), a variation (or perturbation) of the local entropy, deformation state or composition results in a change in internal energy given by:

$$\delta e_{v'} = \frac{\partial e_{v'}}{\partial s_{v'}} \delta s_{v'} + \frac{\partial e_{v'}}{\partial F_{ij}} \delta F_{ij} + \sum_{i=1}^{N_S} \frac{\partial e_{v'}}{\partial \rho_i^{S'}} \delta \rho_i^{S'} + \sum_{i=1}^{N_I} \frac{\partial e_{v'}}{\partial \rho_i^{I'}} \delta \rho_i^{I'} \tag{197}$$

From the second law of thermodynamics, we associate:

$$\frac{\partial e_{v'}}{\partial s_{v'}} = \theta \tag{198}$$

where θ is the absolute temperature. In our study of elastic deformation, we showed that a small change in the elastic strain, ϵ_{ij}, induced a change in the elastic energy density, de_v, given by $de_v = \sigma_{ij} d\epsilon_{ij}$ where σ_{ij} is the Cauchy stress tensor. In this formulation, the energy density was reckoned per unit volume of the deformed material. A similar expression can be written for the strain energy density when it is reckoned per unit volume of the reference state. In this case one writes:

$$de_{v'} = T_{ji} dF_{ij} \tag{199}$$

where T_{ji} is the first Piola-Kirchhoff stress tensor [21]. This allows the association, in analogy to Eq. (97):

$$\frac{\partial e_{v'}}{\partial F_{ij}} = T_{ji}. \tag{200}$$

The stress tensor T_{ji} expresses the force acting on a surface element in terms of the area of the element as it appears in the reference state.

For simplicity, the following definitions for the compositional derivatives of the internal energy density are used:

$$\frac{\partial e_{v'}}{\partial \rho_i^{S'}} = \gamma_i^S \tag{201}$$

and

$$\frac{\partial e_{v'}}{\partial \rho_i^{I'}} = \gamma_i^I. \tag{202}$$

A variation in the internal energy density is therefore given by:

$$\delta e_{v'} = \theta \delta s_{v'} + T_{ji} \delta F_{ij} + \sum_{i=1}^{N_S} \gamma_i^S \delta \rho_i^{S'} + \sum_{i=1}^{N_I} \gamma_i^I \delta \rho_i^{I'}. \tag{203}$$

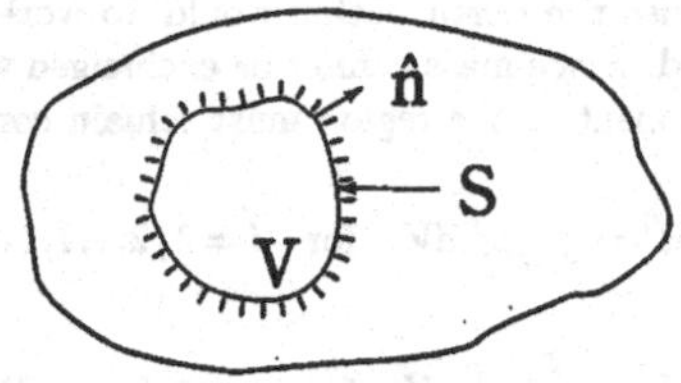

Figure 14: In order to determine the conditions for thermodynamic equilibrium, a region V is imagined to be thermodynamically isolated from the rest of the system. Its thermodynamic state is then perturbed at constant entropy and the conditions for which the internal energy vanishes are established. The requirement that V remain mechanically isolated implies the displacement on the surface S must be held fixed.

The variation of the internal energy of a region V, $\delta\mathcal{E}$, is obtained by integrating the change in the internal energy density, $\delta e_{v'}$, over the reference volume V'. Thus:

$$\delta\mathcal{E} = \int_{V'} \delta e_{v'} dV = \int_{V'} \left\{ \theta \delta s_{v'} + T_{ji} \delta F_{ij} + \sum_{i=1}^{N_S} \gamma_i^S \delta \rho_i^{S'} + \sum_{i=1}^{N_I} \gamma_i^I \delta \rho_i^{I'} \right\} dV. \tag{204}$$

3.2 Single-phase system

In this subsection, the conditions for thermodynamic equilibrium in a multicomponent, single-phase system are derived using Gibbs statement of the first two laws of thermodynamics.

Figure 14 shows a single crystal of material that is presumed to be in equilibrium. An arbitrary region of the crystal of volume V and surface S is chosen and imagined to be isolated from the rest of the crystal. If the crystal as a whole is in equilibrium, then the region V must also be in equilibrium, and Gibbs' criterion for the determination of equilibrium can be applied to it. The region is imagined to be perturbed from its current state. If the perturbation is performed at constant entropy and the internal energy of the region remains unchanged (or increases) for all possible or admissible perturbations, then the internal energy of the isolated region is an extremum (minimum) and the region is in equilibrium.

Three kinds of perturbations in the thermodynamic state of the system are allowed. First, changes in the entropy density of the volume elements comprising V can be made. Such entropy changes can result, for example, from the flow of heat between neighboring volume elements. Second, changes in the compositions of the volume elements resulting from possible diffusion between the volume elements are allowed and, third, the local state of deformation can be varied.

In keeping with the equilibrium principle, the perturbation in the local entropy densities must be accomplished holding the total entropy of the system, S, fixed. Since:

$$S = \int_{V'} s_{v'} dV, \tag{205}$$

constant entropy requires:

$$\delta S = \int_{V'} \delta s_{v'} dV = 0. \tag{206}$$

Since the system is imagined to be isolated, the perturbations in the compositions and displacement field of the volume elements are not arbitrary. For example, the displacement along the surface of the region must vanish, otherwise the perturbation would do work against an external force and the system would not be isolated. Since mass cannot be exchanged with the surroundings, the total number of atoms of each component in the region must remain constant. This mass constraint is expressed by:

$$\mathcal{N}_i^S = \int_{V'} \rho_i^{S'} dV \quad \text{for} \quad i = 1, 2, \ldots, N_S \tag{207}$$

and

$$\mathcal{N}_i^I = \int_{V'} \rho_i^{I'} dV \quad \text{for} \quad i = 1, 2, \ldots, N_I \tag{208}$$

which give the following constraints on the perturbations:

$$\delta\mathcal{N}_i^S = \int_{V'} \delta\rho_i^{S'} dV = 0 \quad \text{for} \quad i = 1, 2, \ldots, N_S \tag{209}$$

and

$$\delta\mathcal{N}_i^I = \int_{V'} \delta\rho_i^{I'} dV = 0 \quad \text{for} \quad i = 1, 2, \ldots, N_I. \tag{210}$$

The constraints imposed on the perturbation of the internal energy can be included in the perturbation procedure by means of Lagrange multipliers [1, 5]. Introducing the multipliers θ_L for the constraint on the entropy, λ_i^S for each of the N_S substitutional components, and λ_i^I for each of the N_I interstitial components, a free energy $\mathcal{E}^*$ can be defined as:

$$\mathcal{E}^* = \mathcal{E} - \theta_L \mathcal{S} - \sum_{i=1}^{N_S} \lambda_i^S \mathcal{N}_i^S - \sum_{i=1}^{N_I} \lambda_i^I \mathcal{N}_i^I. \tag{211}$$

The minimization of $\mathcal{E}^*$ is equivalent to the minimization of $\mathcal{E}$ subject to the constraints of constant entropy and mass. The first variation, or perturbation, of $\mathcal{E}^*$ is:

$$\delta\mathcal{E}^* = \delta\mathcal{E} - \theta_L \delta\mathcal{S} - \sum_{i=1}^{N_S} \lambda_i^S \delta\mathcal{N}_i^S - \sum_{i=1}^{N_I} \lambda_i^I \delta\mathcal{N}_i^I. \tag{212}$$

Substituting Eqs. (204), (206), (209) and (210) into Eq. (212) yields:

$$\delta\mathcal{E}^* = \int_{V'} \left\{ (\theta - \theta_L)\delta s_{v'} + T_{ji}\delta F_{ij} + \sum_{i=1}^{N_S} (\gamma_i^S - \lambda_i^S)\delta\rho_i^{S'} + \sum_{i=1}^{N_I} (\gamma_i^I - \lambda_i^I)\delta\rho_i^{I'} \right\} dV. \tag{213}$$

The variations or perturbations appearing in Eq. (213) are still not independent. For example, consider the perturbation of the displacement gradient tensor:

$$\delta F_{ij} = \frac{\partial}{\partial x_j'}(\delta x_i). \tag{214}$$

Using Eq. (40) to substitute for x_i and recognizing that a perturbation of a reference point is identically zero gives:

$$\delta F_{ij} = \frac{\partial}{\partial x_j'}(\delta x_i' + \delta u_i) = \frac{\partial}{\partial x_j'}(\delta u_i). \tag{215}$$

Making use of the identity:

$$\frac{\partial}{\partial x'_j}(T_{ji}\delta u_i) = T_{ji}\delta u_{i,j} + T_{ji,j}\delta u_i, \tag{216}$$

the volume integral containing the stress term appearing in Eq. (213) becomes, after using Eq. (215):

$$\int_{V'} T_{ji}\delta F_{ij} dV = \int_{V'} \frac{\partial}{\partial x'_j}[T_{ji}\delta u_i]\, dV - \int_{V'} T_{ji,j}\delta u_i dV. \tag{217}$$

Applying the divergence theorem, Eq. (87), to Eq. (217) gives:

$$\int_{V'} T_{ji}\delta F_{ij} dV = \int_{S'} T_{ji}\delta u_i n'_j dS - \int_{V'} T_{ji,j}\delta u_i dV. \tag{218}$$

Equation (218) shows that the nine variations corresponding to δF_{ij} are not independent, but rather can be expressed in terms of the three variations δu_i.

The variations in the local number densities of the components are also not independent on account of the crystalline structure of the material (or network constraint). From Eq. (209), the perturbation in the number density of substitutional component 1 can be expressed in terms of the variations of the densities of the other substitutional components as:

$$\delta\rho_1^{S'} = -\sum_{i=2}^{N_S} \delta\rho_i^{S'} \tag{219}$$

Likewise, Eq. (210) can be used to express the variation of interstitial component 1 in terms of the variations of the densities of the other interstitial components as:

$$\delta\rho_1^{I'} = -\sum_{i=2}^{N_I} \delta\rho_i^{I'}. \tag{220}$$

From Eqs. (219) and (220), it follows that:

$$\sum_{i=1}^{N_S}\left(\gamma_i^S - \lambda_i^S\right)\delta\rho_i^{S'} = \sum_{i=2}^{N_S}\left[\gamma_i^S - \gamma_1^S - \lambda_i^S + \lambda_1^S\right]\delta\rho_i^{S'} \tag{221}$$

and

$$\sum_{i=1}^{N_I}\left(\gamma_i^I - \lambda_i^I\right)\delta\rho_i^{I'} = \sum_{i=2}^{N_I}\left[\gamma_i^I - \gamma_1^I - \lambda_i^I + \lambda_1^I\right]\delta\rho_i^{I'}. \tag{222}$$

Substituting Eqs. (218), (221) and (222) into Eq. (213) gives for the variation in the free energy $\mathcal{E}^*$:

$$\begin{aligned}\delta\mathcal{E}^* &= \int_{V'}\left\{(\theta-\theta_L)\delta s_{v'} - T_{ji,j}\delta u_i - \sum_{i=2}^{N_S}(M_{i1}^S - \lambda_i^S + \lambda_1^S)\delta\rho_i^{S'} + \sum_{i=2}^{N_I}(M_{i1}^I - \lambda_i^I + \lambda_1^I)\delta\rho_i^{I'}\right\} dV \\ &+ \int_{S'} T_{ji}n'_j\delta u_i dS \end{aligned} \tag{223}$$

where the following notation has been introduced:

$$M_{i1}^S = \gamma_i^S - \gamma_1^S = \frac{\partial e_{v'}}{\partial\rho_i^{S'}} - \frac{\partial e_{v'}}{\partial\rho_1^{S'}} \tag{224}$$

and

$$M_{i1}^I = \gamma_i^I - \gamma_1^I = \frac{\partial e_{v'}}{\partial \rho_i^{I'}} - \frac{\partial e_{v'}}{\partial \rho_1^{I'}}. \tag{225}$$

In order to determine the equilibrium conditions, the variation in the system free energy, $\delta\mathcal{E}^*$, is set equal to zero in Eq. (223). In keeping with the equilibrium criterion that the region of the system under consideration must be isolated from its surroundings, the surface integral in Eq. (223) must vanish as the perturbation in the displacement on the surface of the region must be zero at every point. Since all variations appearing in the volume integral of Eq. (223) are now independent, $\delta\mathcal{E}^* = 0$ for an arbitrary region of the system only when the coefficients of each term appearing in the volume integrand of Eq. (223) vanishes. This yields the equilibrium conditions:

$$\theta = \theta_L, \tag{226}$$

$$T_{ji,j} = 0, \tag{227}$$

$$M_{i1}^S = \lambda_i^S - \lambda_1^S \tag{228}$$

and

$$M_{i1}^I = \lambda_i^I - \lambda_1^I. \tag{229}$$

The first condition, Eq. (226), is the condition for thermal equilibrium requiring that the temperature field, $\theta(\bar{x})$, be constant and uniform everywhere in the system. If the temperature is not everywhere uniform, then the system is not in equilibrium and heat flow will occur. The second condition, Eq. (227), is the condition for mechanical equilibrium which states that the divergence of the first Piola-Kirchhoff stress tensor vanishes in the absence of body forces. This condition is analogous to the mechanical equilibrium condition derived in the previous section, Eq. (89). However, instead of being expressed in terms of the Cauchy stress tensor, the mechanical equilibrium condition is expressed in terms of the Piola-Kirchhoff stress tensor.

Equations (228) and (229) are the conditions for chemical equilibrium for substitutional and interstitial components, respectively. The quantities M_{i1}^S and M_{i1}^I are called diffusion potentials [5]. Since the λ_i are Lagrange multipliers (constants), the diffusion potentials of each component must be constant and equal everywhere in the system for chemical equilibrium to obtain. If this condition is not satisfied, mass diffusion can occur. (Hence the name diffusion potential.) Physically, the diffusion potential expresses the change in energy associated with an exchange process performed under certain thermodynamic conditions. M_{i1}^S gives the change in energy when substitutional component 1 is removed from the system and replaced by substitutional component i. Likewise, M_{i1}^I gives the change in energy when interstitial component 1 is removed from the system and replaced by interstitial component i. The choice of component 1 is arbitrary and can be a vacancy. Its choice should facilitate solution of the problem at hand. For a sublattice that is sparsely populated, as in the case of certain interstitial atoms, an interstitial vacancy can be chosen as component 1. In such cases, the diffusion potential M_{iv}^I is usually referred to as the chemical potential of interstitial component i. For many substitutional systems, however, it is often most convenient to choose the predominant (solute) atom as the dependent species.

3.3 Stress-dependence of Diffusion Potential

General relationships for the stress (and composition) dependence of the diffusion potential can be determined formally from the first two laws of thermodynamics. In order to illustrate the procedure, a substitutional binary alloy consisting of components A and B will be used. If component A is chosen as component 1, then, at equilibrium, the first two laws of thermodynamics yield[11]

$$de_{v'} = \theta ds_{v'} + T_{ji}dF_{ij} + \rho'_o M_{BA}dc \tag{230}$$

where c is the composition (mole fraction) of component B: $c = c_B = \rho_B^{S'}/\rho_o^{S'} = \rho_B/\rho_o$. ρ_o is the number density of lattice sites in the actual state and the superscript S designating the lattice has been omitted since it is the only lattice of interest in the present example. As written, the variables θ, T_{ji} and M_{BA} are functions of the independent variables $s_{v'}$, F_{ij} and c. A change in variable is obtained by defining a new function, $g_{v'}$ (using a Legendre transform), as [5]:

$$g_{v'} = e_{v'} - \theta s_{v'} - T_{ji}F_{ij}. \tag{231}$$

$g_{v'}$ is a state function whose value depends only on the current thermodynamic state and is independent of the path taken to reach this state. Differentiating Eq. (231):

$$dg_{v'} = de_{v'} - \theta ds_{v'} - s_{v'}d\theta - T_{ji}dF_{ij} - F_{ij}dT_{ji} \tag{232}$$

and using Eq. (230) gives:

$$dg_{v'} = -s_{v'}d\theta - F_{ij}dT_{ji} + \rho'_o M_{BA}dc. \tag{233}$$

As written, $g_{v'}$ can be considered a function of θ, T_{ji} and c or $g_{v'} = g_{v'}(\theta, T_{ji}, c)$. If the small-strain approximation is used, then:

$$dg_{v'} = -s_{v'}d\theta - \epsilon_{ij}d\sigma_{ij} + \rho'_o M_{BA}dc \tag{234}$$

and $g_{v'} = g_{v'}(\theta, \sigma_{ij}, c)$.

Since $g_{v'}$ is a state function of the variables θ, σ_{ij} and c, $dg_{v'}$ is an exact differential and one can write:

$$dg_{v'}(\theta, \sigma_{ij}, c) = \left(\frac{\partial g_{v'}}{\partial \theta}\right) d\theta + \left(\frac{\partial g_{v'}}{\partial \sigma_{ij}}\right) d\sigma_{ij} + \left(\frac{\partial g_{v'}}{\partial c}\right) dc. \tag{235}$$

In performing the partial differentiation, it is assumed, as is standard practice, that the independent variables not involved in the differentiation are held constant. Comparing Eq. (235) with Eq. (234) allows the following associations to be made:

$$\left(\frac{\partial g_{v'}}{\partial \theta}\right) = -s_{v'} \quad \left(\frac{\partial g_{v'}}{\partial \sigma_{ij}}\right) = -\epsilon_{ij} \quad \left(\frac{\partial g_{v'}}{\partial c}\right) = \rho'_o M_{BA}. \tag{236}$$

Since the order of differentiation can be exchanged for an exact differential, the following equality must hold:

$$\frac{\partial}{\partial c}\left(\frac{\partial g_{v'}}{\partial \sigma_{ij}}\right) = \frac{\partial}{\partial \sigma_{ij}}\left(\frac{\partial g_{v'}}{\partial c}\right). \tag{237}$$

Substituting the relationships of Eq. (236) into Eq. (237) gives the following Maxwell relation:

$$-\left(\frac{\partial \epsilon_{ij}}{\partial c}\right)_{\theta,\sigma_{ij}} = \rho'_o \left(\frac{\partial M_{BA}}{\partial \sigma_{ij}}\right)_{\theta,c,\sigma_{kl\neq ij}} \tag{238}$$

[11]Eq. (230) follows directly from Eq. (203) once equilibrium has been assumed and the constraints on the variation of atomic species have been accounted for.

The subscripts appearing in Eq. (238) indicate those quantities held constant during the partial differentiation. The term $\sigma_{kl \neq ij}$ means that all components of the stress tensor are held constant except the component σ_{ij}. Equation (238) is quite general with no particular assumptions on material behavior other than the validity of employing the small-strain approximation. To proceed further, a constitutive law connecting the stress and strain must be invoked. If the temperature is assumed to remain constant, stresses are induced when the material is deformed or a composition change occurs, and Eq. (144) can be used as a stress-strain constitutive law. Therefore:

$$\sigma_{ij} = C_{ijkl}[\epsilon_{kl} - \epsilon^c_{kl}]. \tag{239}$$

In order to combine Eq. (239) with the Maxwell relation, Eq. (238), it is first necessary to solve for the strain in terms of the stress. This is accomplished by contracting each side of Eq. (239) with the elastic compliance tensor, S_{ijkl}:

$$S_{mnij}\sigma_{ij} = S_{mnij}C_{ijkl}[\epsilon_{kl} - \epsilon^c_{kl}]. \tag{240}$$

Using Eq. (103), Eq. (240) can be simplified to:

$$S_{mnij}\sigma_{ij} = \delta_{mk}\delta_{nl}[\epsilon_{kl} - \epsilon^c_{kl}] = \epsilon_{mn} - \epsilon^c_{mn}. \tag{241}$$

Solving Eq. (241) for the strain gives:

$$\epsilon_{mn} = \epsilon^c_{mn} + S_{mnij}\sigma_{ij}. \tag{242}$$

The compositional derivative of the strain at constant temperature and stress becomes:

$$\left(\frac{\partial \epsilon_{mn}}{\partial c}\right)_{\theta,\sigma_{ij}} = \left(\frac{\partial \epsilon^c_{mn}}{\partial c}\right)_{\theta,\sigma_{ij}} + \left(\frac{\partial S_{mnij}}{\partial c}\right)_{\theta,\sigma_{ij}} \sigma_{ij}. \tag{243}$$

If we assume that, for simplicity, the elastic constants are independent of composition, then:

$$\left(\frac{\partial S_{mnij}}{\partial c}\right)_{\theta,\sigma_{ij}} = 0, \tag{244}$$

and Eq. (243) becomes:

$$\left(\frac{\partial \epsilon_{mn}}{\partial c}\right)_{\theta,\sigma_{ij}} = \left(\frac{\partial \epsilon^c_{mn}}{\partial c}\right)_{\theta,\sigma_{ij}}. \tag{245}$$

If the discussion is further limited to isotropic or cubic materials, the compositional strain is given by Eq. (145), to first order in the composition. The compositional derivative of the strain then becomes:

$$\left(\frac{\partial \epsilon_{ij}}{\partial c}\right)_{\theta,\sigma_{ij}} = \left(\frac{\partial \epsilon^c_{ij}}{\partial c}\right)_{\theta,\sigma_{ij}} = \left(\frac{\partial\left[\epsilon^c(c - c_r)\delta_{ij}\right]}{\partial c}\right)_{\theta,\sigma_{ij}} = \epsilon^c\delta_{ij} \tag{246}$$

where c_r is a reference state composition and the free indices mn have been replaced by ij. Substituting Eq. (246) into Eq. (238) gives:

$$-\epsilon^c\delta_{ij} = \rho'_o\left(\frac{\partial M_{BA}}{\partial \sigma_{ij}}\right)_{\theta,c,\sigma_{kl\neq ij}} \tag{247}$$

which can be rewritten as:

$$\rho_o' dM_{BA} = -\epsilon^c \delta_{ij} d\sigma_{ij} = -\epsilon^c d\sigma_{jj}. \tag{248}$$

For the material parameters employed, Eq. (248) indicates that the diffusion potential depends only on the trace of the stress tensor, σ_{kk}. This is not unexpected, as the compositional strain is dilatational and performs work only against the diagonal components of the stress tensor in a cubic material.

Equation (248) can be integrated directly along a path from the unstressed state to the stressed state as:

$$\int_{M_{BA}(\theta,c)}^{M_{BA}(\theta,c,\sigma_{ij})} \rho_o' dM_{BA} = -\int_0^{\sigma_{jj}} \epsilon^c d\sigma_{jj}. \tag{249}$$

Since ρ_o' and ϵ^c are independent of the state of stress, integration yields:

$$M_{BA}(\theta, c, \sigma_{ij}) = M_{BA}(\theta, c) - \frac{\epsilon^c}{\rho_o'} \sigma_{kk}. \tag{250}$$

$M_{BA}(\theta, c)$ is the diffusion potential at the temperature θ and composition c in the absence of all stress effects. As such, it specifies the change in energy owing to the simultaneous addition of a B atom to and removal of an A atom from the system in the absence of stress and can be represented in terms of the usual chemical potentials, μ_A and μ_B by:

$$M_{BA}(\theta, c) = \mu_B(\theta, c) - \mu_A(\theta, c) \tag{251}$$

where, as defined, the chemical potentials are determined at zero pressure. Substituting Eq. (251) into Eq. (250) yields the following expression for the diffusion potential:

$$M_{BA}(\theta, c, \sigma_{ij}) = \mu_B(\theta, c) - \mu_A(\theta, c) - \frac{\epsilon^c}{\rho_o'} \sigma_{kk}. \tag{252}$$

3.3.1 Example 1: Stress-induced composition change

As a simple example of how application of an elastic stress can induce a change in the equilibrium composition of an alloy, consider an unstressed $(A - B)$ binary crystal in equilibrium with a fluid containing the two components A and B (Fig. 15). The chemical potentials of the two components in the fluid are μ_A^f and μ_B^f and the composition of the unstressed crystal in equilibrium with the fluid is $c_B = c_o$. Chemical equilibrium between the unstressed crystal and fluid requires [5]:

$$M_{BA}(\theta, c_o) = \mu_B^f - \mu_A^f. \tag{253}$$

Equation (253) states that the simultaneous addition of a B atom to and the removal of an A from the crystal produces the same change in free energy as the addition of a B to and the removal of an A from the fluid. If this were not the case, mass exchange between the crystal and fluid would occur.

If a uniaxial stress, $\sigma_{app} = \sigma_{33}$, is applied to the crystal in the x_3 direction while the chemical potentials of A and B in the fluid remain fixed, a change in composition of the crystal will result in order to bring the crystal back into chemical equilibrium with the fluid. The applied stress changes the diffusion potential M_{BA} in the crystal such that the condition for chemical equilibrium, Eq. (253), is no longer satisfied and the resulting gradient in chemical (diffusion) potential will

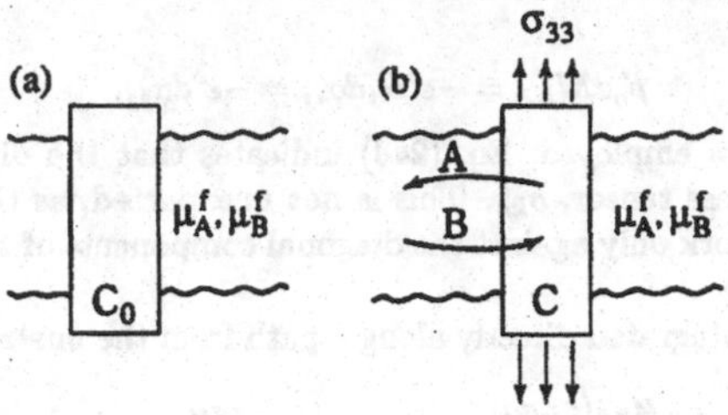

Figure 15: (a) The unstressed crystal, initially of composition c_o, is in equilibrium with a large fluid reservoir of chemical potential μ^f_A and μ^f_B. (b) Application of a uniaxial stress ($\sigma_{app} = \sigma_{33}$) to the crystal changes the diffusion potential of the crystal and induces mass flow between the crystal and fluid reservoir. The equilibrium composition of the crystal under stress is c.

induce mass flow between the crystal and fluid reservoir. In the presence of the applied stress, chemical equilibrium requires:

$$M_{BA}(\theta, c, \sigma_{ij}) = \mu_B(\theta, c) - \mu_A(\theta, c) - \frac{\epsilon^c}{\rho'_o}\sigma_{kk} = \mu^f_B - \mu^f_A \tag{254}$$

where the composition c which satisfies Eq. (254) is different from the initial composition c_o. Comparing Eqs. (253) and (254) provides the following expression to be solved for the equilibrium composition of the stressed crystal:

$$\mu_B(\theta, c) - \mu_A(\theta, c) = \mu_B(\theta, c_o) - \mu_A(\theta, c_o) + \frac{\epsilon^c}{\rho'_o}\sigma_{kk}. \tag{255}$$

An approximate expression for the composition shift, $c - c_o$, can be obtained from Eq. (255) by expanding in a Taylor's series the chemical potentials in terms of the composition about the composition c_o. We first note that the difference in chemical potentials is given by:

$$\mu_B(\theta, c) - \mu_A(\theta, c) = \mu^o_B(\theta) - \mu^o_A(\theta) + k\theta \ln(\gamma_B c) - k\theta \ln[\gamma_A(1 - c)] \tag{256}$$

where γ_A and γ_B are the activity coefficients of A and B, respectively, and μ^o is the composition-independent reference state value of the indicated chemical potential. A Taylor expansion to first-order in the composition gives:

$$\mu_B(\theta, c) - \mu_A(\theta, c) \approx \mu_B(\theta, c_o) - \mu_A(\theta, c_o) + k\theta\frac{\partial \ln(\gamma_B c)}{\partial c}(c - c_o) - k\theta\frac{\partial \ln[\gamma_A(1 - c)]}{\partial c}(c - c_o). \tag{257}$$

Using the Gibbs-Duhem relation:

$$\frac{\partial \ln[\gamma_A(1 - c)]}{\partial c} = -\frac{c}{(1 - c)}\frac{\partial \ln(\gamma_B c)}{\partial c}, \tag{258}$$

Eq. (257) can be simplified to:

$$\mu_B(\theta, c) - \mu_A(\theta, c) \approx \mu_B(\theta, c_o) - \mu_A(\theta, c_o) + \frac{k\theta}{c_o(1 - c_o)}\left[1 + \frac{\partial \ln \gamma_B}{\partial \ln c}\right](c - c_o). \tag{259}$$

Substituting Eq. (259) into Eq. (255) and solving for the composition change yields:

$$c - c_o = \frac{\epsilon^c c_o(1 - c_o)}{\rho'_o k\theta\chi}\sigma_{kk} = \frac{\epsilon^c \mathcal{N}_A}{\rho'_o G''_m}\sigma_{kk}. \tag{260}$$

Recognizing that, in the present example, the only source of stress is the applied uniaxial stress σ_{app}, then $\sigma_{kk} = \sigma_{11} + \sigma_{22} + \sigma_{33} = \sigma_{app}$, and Eq. (260) becomes:

$$c - c_o = \frac{\epsilon^c c_o(1-c_o)}{\rho'_o k\theta\chi}\sigma_{app} = \frac{\epsilon^c \mathcal{N}_A}{\rho'_o G''_m}\sigma_{app} \tag{261}$$

where $\mathcal{N}_A$ is Avogadro's number, G''_m is the second derivative of the molar free energy with respect to composition and χ is the thermodynamic factor defined by:

$$G''_m = \frac{R\theta}{c_o(1-c_o)}\left[1 + \frac{\partial \ln \gamma_B}{\partial \ln c}\right] = \frac{R\theta\chi}{c_o(1-c_o)} \tag{262}$$

and

$$\chi = \left[1 + \frac{\partial \ln \gamma_B}{\partial \ln c}\right]. \tag{263}$$

$R = \mathcal{N}_A k$ is the gas constant.

The magnitude of the composition change induced by application of an external stress, Eq. (261), depends on three terms; the compositional strain (ϵ^c), the applied stress (σ_{app}), and the curvature of the free energy with respect to composition in the absence of stress (G''_m). The product $\epsilon^c\sigma_{app}$ is a measure of the elastic work term. When ϵ^c is large, more elastic energy can be relieved when one type of atom is replaced by another at a given applied stress. Changing the composition, however, engenders a change in the chemical free energy and the elasticity-induced composition change can only proceed until the decrease in elastic energy is exactly balanced by the increase in chemical energy resulting from the composition change. G''_m is an inverse measure of the allowed composition change. If G''_m is large, corresponding to a strong curvature of the free energy, small changes in the composition result in large changes in the chemical free energy. If G''_m is small and the free energy curve is a relatively weak function of the composition, larger changes in composition give rise to progressively smaller increases in the free energy. The magnitude of the elasticity-induced composition changes always depends on the trade-off between the elastic and chemical energies and can range from essentially zero up to about 10 $at\%$.

3.3.2 Example 2: Stress-induced solute redistribution

Another example illustrating the interaction between elastic stress and composition concerns solute redistribution around such stress centers as dislocations and second-phase precipitates [28, 29, 30]. Consider an infinite, and initially homogeneous, binary crystal of composition c_o. A dislocation or second-phase particle is introduced into the crystal and its stress field induces solute redistribution. As observed in the previous example, it is expected that larger solute atoms would segregate to regions of tension in the crystal while smaller solute atoms would segregate to regions of compression, both in order to reduce the elastic energy of the system. Solute redistribution results in a concomitant increase in the chemical energy of the system, however, and will continue until the decrease in elastic energy is exactly offset by the increase in chemical energy. Equilibrium is achieved once the diffusion potential, M_{BA}, is uniform throughout the crystal. Since the elastic field is position dependent, the equilibrium composition will also be position dependent. The magnitude of solute redistribution cannot be determined by direct application of Eq. (260) with the stress field of the stress center substituted for the stress term on the right-hand-side of the equation, however, since a change in composition (from c_o) induces a compositional strain that partially offsets the stress field of the stress center. Stresses thus arise from two sources in this case, the stress center and the heterogeneous composition field[12].

The stress and composition are coupled in two terms: the expression for the stress dependence of the diffusion potential, Eq. (252), and the stress-strain constitutive equation of Eq. (143):

$$\sigma_{ij} = C_{ijkl} \left[\epsilon_{kl} - \epsilon^c(c - c_o)\delta_{kl}\right], \tag{264}$$

where c is the local composition and ϵ^c is the compositional strain. In writing Eq. (264), it has been assumed that the stress field generated by the solute atom is similar to that generated by a very small misfitting sphere [31].

Far from the source of stress, the crystal composition remains unperturbed at c_o and solute redistribution can be considered to occur at constant diffusion potential or under open-system conditions[13]. Equilibrium is obtained when the diffusion potential at all points near the stress center assumes a value equal to the diffusion potential far from the stress center. For an isotropic system in which the elastic constants do not depend upon composition, this requires:

$$M_{BA}(\theta, c_o) = \mu_B(\theta, c_o) - \mu_A(\theta, c_o) = \mu_B(\theta, c) - \mu_A(\theta, c) - \frac{\epsilon^c}{\rho'_o}\sigma_{kk}. \tag{265}$$

If the chemical potentials are expanded in a Taylor series in the composition about the far-field composition c_o as in Eq. (259), Eq. (265) can be written in a form identical to Eq. (260).

Equations (260) and (264) provide two equations for the stress and composition fields. Substituting Eq. (260) into Eq. (264) gives:

$$\sigma_{ij} = C_{ijkl} \left[\epsilon_{kl} - \frac{(\epsilon^c)^2 \mathcal{N}_A}{G''_m \rho'_o}\sigma_{nn}\delta_{kl}\right]. \tag{266}$$

Contracting each side of Eq. (266) with the elastic compliance tensor, S_{mnij}, yields:

$$S_{mnij}\sigma_{ij} = S_{mnij}C_{ijkl} \left[\epsilon_{kl} - \frac{(\epsilon^c)^2 \mathcal{N}_A}{G''_m \rho'_o}\sigma_{nn}\delta_{kl}\right]. \tag{267}$$

Making use of Eq. (103) for the inverse relationship between S_{ijkl} and C_{ijkl} allows Eq. (267) to be simplified to:

$$S_{mnij}\sigma_{ij} = \delta_{km}\delta_{ln} \left[\epsilon_{kl} - \frac{(\epsilon^c)^2 \mathcal{N}_A}{G''_m \rho'_o}\sigma_{nn}\delta_{kl}\right] = \epsilon_{mn} - \frac{(\epsilon^c)^2 \mathcal{N}_A}{G''_m \rho'_o}\sigma_{nn}\delta_{mn}. \tag{268}$$

Solving for the strain field and using $\sigma_{nn} = \sigma_{jj} = \sigma_{ij}\delta_{ij}$ gives:

$$\epsilon_{mn} = \left[S_{mnij} + \frac{(\epsilon^c)^2 \mathcal{N}_A}{G''_m \rho'_o}\delta_{ij}\delta_{mn}\right]\sigma_{ij} = S^o_{mnij}\sigma_{ij} \tag{269}$$

[12] In the previous example, a change in composition did not induce a stress, since the crystal was mechanically unconstrained; i.e., the crystal was free to change its dimensions.

[13] Here, the crystal is assumed to be sufficiently large so as to act as a chemical potential reservoir. Mass is considered to flow towards or away from the source of stress.

where a modified, or open-system compliance tensor, S^o_{mnij}, has been defined as:

$$S^o_{mnij} = S_{mnij} + \frac{(\epsilon^c)^2 N_A}{G''_m \rho'_o} \delta_{ij}\delta_{mn} \tag{270}$$

Equation (269) is a modified constitutive law connecting the stress and strain. The modified compliance tensor, S^o_{mnij}, contains both elastic and chemical information about the system. It permits the actual stress field in the system, arising from both the stress center (e.g., dislocation) and solute redistribution, to be determined to first-order in composition change. The elastic solution pertaining to the system without solute redistribution is used, but the actual elastic constants are replaced by the modified elastic compliances of Eq. (270). A set of modified elastic constants, C^o_{ijkl}, connecting the strain to the stress tensor can also be defined by the relation:

$$\sigma_{ij} = C^o_{ijkl}\epsilon_{kl}. \tag{271}$$

The components of C^o_{ijkl} are found by solving $C^o_{ijkl}S^o_{klmn} = \delta_{im}\delta_{jn}$ as in Eq. (103). The modified elastic constants for an isotropic system, μ^o (shear modulus), ν^o (Poisson's ratio) and E^o (Young's modulus) are defined by:

$$\mu^o = \mu, \tag{272}$$

$$\nu^o = \frac{\nu - 2\mu(1+\nu)(\epsilon^c)^2 N_A / G''_m \rho'_o}{1 + 2\mu(1+\nu)(\epsilon^c)^2 N_A / G''_m \rho'_o}, \tag{273}$$

and

$$E^o = \frac{E}{1 + (\epsilon^c)^2 N_A E / G''_m \rho'_o}. \tag{274}$$

Since solute redistribution occurs under open-system conditions (constant diffusion potential) the modified elastic constants are referred to as open-system elastic constants [5].

In order to obtain the solute redistribution around an edge dislocation, a straight dislocation lying along the positive x_3 axis with Burger's vector b directed along the x_2 axis is considered. The trace of the stress field owing to the edge dislocation is given by [32]:

$$\sigma_{kk} = \frac{-\mu b(1+\nu)\sin\phi}{\pi(1-\nu)r} \tag{275}$$

where ϕ is the angle measured from the x_2 axis in the direction of the x_1 axis and $r^2 = x_1^2 + x_2^2$. The actual stress field in the presence of solute redistribution is, according to Eq. (271), obtained by replacing the actual elastic constants by the open system elastic constants of Eqs. (272) and (273). The trace of the stress is then given by:

$$\sigma_{kk} = \frac{-\mu^o b(1+\nu^o)\sin\phi}{\pi(1-\nu^o)r}. \tag{276}$$

Substituting Eq. (276) into Eq. (260) gives the composition field in the vicinity of the dislocation:

$$c(r,\phi) = c_o - \frac{\mu^o b(1+\nu^o)c_o(1-c_o)\epsilon^c \sin\phi}{\rho'_o k\theta\chi\pi(1-\nu^o)r} \tag{277}$$

where c_o is the initial (uniform) composition of component B in the crystal. From Eq. (276), the trace of the stress is negative (the region is under compression) when $\sin\phi > 0$, corresponding to a point for which $x_1 > 0$. If $\epsilon^c > 0$; i.e., the partial molar volume of component B is greater than that of component A, then Eq. (277) predicts that regions of the crystal under compression will be depleted of component B while regions under tension will be enhanced in component B. This result is expected intuitively as the strain energy can be partially relieved by replacing a larger atom with a smaller atom in a region of compression.

The net solute enhancement in the vicinity of the dislocation is obtained by integrating the change in composition over the volume surrounding the dislocation. Solute enhancement per unit length of dislocation is thus:

$$\int_0^\infty \int_0^{2\pi} (c - c_o) r d\phi dr = \int_0^\infty \frac{-\mu^o b(1+\nu^o) c_o (1-c_o)\epsilon^c}{\rho'_o k\theta\chi\pi(1-\nu^o) r} r dr \int_0^{2\pi} \sin\phi \, d\phi = 0 \tag{278}$$

Thus, for the case of a straight edge dislocation in an isotropic matrix, there is no *net* solute enhancement around the dislocation. Regions enriched in solute are compensated by regions depleted in solute.

Precipitates, or second-phase particles, are sources of stress that can also lead to solute redistribution in the surrounding matrix phase. For the case of an isolated misfitting spherical precipitate in an isotropic matrix, we showed in Eq. (179) that the trace of the stress field in the matrix vanishes, $\sigma_{kk} = 0$. This means that, for an isotropic matrix, there is no solute enhancement in the vicinity of the precipitate. For an anisotropic matrix, the trace of the stress in the matrix does not vanish, and solute redistribution in the vicinity of the precipitate would be expected [33].

Figure 16 shows the solute redistribution arising from the interaction between two elastically inhomogeneous spherical precipitates subjected to a uniaxial stress field, σ_{app} applied in the x_3 direction [28]. Solute redistribution was calculated using the open-system elastic constants[14] developed above and, therefore, accounts for the stress fields from the precipitates as well as the compositional inhomogeneity. Contour lines are lines of constant concentration determined by solving the elasticity problem using the open-system elastic constants. The isoconcentrates shown have been normalized by Z_a where:

$$Z_a = \frac{\epsilon^c \sigma_{app}(1+\nu^o)(C_{44}^\beta - C_{44}^\alpha)\mathcal{N}_A}{G''_m C_{44}^\alpha (1-\nu^o)\rho_o}. \tag{279}$$

Solute enhancement occurs when $Z_a > 0$ in the regions of the solid lines and depletion in the regions of dashed lines. Changing the sign of ϵ^c, the applied stress or the relative hardness of the phases changes the sign of Z_a. The contour lines are in intervals of 0.062.

The approach taken above, dependent upon concentration changes occurring at constant chemical potential, can be applied to other problems of solute redistribution, as long as the system can be considered to be in contact with a chemical potential reservoir.

3.4 Two-Phase Equilibria

In this section, the conditions for thermodynamic equilibrium in a two-phase crystal are derived for the case when the interface between the crystals is coherent. The approach employed is identical to that used in the single-phase case; Gibbs' equilibrium criterion is applied to an isolated region

[14] Once again, the crystal is treated as a chemical potential reservoir allowing the open-system elastic constants to be employed.

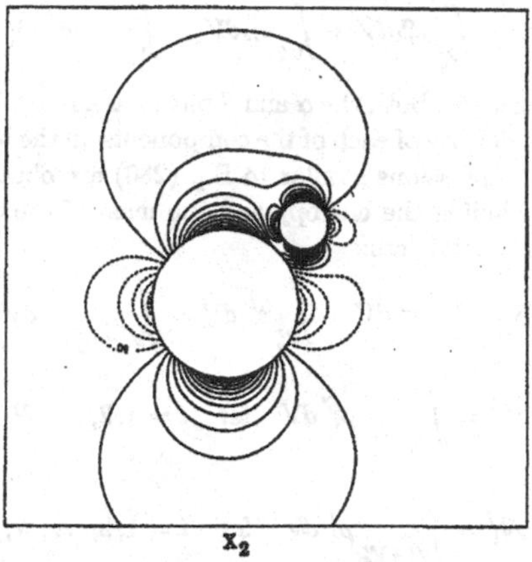

Figure 16: Solute redistribution in the vicinity of two elastically heterogeneous spherical precipitates subjected to a uniaxial stress field in the x_3 direction is depicted for an isotropic matrix. The solid lines represent regions of solute enhancement when $Z_a > 0$. The contour lines are in intervals of 0.062.

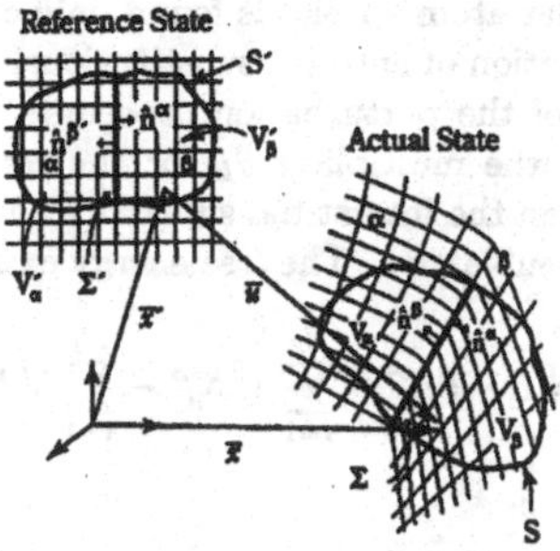

Figure 17: A region V containing both the α and β phases is imagined to be thermodynamically isolated from the rest of the system in its actual state. For a coherent interface, all points in V can be mapped back to a continuous reference state. In order to determine the equilibrium conditions, the thermodynamic state of the isolated region is perturbed at constant entropy and the conditions for which the internal energy vanishes are established.

of material containing both phases and the conditions for which the internal energy is extremized at constant entropy are determined. The equilibrium conditions thus obtained are used to examine phase equilibria in two-phase binary systems in the next section.

The coherency condition implies that each of the crystalline phases, α and β, can be mapped back into an identical reference state, as shown in Fig. 17. Physically, this means there is a continuity of the stacking sequence across the interface, Σ, from one phase into the other. Although the reference state is arbitrary, it is often simplest to choose the unstressed lattice of one of the phases, at a given temperature and composition, as the reference state. The isolated region and its surface in the actual state are denoted by V and S, respectively, and by V' and S' in the reference state.

As for a single-phase system, the total internal energy $\mathcal{E}$ of the region defined by $V = V_\alpha + V_\beta$

is:

$$\mathcal{E} = \int_{V'_\alpha} e^{\alpha}_{v'} dV + \int_{V'_\beta} e^{\beta}_{v'} dV = \int_{V'_\alpha + V'_\beta} e_{v'} dV \tag{280}$$

where the internal energy density for both the α and β phases is a function of the entropy density, the deformation, and the number density of each of the components on the interstitial and substitutional sublattices (see Eq. (194).) Expressions similar to Eq. (280) are obtained for the conservation of other extensive quantities including the entropy and the mass of the different components on the interstitial and substitutional sublattices:

$$\mathcal{S} = \int_{V'_\alpha} s^{\alpha}_{v'} dV + \int_{V'_\beta} s^{\beta}_{v'} dV = \int_{V'_\alpha + V'_\beta} s_{v'} dV \tag{281}$$

$$\mathcal{N}^S_i = \int_{V'_\alpha + V'_\beta} \rho^{S'}_i dV \quad \text{for} \quad i = 1, 2, \ldots, N_S \tag{282}$$

and

$$\mathcal{N}^I_i = \int_{V'_\alpha + V'_\beta} \rho^{I'}_i dV \quad \text{for} \quad i = 1, 2, \ldots, N_I \tag{283}$$

where $\mathcal{N}^S_i$ and $\mathcal{N}^I_i$ are the total number of atoms of species i on the substitutional and interstitial sublattices, respectively. N_I and N_S are the number of different component species found on the interstitial and substitutional sublattices, respectively. As for the treatment of a single-phase system, we assume that a particular atom species is found only on one sublattice.

The constraints on the conservation of entropy, substitutional components and interstitial components required for application of the perturbation procedure are treated using Lagrange multipliers. As before, we introduce the multipliers θ_L for the conservation of entropy, λ^I_i for the conservation of mass components on the interstitial sublattice and λ^S_i for the conservation of mass components on the substitutional sublattice. The free energy extremized at equilibrium is thus:

$$\mathcal{E}^* = \mathcal{E} - \theta_L \mathcal{S} - \sum_{i=1}^{N_S} \lambda^S_i \mathcal{N}^S_i - \sum_{i=1}^{N_I} \lambda^I_i \mathcal{N}^I_i. \tag{284}$$

The first variation in $\mathcal{E}^*$, $\delta\mathcal{E}^*$, is:

$$\delta\mathcal{E}^* = \delta\mathcal{E} - \theta_L \delta\mathcal{S} - \sum_{i=1}^{N_S} \lambda^S_i \delta\mathcal{N}^S_i - \sum_{i=1}^{N_I} \lambda^I_i \delta\mathcal{N}^I_i. \tag{285}$$

The perturbation in the internal energy of the system must account for the possibility of a phase transition between α and β as well as variations in the thermodynamic state of the volume elements as given by Eq. (203). Consequently:

$$\delta\mathcal{E} = \int_{V'_\alpha + V'_\beta} \delta e_{v'} dV + \int_{\Sigma'_\alpha} e^{\alpha}_{v'} \{\delta\vec{a}^{\alpha'} \cdot \hat{n}^{\alpha'}\} dS + \int_{\Sigma'_\beta} e^{\beta}_{v'} \{\delta\vec{a}^{\beta'} \cdot \hat{n}^{\beta'}\} dS. \tag{286}$$

The surface integrals arise from the perturbations that permit the transformation of one phase into the other, which can only occur at the interface between the two phases, Σ. The α or β superscript appearing in the surface integrals identifies the phase in which the interfacial quantity is evaluated. Since both crystalline phases can be referred to the same reference state when they

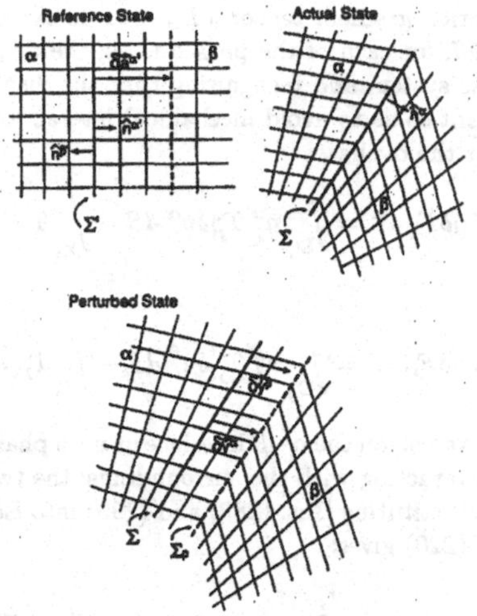

Figure 18: **The interpretation of the perturbation in the position of the coherent interface, Σ, is shown. Σ and Σ_p show the actual and varied positions of the interface respectively. In this illustration, a portion of the β phase has transformed into α.**

are coherent, $\Sigma'_\alpha = \Sigma'_\beta = \Sigma'$. These terms give the perturbation in the internal energy when a small portion of one phase transforms into the other.

The vectors $\delta\vec{a}^{\alpha'}$ and $\delta\vec{a}^{\beta'}$ are measures of the virtual accretion to the α and β phases, respectively, as measured with respect to the reference state. (See Fig. 18.) For example, $\delta\vec{a}^{\alpha'}$ gives the amount and direction for which α phase is added at the interface. If interfacial (capillary) effects are ignored and the interface is assumed to be planar, then it is simplest to consider that the transformation of one phase into the other (the accretion term) occurs perpendicular to the interface in the reference state[15]. The magnitude of the accretion can be defined then by the scalars $\delta a^{\alpha'}$ and $\delta a^{\beta'}$ where $\delta a^{\alpha'} = \delta\vec{a}^{\alpha'} \cdot \hat{n}^{\alpha'}$ and $\delta a^{\beta'} = \delta\vec{a}^{\beta'} \cdot \hat{n}^{\beta'}$. Since the interface is coherent, $\delta\vec{a}^{\alpha'} = \delta\vec{a}^{\beta'}$ and $\delta a^{\alpha'} = \delta a^{\beta'}$. These relationships are developed more completely in the appendix.

Expressions similar to Eq. (286) exist for variations in the other extensive quantities. For example, the variation in the system entropy is given by:

$$\delta S = \int_{V'_\alpha + V'_\beta} \delta s_{v'} dV + \int_{\Sigma'_\alpha} s^\alpha_{v'} \{\delta\vec{a}^{\alpha'} \cdot \hat{n}^{\alpha'}\} dS + \int_{\Sigma'_\beta} s^\beta_{v'} \{\delta\vec{a}^{\beta'} \cdot \hat{n}^{\beta'}\} dS. \tag{287}$$

Using Eq. (204), the change in the free energy $\mathcal{E}^*$ owing to perturbations in the thermodynamic state can be written:

$$\begin{aligned}
\delta\mathcal{E}^* \;=\; & \int_{V'} \left\{ (\theta - \theta_L)\delta s_{v'} + T_{ji}\delta F_{ij} + \sum_{i=1}^{N_S} (\gamma_i^S - \lambda_i^S)\delta\rho_i^{S'} + \sum_{i=1}^{N_I} (\gamma_i^I - \lambda_i^I)\delta\rho_i^{I'} \right\} dV \\
& + \int_{\Sigma'_\alpha} \left\{ e^\alpha_{v'} - \theta_L s^\alpha_{v'} - \sum_{i=1}^{N_S} \lambda_i^S \rho_i^{S\alpha'} - \sum_{i=1}^{N_I} \lambda_i^I \rho_i^{I\alpha'} \right\} \left\{ \delta\vec{a}^{\alpha'} \cdot \hat{n}^{\alpha'} \right\} dS \\
& + \int_{\Sigma'_\beta} \left\{ e^\beta_{v'} - \theta_L s^\beta_{v'} - \sum_{i=1}^{N_S} \lambda_i^S \rho_i^{S\beta'} - \sum_{i=1}^{N_I} \lambda_i^I \rho_i^{I\beta'} \right\} \left\{ \delta\vec{a}^{\beta'} \cdot \hat{n}^{\beta'} \right\} dS.
\end{aligned} \tag{288}$$

[15]**When the accretion term is referred to the reference state, accretion is nothing other than the addition of *undeformed* lattice planes to the surface of one phase and the removal of those planes from the other phase.**

The variation in the deformation gradient tensor, δF_{ij}, can be expressed in terms of the variation of the displacement vector, $\delta \vec{u}$, for each of the phases using the divergence theorem. Referring to Eq. (218) for the single-phase system and recognizing that the displacement at the surface of the region V must remain constant so as to avoid mechanical interaction with the surroundings (thus $\delta \vec{u} = 0$ on S), one obtains for the α phase:

$$\int_{V'_\alpha} T_{ji}\delta F_{ij} dV = \int_{\Sigma'_\alpha} n_j^{\alpha'} T_{ji}^{\alpha} \delta u_i^{\alpha'} dS - \int_{V'_\alpha} T_{ji,j}\delta u_i dV \tag{289}$$

and for the β phase:

$$\int_{V'_\beta} T_{ji}\delta F_{ij} dV = \int_{\Sigma'_\beta} n_j^{\beta'} T_{ji}^{\beta} \delta u_i^{\beta'} dS - \int_{V'_\beta} T_{ji,j}\delta u_i dV. \tag{290}$$

The surface integral does not vanish on the $\alpha-\beta$ interface as each phase is free to perform mechanical work on the other without interacting with the surroundings; the two phases are not mechanically isolated from one another. Substituting Eqs. (289) and (290) into Eq. (288) and using the network constraint of Eqs. (219) and (220) gives:

$$\begin{aligned}
\delta \mathcal{E}^* &= \int_{V'_\alpha + V'_\beta} \left\{ (\theta - \theta_L)\delta s_{v'} + \sum_{i=2}^{N_S} (M_{i1}^S - \lambda_i^S + \lambda_1^S)\delta \rho_i^{S'} + \sum_{i=2}^{N_I} (M_{i1}^I - \lambda_i^I + \lambda_1^I)\delta \rho_i^{I'} \right\} dV \\
&- \int_{V'_\alpha + V'_\beta} T_{ji,j}\delta u_i dV + \int_{\Sigma'_\alpha} n_j^{\alpha'} T_{ji}^{\alpha} \delta u_i^{\alpha} dS + \int_{\Sigma'_\beta} n_j^{\beta'} T_{ji}^{\beta} \delta u_i^{\beta} dS \\
&+ \int_{\Sigma'_\alpha} \{\omega_{v'}^{\alpha}\}\{\delta \vec{a}^{\alpha'} \cdot \hat{n}^{\alpha'}\} dS + \int_{\Sigma'_\beta} \{\omega_{v'}^{\beta}\}\{\delta \vec{a}^{\beta'} \cdot \hat{n}^{\beta'}\} dS
\end{aligned} \tag{291}$$

where the diffusion potentials, M_{i1}^S and M_{i1}^I are defined by Eqs. (224) and (225), respectively, and the thermodynamic potential, $\omega_{v'}$, is defined as:

$$\omega_{v'} = e_{v'} - \theta_L s_{v'} - \sum_{i=1}^{N_S} \lambda_i^S \rho_i^{S'} - \sum_{i=1}^{N_I} \lambda_i^I \rho_i^{I'}. \tag{292}$$

Although the variations in the volume integral are now independent, the variations appearing in the surface integrals are not, owing to the coherency constraint and material continuity. Using Eq. (362) derived in the appendix, Eq. (291) can be expressed as:

$$\begin{aligned}
\delta \mathcal{E}^* &= \int_{V'_\alpha + V'_\beta} \left\{ (\theta - \theta_L)\delta s_{v'} + \sum_{i=2}^{N_S} (M_{i1}^S - \lambda_i^S + \lambda_1^S)\delta \rho_i^{S'} + \sum_{i=2}^{N_I} (M_{i1}^I - \lambda_i^I + \lambda_1^I)\delta \rho_i^{I'} \right\} dV \\
&- \int_{V'_\alpha + V'_\beta} T_{ji,j}\delta u_i dV \\
&+ \int_{\Sigma'_\alpha} \left\{ \left[\omega_{v'}^{\alpha} - \omega_{v'}^{\beta} + n_k^{\alpha'} T_{ki}^{\beta} \left(F_{ij}^{\beta} - F_{ij}^{\alpha} \right) n_j^{\alpha'} \right] \delta a^{\alpha'} + \left[n_j^{\alpha'} T_{ji}^{\alpha} + n_j^{\beta'} T_{ji}^{\beta} \right] \delta u_i^{\alpha} \right\} dS.
\end{aligned} \tag{293}$$

The variations appearing in both the surface and volume integrals are now independent of one another. The free energy $\mathcal{E}^*$ is extremized when the variation in $\mathcal{E}^*$ is zero ($\delta \mathcal{E}^* = 0$.) This condition is satisfied when the term multiplying each individual variation in Eq. (293) vanishes.

The resulting equilibrium conditions for the two crystals are uniform temperature and diffusion potentials given by:

$$\theta^{\alpha} = \theta^{\beta} = \theta_L, \tag{294}$$

$$M_{i1}^{S\alpha} = M_{i1}^{S\beta} = \lambda_i^S - \lambda_1^S \quad \text{for } i = 2, \ldots, N_S, \tag{295}$$

and

$$M_{i1}^{I\alpha} = M_{i1}^{I\beta} = \lambda_i^I - \lambda_1^I \quad \text{for } i = 2, \ldots, N_I. \tag{296}$$

Using these equilibrium conditions and the network constraints of Eqs. (195) and (196), the thermodynamic potential of Eq. (292) can be written:

$$\omega_{v'} = e_{v'} - \theta_L s_{v'} - \sum_{i=2}^{N_S} M_{i1}^S \rho_i^{S'} - \sum_{i=2}^{N_I} M_{i1}^I \rho_i^{I'} - \lambda_1^S \rho_o^{S'} - \lambda_1^I \rho_o^{I'}. \tag{297}$$

Mechanical equilibrium in each crystal requires:

$$T_{ji,j}^{\alpha} = 0 \text{ and } T_{ji,j}^{\beta} = 0. \tag{298}$$

At the interface, mechanical equilibrium obtains when the tractions across the interface are balanced:

$$n_j^{\alpha'} T_{ji}^{\alpha} + n_j^{\beta'} T_{ji}^{\beta} = 0 \tag{299}$$

which, in the small strain approximation, becomes:

$$n_j^{\alpha} \sigma_{ji}^{\alpha} + n_j^{\beta} \sigma_{ji}^{\beta} = 0. \tag{300}$$

Equation (299) or (300) requires the sum of the mechanical forces to vanish at each point along the interface.

The final equilibrium condition at the interface arises from the virtual phase transformation and is given by:

$$\omega_{v'}^{\alpha} - \omega_{v'}^{\beta} + n_k^{\alpha'} T_{ki}^{\beta} \left(F_{ij}^{\beta} - F_{ij}^{\alpha}\right) n_j^{\alpha'} = 0. \tag{301}$$

Equation (301) can also be expressed in terms of the small strain and Cauchy stress tensors:

$$\omega_{v'}^{\alpha} - \omega_{v'}^{\beta} + \sigma_{ki}^{\alpha} \left(\epsilon_{ij}^{\beta} - \epsilon_{ij}^{\alpha}\right) = 0. \tag{302}$$

Equation (301) (or Eq. (302)) must be satisfied at each point along the interface for the system to be in equilibrium. When Eq. (301) is not satisfied, there exists a thermodynamic driving force for the transformation of one phase into the other. This equation can be used to define kinetic equations for the local velocity of the interphase interface [34, 35, 36] and will be used in the next section to examine thermodynamic equilibrium in two-phase stressed crystals.

4 Phase Equilibria

During many diffusional phase transformations, the atoms comprising the crystal will rearrange themselves on the lattice to form new phases of different composition. During this process the lattice often remains intact, even though it is elastically distorted when the partial molar volumes (strains) of the component species differ. If the lattice remains intact and the individual

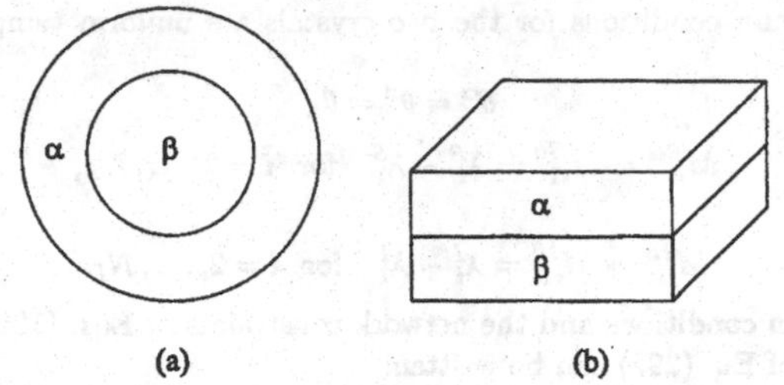

Figure 19: **Two idealized system geometries are depicted that allow the conditions for thermodynamic equilibrium to be satisfied simultaneously. In (a) the α and β phases appear as concentric spheres with the stress state of the α phase being a function of position. The elastic constants of both phases must be isotropic in this configuration. In (b) the phases appear as parallel plates. If the system is not allowed to bend, the stress state will be uniform in each phase. If no external stresses are applied and the system (alloy) composition and temperatures are the same, the equilibrium compositions and volume fractions could differ between the geometries because of the different stress states.**

lattice sites can be associated with the sites in a perfect lattice (or reference state) then the phase transformation is coherent and the equilibrium that is established in this case is termed coherent equilibrium [37]. Coherent equilibrium is not restricted to systems in which the two phases possess the same crystal structure. For example, a diffusional transformation can result in a coherent tetragonal or orthorhombic precipitate in a cubic matrix.

Just as a change in pressure will alter the relative stability of two phases and thereby affect the equilibrium compositions and volume fraction in a multiphase system, the application of an applied stress or the presence of misfit strains will influence phase equilibria in multiphase crystals [37-57]. Sometimes the changes induced by elastic deformation are qualitatively similar to those induced by a pressure, such as when the phases are constrained to be epitaxial thin films on a thick or rigid substrate [37, 45, 49]. In other cases, qualitatively different phenomena are observed, including the existence of more than one thermodynamically stable state and discontinuous jumps in the equilibrium phase fraction with a continuous change in temperature or bulk composition [38, 43, 55].

The difference in the predicted behavior between elastically stressed coherent systems and incoherent (or fluid) systems is thought to be a result of the long-range elastic interaction between phases in a coherent system. This elastic interaction renders the state of deformation and, therefore, the thermodynamic state of one phase dependent on material properties and volume fraction of the other phases present in the system [51]. In addition, the thermodynamic state of a phase will depend on the morphology and spatial distribution of the coexisting phases, see Fig. 19. This is in contrast to a fluid system subjected to a pressure in that the pressure in each of the phases and individual phase domains at equilibrium, in the absence of capillarity effects, is equal to the applied pressure and is independent of the domain morphology [1].

In this section, the conditions for thermodynamic equilibrium will be used to illustrate some of the qualitative changes in phase equilibria and phase diagrams engendered by elastic stress. In order to do so, it is important first to define carefully what is meant by a phase and to classify appropriately the various thermodynamic variables.

4.1 Thermodynamic Relationships

4.1.1 Thermodynamic fields and densities

Following Griffiths and Wheeler [58], thermodynamic quantities can be divided into two categories, thermodynamic *fields* and *densities*. Thermodynamic fields have the property that they assume identical values within and between all homogeneous phases at equilibrium. As such, they are equivalent to thermodynamic intensive variables. Common examples of thermodynamic fields in unstressed systems include the temperature and chemical potentials. Thermodynamic densities are those thermodynamic variables which, at equilibrium, are constant throughout a given phase but do not assume identical values in the different phases at equilibrium. In addition, thermodynamic densities must remain unchanged by the combination of two identical systems. The entropy per unit volume or the phase compositions are examples of thermodynamic densities.

In addition to the thermodynamic fields and densities, we will consider externally controlled variables. These variables are problem-dependent and are fixed by experiment. The externally controlled variables are usually a combination of thermodynamic fields and thermodynamic densities. Since thermodynamic densities are not equal between phases at equilibrium, we will refer to thermodynamic densities that relate to the system in its entirety as system densities. A system density commonly employed as an externally controlled variable is the alloy composition. For example, in a single-component fluid system that reaches equilibrium under conditions of imposed constant temperature and pressure, the externally controlled variables are the two thermodynamic fields, temperature and pressure. For a binary system at constant temperature, pressure and composition, the externally controlled variables would correspond to two thermodynamic fields (temperature and pressure) and one system density (bulk composition). The correct recognition of the appropriate thermodynamic fields, thermodynamic densities, and externally controlled variables is crucial to the construction and understanding of phase diagrams in coherent solids.

A system with n independent variables can be characterized by $n+1$ thermodynamic fields, h_i, $i = 0, 1, ..., n$ [1]. One thermodynamic field can be treated as a function of the other n thermodynamic fields[16] and assumes equal values in coexisting, homogeneous phases at equilibrium. A particularly simple example is a one-component fluid for which the temperature and pressure can be varied. The chemical potential or molar Gibbs free energy assumes the role of the dependent thermodynamic field as it is equal within and between phases at equilibrium and is a function of the temperature and pressure. The dependent thermodynamic field is designated as h_0 or ϕ[17]. Any of the thermodynamic fields can be chosen as the dependent thermodynamic field ϕ. However, a particularly convenient choice can be identified from the condition of thermodynamic equilibrium that applies at a two-phase interface; i.e., that condition relating to the transformation of one phase into the other[18]. In an unstressed or fluid system, this equilibrium condition is the equality of the pressures, $P^\alpha = P^\beta$, across the interface. This is equivalent to the equality of the density of the grand canonical or thermodynamic potential across the interface, $\omega_v^\alpha = \omega_v^\beta$, measured in energy per unit volume. For example, in a binary fluid system:

$$-P = \omega_v = \omega_v(\theta, \mu_1, \mu_2) = e_v - \theta s_v - \mu_1 \rho_1 - \mu_2 \rho_2 \tag{303}$$

[16] Since only n thermodynamic fields are independent, one of the thermodynamic fields must be dependent.

[17] As will be shown, the dependent thermodynamic field ϕ is often a density, but not a thermodynamic density, in that it has units of energy per unit volume. In such cases it will be identified by ϕ_v.

[18] This equilibrium condition is an interfacial condition. For the two-phase coherent solid, it derives from the perturbation $\delta a^{\alpha'}$ appearing in Eq. (293).

where e_v and s_v are the internal energy and entropy densities, respectively, θ is the absolute temperature, μ is the chemical potential, and ρ is the mass density of the indicated component. The free energy and thermodynamic field, ω_v, is a function of the three independent thermodynamic fields θ, μ_1 and μ_2 in the binary fluid. Furthermore, it follows from Eq. (303) that:

$$-dP = d\omega_v = -s_v d\theta - \rho_1 d\mu_1 - \rho_2 d\mu_2. \tag{304}$$

The thermodynamic fields are obtained by differentiation of ω_v:

$$-s_v = \left(\frac{\partial \omega_v}{\partial \theta}\right)_{\mu_1,\mu_2} \quad -\rho_1 = \left(\frac{\partial \omega_v}{\partial \mu_1}\right)_{\theta,\mu_2} \quad -\rho_2 = \left(\frac{\partial \omega_v}{\partial \mu_2}\right)_{\theta,\mu_1}. \tag{305}$$

In general, the n thermodynamic densities are given by:

$$d_i = \frac{\partial \phi}{\partial h_i} \quad \text{for} \quad i = 1, ..., n \tag{306}$$

where the thermodynamic fields other than h_i are held constant in the differentiation of Eq. (306). Equation (304) is a fundamental equation and the thermodynamic densities are the conjugate thermodynamic variables to the thermodynamic fields. This choice of thermodynamic fields is particularly useful as ω is the free energy that is minimized at equilibrium for the fixed set of fields θ, μ_1, and μ_2.

4.1.2 Field diagrams and phase diagrams

The n independent thermodynamic fields, h_i, can be considered to form an n-dimensional thermodynamic space, which shall be referred to as the thermodynamic field space. Two systems which occupy the same point in the thermodynamic field space are the same phase if the corresponding thermodynamic densities of the phases are equal and are different phases when at least one of the thermodynamic densities is different [58]. Two phases, α and β, cannot coexist unless they occupy the same point in the thermodynamic field space since, if they are in equilibrium, the following conditions for the independent thermodynamic fields must be satisfied:

$$h_i^\alpha = h_i^\beta. \tag{307}$$

The n equalities of Eq. (307), in connection with the equilibrium condition $\phi^\alpha = \phi^\beta$, define a hypersurface of dimension $n-1$ in the n-dimensional field space[19]. This hyper- or coexistence surface represents a first-order phase transformation between the two phases when at least one of the thermodynamic densities is discontinuous between the two phases [58]. Three-phase coexistence occurs along a hypersurface of dimension $n-2$ (the intersection of two hypersurfaces of dimension $n-1$). Hypersurfaces of dimension $n-2$ also result from the termination of a hypersurface of dimension $n-1$ and give a surface of critical points. The depiction of phase equilibria in the thermodynamic field space is referred to as a field-space diagram or, simply, a field diagram. The depiction of the phase boundaries in a thermodynamic space that is spanned by at least one of the thermodynamic densities is referred to as a phase diagram.

[19] As an example, consider the temperature-pressure phase diagram of a single-component system. There are two independent thermodynamic fields (temperature and pressure) and the dependent field is the molar Gibbs free energy. In the two-dimensional space of temperature and pressure, single phases are found in areas (two dimensions), regions of two-phase coexistence along lines (one dimension) and three-phase coexistence at points.

4.1.3 Phase rule

The Gibbs phase rule relates the number of homogeneous phases present in a system to the number of thermodynamic fields that can be varied independently without changing the number of phases in the system. In the previous paragraph it was argued that, for a system with n independent thermodynamic fields, single-phase equilibria exists in a region of dimension n in the thermodynamic field space, two-phase equilibria is found along a hypersurface of dimension $n-1$ and three-phase equilibria exists along a hypersurface of dimension $n-2$. In general, the coexistence of p phases in an n-dimensional field space corresponds to a hypersurface of dimension $n-p+1$. Recognizing that a hypersurface of dimension $n-p+1$ allows $n-p+1$ thermodynamic fields to be independently varied while still remaining on the hypersurface, the Gibbs phase rule follows directly as:

$$f = n - p + 1 \tag{308}$$

where f is the number of thermodynamic fields that can be changed independently leaving the number of phases in the system constant and is referred to as the number of degrees of freedom. (In fluids, the thermodynamic fields are equivalent to the potentials [47].) This development applies equally well to nonhydrostatically stressed coherent systems so long as the thermodynamic fields are correctly identified and the phases are homogeneously deformed. It is important to remember that the number of degrees of freedom in the system is different from the number of quantities that must be specified in order to define a system uniquely[20].

If all the externally controlled thermodynamic variables are thermodynamic fields, then f is also the number of those externally controlled thermodynamic variables that can be changed independently, keeping the number of phases in the system constant. However, if one or more of the externally controlled variables is a thermodynamic density, the degrees of freedom in the system (given by Eq. (308)), might no longer be equal to the number of externally controlled thermodynamic variables that can be changed while leaving the number of phases present in the system fixed. This is not to say that the Gibbs phase rule (which applies to the fields) is no longer applicable, as it is valid regardless of the choice of the externally controlled variables, but rather that it is possible to change the value of a system density which is an externally controlled variable, holding the other fields fixed, without changing the number of phases present in the system. This is accomplished by a change in the volume fraction of the system as discussed below.

A simple example illustrating this point is a binary fluid system (assume the independent thermodynamic fields are θ, μ_2 and P, and the dependent thermodynamic field is μ_1) in which the externally controlled variables are chosen as θ, P, and c_o where c_o is the alloy (bulk) composition measured in terms of the mole fraction of component 2. Within a single-phase field of a temperature-pressure-composition phase diagram, θ, P and c_0 can all be changed independently without changing the number of phases (the number of degrees of freedom is equal to the number of independent thermodynamic fields.) However, at a point within a two-phase field of the same phase diagram that is not contiguous to a phase boundary, the externally controlled variables θ, P and c_0 can again all be changed independently leaving the two phases in equilibrium. If one of the thermodynamic field variables (θ or P) is changed, all the thermodynamic densities change, as the phases would now occupy a new point in the field space. (In other words, a change in θ or P results in a change in the equilibrium composition, entropy density and molar volume of each

[20] A unique description of the system not only requires knowledge of which phases are present but also the relative amounts of each phase.

phase.) In contrast, if the alloy composition (a system density) is changed at constant θ and P and the two-phase system remains in equilibrium, the thermodynamic densities remain constant; i.e., the equilibrium composition, molar volume and entropy density of each phase does not change. If this were not the case, the system would occupy a new point in the thermodynamic field space which, in general, would not permit two-phase equilibrium. Although the number of externally controlled variables that can be changed independently in the two-phase system, without changing the number of phases present in the system, is again equal to three, the same number as for the single-phase field, there are only two degrees of freedom for the two-phase system. These arguments are also applicable to systems with more than two phases and more than one density as externally controlled variables. The correct identification of the thermodynamic fields and externally controlled variables is important in discussing the applicability of the Gibbs phase rule to coherent systems.

4.1.4 Common-tangent construction

The common-tangent construction [1, 59, 60] is a graphical means of satisfying the thermodynamic equilibrium conditions of systems in which one or more of the externally controlled variables is a system density. Implementing the common-tangent construction for coherent systems requires the appropriate thermodynamic free energy to be identified and plotted as a function of those densities that are experimentally controlled. Phase equilibria between two or more phases are possible when hyperplanes tangent to these surfaces coincide. The densities of the phases at equilibrium are established by the point of tangency of the hyperplane and the free energy curve. The common-tangent construction of coherent systems follows directly from the thermodynamic equilibrium conditions as shown below.

Assume that the externally controlled thermodynamic variables of a two-phase homogeneous system consist of $n-r$ thermodynamic fields h_i $(i = r+1, ..., n)$ and r system densities D_i $(i = 1, ..., r)$. The system densities are the weighted sums of the thermodynamic densities of the individual phases:

$$D_i = (1-z)d_i^{\alpha} + z d_i^{\beta} \quad i = 1, ..., n \tag{309}$$

where z is the volume fraction of the β phase. If ϕ is the free energy which is extremized when all the thermodynamic fields are specified experimentally, the free energy extremized at equilibrium under the above experimental conditions, K, is;;

$$K = [(1-z)k_v^{\alpha} + z k_v^{\beta}]V \tag{310}$$

with;

$$k_v = \phi_v - \sum_{i=1}^{r} d_i h_i. \tag{311}$$

V is the volume of the system, referred to an appropriate reference state, and k_v and ϕ_v are the volume density of the respective free energies.

When the $n-r$ thermodynamic fields h_i $(i = r+1, ..., n)$ are held constant, the free energy density of each phase, k_v, can be plotted as a function of the r thermodynamic densities d_i $(i = 1, ..., r)$. For simplicity, we denote a point in the space spanned by the r thermodynamic densities as $\vec{d}$. The components of the vector $\vec{d}$ are the r thermodynamic densities d_i $(i = 1, ..., r)$. Thus $\vec{d} = (d_1, d_2, ..., d_r)$. If a similar vector $\vec{h}$ is introduced for the thermodynamic fields considered to be established by experiment, that is $\vec{h} = (h_{r+1}, h_{r+2}, ..., h_r)$, then the free energy density, k_v, is a function of $\vec{h}$ and $\vec{d}$: $k_v = k_v(\vec{d}, \vec{h})$. The free energy density, k_v, will be different for the α and β phases.

The equation of the hyperplane tangent to the surface of $k_v^\alpha(\vec{d}, \vec{h})$ at the specific point $\vec{d}^\alpha$ in the thermodynamic space spanned by the free energy k_v and the thermodynamic densities d_i $(i = 1, ..., r)$ is:

$$k_T^\alpha = k_v^\alpha(\vec{d}^\alpha, \vec{h}) + \sum_{i=1}^{r}(d_i - d_i^\alpha)\left(\frac{\partial k_v^\alpha}{\partial d_i}\right)_{\vec{d}=\vec{d}^\alpha} \tag{312}$$

where k_T^α designates the tangent plane to the α phase. The partial derivatives are evaluated at the point of tangency $\vec{d} = \vec{d}^\alpha$.

If a common-tangent construction exists for identifying the equilibrium thermodynamic densities of the phases in equilibrium, then the tangent plane to the α phase at point $\vec{d}^\alpha$ must coincide with the tangent plane to the β phase at the point of tangency $\vec{d}^\beta$. This requires:

$$k_T^\alpha = k_T^\beta. \tag{313}$$

Using Eq. (312), Eq. (313) can be written:

$$k_v^\alpha(\vec{d}^\alpha, \vec{h}) + \sum_{i=1}^{r}(d_i - d_i^\alpha)\left(\frac{\partial k_v^\alpha}{\partial d_i}\right)_{\vec{d}=\vec{d}^\alpha} = k_v^\beta(\vec{d}^\beta, \vec{h}) + \sum_{i=1}^{r}(d_i - d_i^\beta)\left(\frac{\partial k_v^\beta}{\partial d_i}\right)_{\vec{d}=\vec{d}^\beta}. \tag{314}$$

Rearranging Eq. (314) gives:

$$\begin{aligned} 0 \;=\; & k_v^\alpha(\vec{d}, \vec{h}) - \sum_{i=1}^{r} d_i^\alpha \left(\frac{\partial k_v^\alpha}{\partial d_i}\right)_{\vec{d}=\vec{d}^\alpha} - k_v^\beta(\vec{d}, \vec{h}) + \sum_{i=1}^{r} d_i^\beta \left(\frac{\partial k_v^\beta}{\partial d_i}\right)_{\vec{d}=\vec{d}^\beta} \\ & + \sum_{i=1}^{r} d_i \left[\left(\frac{\partial k_v^\alpha}{\partial d_i}\right)_{\vec{d}=\vec{d}^\alpha} - \left(\frac{\partial k_v^\beta}{\partial d_i}\right)_{\vec{d}=\vec{d}^\beta}\right]. \end{aligned} \tag{315}$$

Equation (315) is satisfied identically for all $\vec{d}$ (that is, arbitrary d_i) only when each term appearing in square brackets is identically zero:

$$\left(\frac{\partial k_v^\alpha}{\partial d_i}\right)_{\vec{d}=\vec{d}^\alpha} = \left(\frac{\partial k_v^\beta}{\partial d_i}\right)_{\vec{d}=\vec{d}^\beta} \quad \text{for } i = 1, 2, \ldots, r \tag{316}$$

and when the constant term vanishes:

$$k_v^\alpha(\vec{d}^\alpha, \vec{h}) - \sum_{i=1}^{r} d_i^\alpha\left(\frac{\partial k_v^\alpha}{\partial d_i}\right)_{\vec{d}=\vec{d}^\alpha} = k_v^\beta(\vec{d}^\beta, \vec{h}) - \sum_{i=1}^{r} d_i^\beta\left(\frac{\partial k_v^\beta}{\partial d_i}\right)_{\vec{d}=\vec{d}^\beta}. \tag{317}$$

The set of conditions given by Eq. (316) assures that the tangent planes are parallel. These conditions are equivalent to the equilibrium conditions given by Eq. (307) on the equality of the fields $h_i^\alpha = h_i^\beta$ $(i = 1, \ldots, r)$ since $\partial k_v / \partial d_i = -h_i$ (see Eq. (311)). Equation (317) assures that the parallel tangent planes coincide by intersecting at the origin and is equivalent to the equilibrium condition on the equality of the dependent field in the two phases, $\phi_v^\alpha = \phi_v^\beta$. Thus the common-tangent construction is equivalent to satisfying the conditions for phase equilibria in systems comprised of homogeneous phases. These results are directly applicable to nonhydrostatically stressed coherent systems, provided the phases are homogeneously deformed at equilibrium [51].

The points of tangency to the free energy curves, $\vec{d^{\alpha}}$ and $\vec{d^{\beta}}$, correspond to the equilibrium thermodynamic densities of the phases. It is only when the respective phases possess the thermodynamic densities $\vec{d^{\alpha}}$ and $\vec{d^{\beta}}$ that the tangent planes coincide and all conditions for thermodynamic equilibrium are satisfied. A line drawn connecting the points $\vec{d^{\alpha}}$ and $\vec{d^{\beta}}$ is termed a tie line. The ends of the tie line, by definition, give the equilibrium thermodynamic densities of the phases. (Note that not all thermodynamic densities are given by $\vec{d^{\alpha}}$ and $\vec{d^{\beta}}$, but just those defined by the thermodynamic space $\vec{d}$; i.e., those system densities that were experimentally controlled. The other system densities must be determined from constitutive relations.) The tie lines clearly span the space of the r thermodynamic densities and, when a phase diagram is viewed in any thermodynamic space that does not include all r thermodynamic densities corresponding to those densities controlled experimentally, it will appear as if the common-tangent construction is invalid. In certain stressed systems, the points of tangency are not necessarily equivalent to the phase boundaries but form a set of distinct lines (planes) called *density* lines [51]. The above analysis can be extended directly to systems with more than two phases.

4.2 Example: Parallel Plates

In this subsection, we use the conditions for thermodynamic equilibrium and the results of the previous subsection to examine the influence of elastic stress on two-phase equilibria and the construction of phase diagrams for coherent systems.

The elastic state of a phase depends upon the geometry of the system. Consequently, investigations of phase equilibria in stressed systems usually assume a specific system geometry a priori. The conditions for thermodynamic equilibrium have been applied to at least two different system geometries, concentric spheres and parallel plates (see Fig. 19), in order to examine the characteristics of equilibrium in coherent systems [38-47]. Since the results in these two cases are qualitatively similar and both satisfy the conditions for thermodynamic equilibrium, the parallel plate geometry will be used to illustrate how elastic stress can influence phase diagram construction[21]. Unless stated otherwise, the temperature is held fixed, the binary system is closed with respect to mass and only systems with isotropic or cubic symmetry are considered.

As we saw in the previous subsection, before phase diagrams can be constructed, it is necessary to establish which thermodynamic variables behave as thermodynamic fields, which as thermodynamic densities and which variables are the externally controlled variables. The importance of these concepts are conveyed by considering the two types of mechanical loading conditions applied to the elastically stressed system of parallel plates shown in Fig. 20. In Fig. 20a, a stress, σ^{o}_{33}, is applied in the x_3 (vertical) direction. In the x_1 and x_2 (horizontal) directions, the position of the surfaces have been fixed. By fixing the position of the surface, the strain in the plane of the plates, ϵ_{11} and ϵ_{22}, is established experimentally. In Fig. 20b, stresses are applied to all surfaces. In each case, the deformation in the x_1 and x_2 directions is constrained to remain homogeneous by the presence of rigid clamps along the surface. This configuration results in the stress and strain within each phase being constant and independent of position. It also means that the components of the strain tensor (or deformation gradient tensor F_{ij}) lying within the plane of the interface are continuous

[21]The parallel plate construction has the additional advantage that the phases are homogeneously deformed at equilibrium and the results of the previous section can be applied.

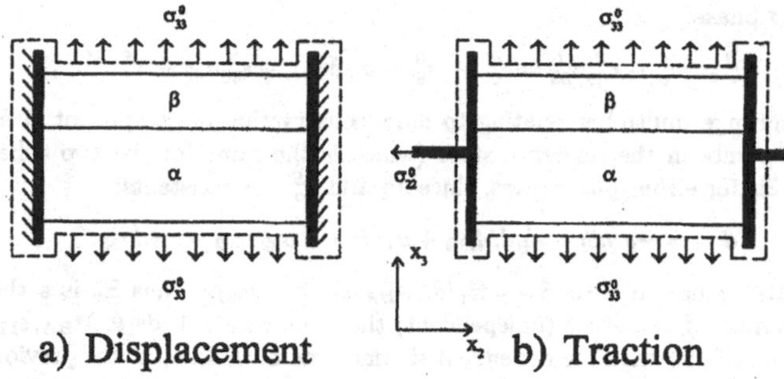

Figure 20: **Two thermodynamic systems for the parallel plate geometry are shown. In (a) displacement boundary conditions have been imposed to hold the dimensions of the system in the $x_1 - x_2$ plane constant. When tractions are imposed as in (b), the dimensions in the $x_1 - x_2$ plane can change, but the stresses must balance. In each case, the stress and strain within each phase are constant and independent of position.**

across the interface: For an isotropic system or a cubic system with principal axes lying along the coordinate axes, this means that $\epsilon_{11}^{\alpha} = \epsilon_{11}^{\beta} = \epsilon_{11}$, $\epsilon_{22}^{\alpha} = \epsilon_{22}^{\beta} = \epsilon_{22}$ and all off-diagonal components of the strain tensor are zero.

As a result of the imposed system geometry, the strain components ϵ_{11} and ϵ_{22} behave as thermodynamic fields: they are constant within each phase and equal between phases at equilibrium. The strain component ϵ_{33}, however, will differ in each of the phases, even when the traction in the x_3 direction vanishes, because the material is free to deform in the x_3 direction in order to lower its elastic energy. The extent of deformation depends on the elastic constants (Poisson's ratio in the isotropic case) and the stress-free lattice parameter of each phase. The conjugate stress term, σ_{33}, is constant within each phase and equal between phases (see Eq. (300)) and is thus classified as a thermodynamic field. Thus the strain components ϵ_{11} and ϵ_{22} and the stress component σ_{33} behave as thermodynamic fields while their conjugate variables, σ_{11}, σ_{22} and ϵ_{33} behave as thermodynamic densities. In order to simplify the following presentation, it is assumed that $\epsilon_{11} = \epsilon_{22}$, from which it follows that $\sigma_{11}^{\alpha} = \sigma_{22}^{\alpha}$ and $\sigma_{11}^{\beta} = \sigma_{22}^{\beta}$ ($\sigma_{11}^{\alpha} \neq \sigma_{11}^{\beta}$). The thermodynamic fields for this problem are temperature (θ), diffusion potential (M_{BA}), two strain components ($\epsilon_{11} = \epsilon_{22}$) and one stress component (σ_{33}).

Finally, for the purpose of constructing phase diagrams and free energy curves for stressed systems, it is necessary to identify the dependent thermodynamic field ϕ_v of the previous subsection. This is accomplished by considering the interfacial equilibrium condition in the small strain approximation, Eq. (302), and using the parallel plate condition for which $\epsilon_{11}^{\alpha} = \epsilon_{11}^{\beta}$ and $\epsilon_{22}^{\alpha} = \epsilon_{22}^{\beta}$. (The normal to the interface lies in the x_3 direction.) Since the off-diagonal components of the strain tensor vanish in this configuration, the only non-zero strain component contributing to Eq. (302) is ϵ_{33}. Equation (302) can, therefore, be written for the assumed parallel plate geometry as:

$$\omega_{v'}^{\alpha} - \epsilon_{33}^{\alpha}\sigma_{33}^{\alpha} = \omega_{v'}^{\beta} - \epsilon_{33}^{\beta}\sigma_{33}^{\beta}. \tag{318}$$

For the present geometry, $\sigma_{33}^{\alpha} = \sigma_{33}^{\beta} = \sigma_{33}^{o} = \sigma_{33}$, and we can define the free energy density, Ξ_v, as:

$$\Xi_v^{\alpha} = \omega_{v'}^{\alpha} - \epsilon_{33}^{\alpha}\sigma_{33}. \tag{319}$$

Using the expression for $\omega_{v'}$ given by Eq. (297) and limiting the analysis to a binary substitutional alloy gives:

$$\Xi_v^{\alpha} = e_{v'}^{\alpha} - \theta s_{v'}^{\alpha} - \rho_B^{\alpha'} M_{BA} - \epsilon_{33}^{\alpha}\sigma_{33} - \lambda_A^S \rho_o^{S'}. \tag{320}$$

Likewise, for the β phase:

$$\Xi_v^\beta = \omega_{v'}^\beta - \epsilon_{33}^\beta \sigma_{33}^\beta = e_{v'}^\beta - \theta s_{v'}^\beta - \rho_B^{\beta'} M_{BA} - \epsilon_{33}^\beta \sigma_{33} - \lambda_A^S \rho_o^{S'}. \tag{321}$$

where λ_A^S is a Lagrange multiplier relating to mass conservation of component A and $\rho_o^{S'}$ is the density of lattice points in the reference state (which is the same for the two coherent phases.) Differentiation of Ξ_v for either phase gives, since λ_A^S and $\rho_o^{S'}$ are constants:

$$d\Xi_v = -s_{v'} d\theta - \rho'_B dM_{BA} + \sigma_{11} d\epsilon_{11} + \sigma_{22} d\epsilon_{22} - \epsilon_{33} d\sigma_{33}. \tag{322}$$

Ξ_v is a fundamental equation with $\Xi_v = \Xi_v(\theta, M_{BA}, \epsilon_{11}, \epsilon_{22}, \sigma_{33})$. Thus Ξ_v is a thermodynamic field that is a function of the other (independent) thermodynamic fields $\theta, M_{BA}, \epsilon_{11}, \epsilon_{22}$ and σ_{33}. As such, we can identify Ξ_v with the dependent thermodynamic field ϕ_v of the previous subsection.

4.2.1 Displacement boundary conditions

Specifying the displacement along the edges of the plate (directions perpendicular to the x_3 axis) establishes the strain components ϵ_{11} and ϵ_{22} in both phases. The externally controlled variables in this case are the thermodynamic fields θ, σ_{33} and ϵ_{11} and the thermodynamic (system) density c_o, where c_o is the bulk alloy composition of component B. If z is the phase fraction of β, then the equilibrium phase compositions, c^α and c^β, must satisfy the mass conservation equation (lever rule):

$$c_o = (1-z)c^\alpha + zc^\beta. \tag{323}$$

Equation (323) is an example of the general relationship for homogeneous phases given by Eq. (309).

In this example, the alloy composition is the only system density that is an externally controlled variable. Thus the free energy extremized at equilibrium can be determined from the free energy density k_v using Eq. (311) as (dropping the prime superscript in keeping with the small-strain approximation):

$$k_v = \Xi_v - (-\rho_B) M_{BA} = e_v - \theta s_v - \sigma_{33}\epsilon_{33}. \tag{324}$$

If the traction in the x_3 direction is furthermore taken to be zero, $\sigma_{33} = 0$, then the free energy extremized at equilibrium is $k_v = e_v - \theta s_v = f_v$, the Helmholtz free energy. The common-tangent construction is valid when the Helmholtz free energy, accounting for the elastic energy of each phase, is plotted as a function of composition. Since tie lines span the thermodynamic space defined by those externally controlled variables that are system densities, tie lines, in this case, are found to lie in any thermodynamic space containing the composition (a thermodynamic density).

A system that is well-modelled by the displacement boundary conditions and the parallel plate geometry is that of a thin film deposited epitaxially on a thick or rigid substrate. We assume that the two phases comprising the film are of uniform thickness and lie parallel to the substrate surface and that no stress is applied in the x_3 direction perpendicular to the substrate. The substrate is assumed to fix the lattice parameter of each phase in the $x_1 - x_2$ plane of the film to the lattice parameter of the substrate throughout the thickness of the film, a good approximation away from the edges. Physically, this corresponds precisely to the displacement boundary conditions and system geometry of Fig. 20a.

The Helmholtz free energy density of each phase is given by the sum of the chemical and elastic energy density (e_v^{el}) contributions as [45]:

$$f_v = \rho_o c \mu_B(\theta, c) + (1-c)\rho_o \mu_A(\theta, c) + e_v^{el} \tag{325}$$

where the chemical potentials are evaluated in the stress-free condition and e_v^{el} is given by Eq. (128). In order to determine the strain energy density, a reference state for the measurement of strain is first defined. Although different reference states are possible, it is simplest here to choose the unstressed lattice lattice parameter of the phase of interest as the reference state. Since the lattice parameter in the plane of the film is constrained by the epitaxy to be equal to the lattice parameter of the substrate, a_s, the strain components $\epsilon_{11} = \epsilon_{22}$ are given by:

$$\epsilon_{11} = \epsilon_{22} = \epsilon = (a_s - a)/a \tag{326}$$

where a is the unstressed lattice parameter of the phase of interest. The stress in the plane of the film $\sigma_{11} = \sigma_{22}$ and the strain component ϵ_{33} must be determined using the stress-strain constitutive relations. For phases with cubic symmetry and the crystallographic directions aligned parallel to the coordinate axes as in Fig. 20, we have from Eq. (129):

$$\sigma_{ij} = \{C_{12}\delta_{ij}\delta_{kl} + C_{44}\left(\delta_{ik}\delta_{jl} + \delta_{il}\delta_{jk}\right) + (C_{11} - C_{12} - 2C_{44})\delta_{ijkl}\}\,\epsilon_{kl}. \tag{327}$$

The first equation for the two unknowns (σ_{11} and ϵ_{33}) is obtained by setting $i = j = 3$ with $\sigma_{33} = 0$:

$$\sigma_{33} = 0 = C_{12}\epsilon_{kk} + 2C_{44}\epsilon_{33} + (C_{11} - C_{12} - 2C_{44})\epsilon_{33}. \tag{328}$$

Since $\epsilon_{kk} = \epsilon_{11} + \epsilon_{22} + \epsilon_{33} = 2\epsilon + \epsilon_{33}$, Eq. (328) can be solved directly for ϵ_{33}:

$$\epsilon_{33} = -2C_{12}\epsilon/C_{11}. \tag{329}$$

The stress component $\sigma_{11} = \sigma_{22}$ is obtained by setting $i = j = 1$ in Eq. (327) to give:

$$\sigma_{11} = C_{12}\epsilon_{kk} + 2C_{44}\epsilon_{11} + (C_{11} - C_{12} - 2C_{44})\epsilon_{11}. \tag{330}$$

Using Eq. (329), Eq. (330) becomes:

$$\sigma_{11} = \sigma_{22} = Y_{100}\epsilon \tag{331}$$

where Y is an effective elastic modulus defined by:

$$Y_{100} = (C_{11} - C_{12})(C_{11} + 2C_{12})/C_{11}. \tag{332}$$

In general, the effective elastic modulus depends on the crystallographic orientation of the phases [61]. Of course, both the misfit strain and elastic modulus depend on the phase of interest.

The elastic energy density of a phase in the film is thus:

$$e_v^{el} = \frac{1}{2}\sigma_{ij}\epsilon_{ij} = \frac{1}{2}(\sigma_{11}\epsilon_{11} + \sigma_{22}\epsilon_{22} + \sigma_{33}\epsilon_{33}) = (\epsilon^T)^2 Y. \tag{333}$$

The Helmholtz free energy density of each phase is thus:

$$f_v = \rho_o c\mu_B(\theta, c) + (1 - c)\rho_o\mu_A(\theta, c) + (\epsilon^T)^2 Y. \tag{334}$$

The total Helmholtz free energy, $\mathcal{F}$, becomes:

$$\mathcal{F} = f_v^\alpha V_\alpha + f_v^\beta V_\beta \tag{335}$$

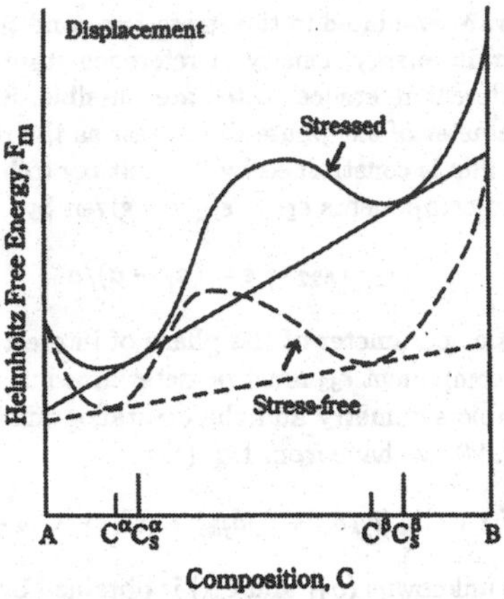

Figure 21: The free energy of a phase in the absence of stress (dashed line) is compared to its free energy when the phase is constrained by displacement boundary conditions for the case where the alloy exhibits a miscibility gap. The common-tangent construction is applied to both sets of free energy curves and the equilibrium phase compositions for the unstressed (c^{α} and c^{β}) and the stressed (c^{α}_s and c^{β}_s) systems are shown. The imposed displacement is taken so that a phase composition of pure A does not engender stress.

where V_α and V_β are the volumes of the α and β phases, respectively.

Figure 21 compares schematically the Helmholtz free energy of a system in the absence of stress (dashed line) with the Helmholtz free energy when the phase is epitaxial to a thick or rigid substrate (corresponding to displacement boundary conditions). The substrate lattice parameter is taken to be that of pure component A, so that no stresses are present for a bulk alloy composition $c = 0$. (This is why the two curves intersect at an alloy composition of pure A.) When the lattice parameter is a function of composition, changes in composition from pure A will induce a stress and the free energy curve will be shifted upwards as shown. In keeping with the development of the previous section, the common-tangent construction can be used in the Helmholtz free energy-composition space in order to obtain the equilibrium compositions of the individual phases graphically. In Fig. 21, the equilibrium phase compositions of the unstressed system are c^{α} and c^{β} whereas the equilibrium phase compositions of the stressed stressed system have shifted to c^{α}_s and c^{β}_s.

Figure 22 depicts how the substrate-induced strains can affect the temperature-composition phase diagram for a ternary III-V semiconductor material [45]. In the absence of all stress effects, these pseudobinaries often exhibit regular solution behavior with a positive regular solution constant, Ω, yielding a free energy curve qualitatively similar to that depicted in Fig. 21. The regular solution constant can be relatively small and the elastic energy can make a significant contribution to the system energy [62]. In Fig. 22, the substrate is taken to be GaP. The broken line represents the miscibility gap in the absence of stress calculated for a regular solution constant, $\Omega = 15.1$kJ [63]. The solid lines depict the calculated miscibility gap for three substrate orientations when the films are epitaxial with the substrate. The curves can be obtained by using the common-tangent construction to the Helmholtz free energy curves as in Eq. (325). A compositional strain of $\epsilon^c = 0.077$ [64] was used (see Eq. (142)). In addition, the elastic constants were assumed to be composition dependent and were obtained using a rule of mixtures between GaP and InP [45, 64].

In the absence of stress, the critical temperature is 910°K. Miscibility gaps appear for all three crystal orientations, although the critical temperature depends on the crystal orientation. The miscibility gap is smaller in the [110] and [111] directions than in [100] owing to the orientation

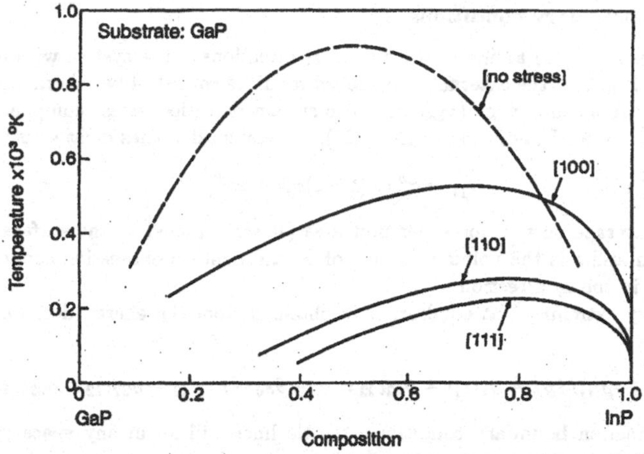

Figure 22: Calculated miscibility gaps for the GaInP pseudobinary for three different substrate orientations; [100], [110] and [111] are shown. The substrate is taken to have the lattice parameter of GaP and is assumed thick or rigid enough to fix the displacement of the phases in the plane normal to the surface (displacement boundary condition.) Tie lines are contained within the plane of the figure.

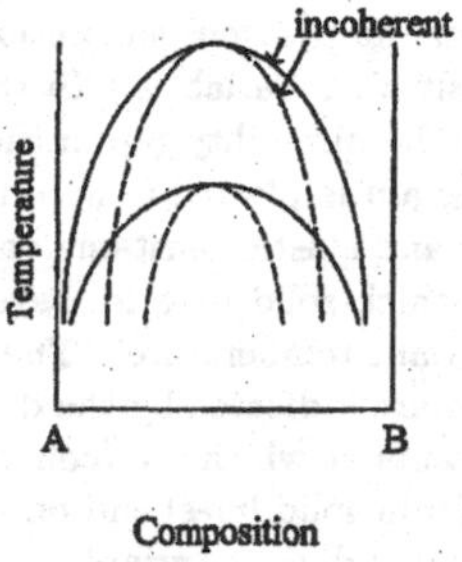

Figure 23: The suppression of the miscibility gap owing to coherency strain is shown as calculated according to Cahn's seminal work [37].

dependence of the elastic energy: the effective elastic moduli for the [110] and [111] substrate orientations, Y_{110} and Y_{111}, are greater than Y_{100} defined by Eq. (332), which results in a greater elastic energy for planes with a [110] and [111] orientation.

The results presented here resemble those of Cahn's early calculations on the stress-induced suppression of the critical point [37]. Figure 23 depicts the change in the phase diagram predicted by Cahn owing to coherency strain and shows the coherent miscibility gap lying completely within the incoherent miscibility gap. This occurs when the displacement boundary conditions are applied to the unstressed matrix phase of the bulk composition. This is equivalent to fixing the substrate lattice parameter to that of the bulk composition before any phase separation has taken place. Thus, with comparison to Fig. 22, Fig. 23 is computed using a different substrate lattice parameter for each bulk alloy composition while Fig. 22 uses the same substrate lattice parameter for all bulk alloy compositions.

4.2.2 Traction boundary conditions

When the tractions (stresses) acting in the x_1 and x_2 directions of the system with parallel plates are specified, as in Fig. 20b, the externally controlled variables consist of two thermodynamic fields, temperature and stress component (σ_{33}), and two system densities, stress component $\sigma_{11} = \sigma_{22}$ and alloy composition c_B. In addition to Eq. (323), the system densities must satisfy:

$$\sigma_{11}^o = \sigma^o = (1-z)\sigma_{11}^\alpha + z\sigma_{11}^\beta \tag{336}$$

where σ_{11}^o is the average normal force per unit area (stress) applied to the surface with normal in the x_1 direction and z is the volume fraction of β. An identical expression connects the stress components (σ_{22}) in the x_2 direction.

The free energy extremized at equilibrium is obtained from the energy density, k_v, using Eq. (311) as:

$$k_v = \Xi_v - (-\rho_B)M_{BA} - \sigma_{11}\epsilon_{11} - \sigma_{22}\epsilon_{22} = e_v - \theta s_v - \sigma_{11}\epsilon_{11} - \sigma_{22}\epsilon_{22} - \sigma_{33}\epsilon_{33}. \tag{337}$$

For the case of traction boundary conditions, the tie lines will lie in any space containing the composition and stress component $\sigma_{11} = \sigma_{22} = \sigma^o$.

Figure 24 illustrates how changing the mechanical loading from displacement to traction boundary conditions can lead to a qualitative change in phase diagram construction and the characteristics of phase equilibrium. In the first column, Figs. 24a-c, the temperature, strain ($\epsilon_{11} = \epsilon_{22}$) and composition have been chosen as the externally controlled variables. (This corresponds to the displacement boundary conditions of the previous subsection with two thermodynamic fields and one system density as externally controlled variables.) In the absence of all stress effects, it is assumed the alloy would exhibit a simple miscibility gap in the composition. The three-dimensional, temperature-strain-composition diagram is plotted in non-dimensional units in part (a) for a case in which both the lattice parameter and elastic constants vary linearly with composition [56]. The phase boundary is depicted by the thick solid lines for several non-dimensional temperatures (at constant strain) and strains (at constant temperature). The outlines of these two-dimensional cuts through the three-dimensional space are indicated by the dashed lines in Fig. 24a. The dotted line AB indicates the locus of critical points at which the (coherent) spinodal coincides with the phase boundary. Representative tie lines (thin solid lines) end on the phase boundary and are parallel to the axis of the only externally controlled density variable, the composition c. The material is taken to be cubic with a [100] orientation.

Figures 24b and c represent two-dimensional cuts through the three-dimensional phase diagram of Fig. 24a: Fig. 24b is a temperature-composition phase diagram at constant strain, $\epsilon = 0$, and Fig. 24c is a strain-composition phase diagram at constant temperature. Since both of these phase diagrams span the composition axis (the only externally controlled variable that is a system density) tie lines lie in the plane of the phase diagram in both cases. The dotted lines indicate the coherent spinodal. The critical point, which identifies a second-order transition in both the temperature-composition and strain-composition phase diagrams, is denoted with C.

Figures 24d-f represent the phase diagrams for the same material system as depicted in Figs. 24a-c except that the applied stress in the x_1 and x_2 directions ($\sigma_{11}^o = \sigma_{22}^o = \sigma^o$) has been chosen as the externally controlled variable in place of ϵ_{11}. Once again the heavy solid lines denote the phase boundary. The dotted line AB and the solid line PQ are the line of consolute critical points and maximum temperatures of the miscibility gap, respectively. Two system densities, composition and stress component σ_{11}^o, are externally controlled variables. Tie lines lie in the $\sigma_{11} - c$ plane and are denoted by the thin solid lines.

Figure 24e is the temperature-composition phase diagram taken at constant applied stress ($\sigma_{11}^o = \sigma_{22}^o = \sigma^o = 0$). The phase diagram spans only one of the system densities which are an externally controlled variable; i.e., the composition c. (The other system density is the stress component σ_{11}.) Therefore, tie lines do not lie in the plane of the phase diagram, even though the applied stress is zero. Unlike the constant *strain*, temperature-composition phase diagram of Fig. 24b, the critical temperature (point C) in the constant *stress* temperature-composition phase diagram does not coincide with the maximum temperature of the miscibility gap (point M)[22]. Figure 24f is the constant temperature, stress-composition phase diagram. Tie lines lie in this thermodynamic space, since the phase diagram spans the space of the two externally controlled variables that are system densities. The tie lines are represented by the thin, downward sloping lines which end on the phase boundaries, e.g., the line segment a_2b_2. The ends of the tie lines give the equilibrium composition and stress component σ_{11} for each of the phases. Like the tie lines of a ternary isotherm in an unstressed ternary alloy, there is no reason to expect the tie lines to be horizontal.

Interpretation of phase diagrams for homogeneously deformed stressed systems is analogous to unstressed systems [51]. Figures 25a and b depict two schematic phase diagrams for a system with the parallel plate geometry in which the temperature, stress component $\sigma_{11}^o = \sigma_{22}^o$ and composition have been chosen as the experimentally controlled variables. The phase boundaries and tie lines are denoted by the thick solid and dashed lines, respectively. In Fig. 25a, assume the applied stress in the x_1 and x_2 directions is zero and the bulk composition is 0.5. This point is given by the solid circle lying on the tie line ending at the points identified by the open circles. The open circles indicate the equilibrium stresses and compositions of the two phases in equilibrium: the composition of the α phase is $c^\alpha = 0.25$ and the nondimensional stress component σ_{11} is $\sigma_{11}^\alpha = 1$ while the corresponding equilibrium thermodynamic densities in the β phase are $c^\beta = 1.25$ and $\sigma_{11}^\beta = -1$. Now assume the bulk composition is increased from 0.5 to about 0.62 holding the applied stress at zero. The solid triangle identifies the point on the phase diagram of the new (imposed) system densities. This point lies on a different tie line from the tie line for the case in which the bulk composition was 0.5. The ends of the tie line are indicated by the open triangles. These open triangles identify the new equilibrium compositions and stresses of the phases in the absence of an applied stress. Since the phases have different lattice parameters in their unstressed states, the individual phases are stressed in the coherent condition.

Figure 25b is the corresponding temperature-composition phase diagram for the case where the applied stress vanishes. The solid lines denote the phase boundaries; i.e., they delineate the single-phase from two-phase regions. Tie lines do not lie in the plane of the temperature-composition phase diagram[23]. Indeed, if the equilibrium phase compositions of Fig. 25a are projected onto the zero-stress plane, as indicated by the dotted arrows of Fig. 25a and then plotted on the temperature-composition phase diagram of Fig. 25b, it appears that the compositions do not lie on the phase boundaries. (The open circles of Fig. 25b give the equilibrium compositions of the phases when the

[22]This behavior is completely analogous to that of unstressed ternary alloys. In a plot of temperature and the composition of just one of the alloy components, the critical temperature does not coincide with the maximum temperature of the miscibility gap.

[23]The plane of the temperature-composition phase diagram corresponds to zero stress. But the individual phases are stressed, even when the applied stress vanishes, owing to the misfit strains.

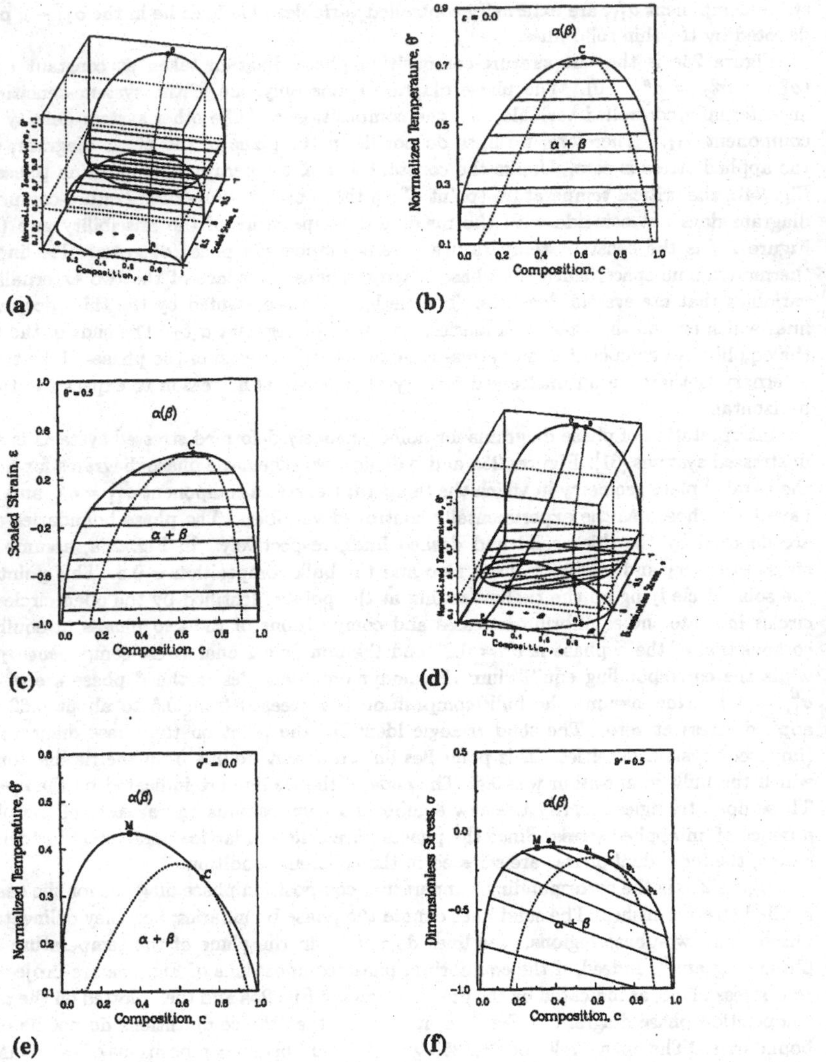

Figure 24: **Two sets of coherent phase diagrams are shown for a system with the parallel plate geometry. In (a)-(c) the temperature, composition and strain component $\epsilon_{11} = \epsilon_{22}$ are the externally controlled variables. The two-dimensional phase diagrams of (b) and (c) are slices through the three-dimensional phase diagram of (a). In (d)-(f), the temperature, composition and stress component $\sigma^o_{11} = \sigma^o_{22} = \sigma^o$ are the externally controlled variables. (e) and (f) are two-dimensional slices through the three-dimensional phase diagram of (d). In all figures the thick solid lines denote the phase boundaries and the thin solid lines the tie lines. For (a)-(c) the tie lines give the equilibrium phase compositions and in (f) the equilibrium compositions and stresses. Dashed lines indicate the coherent spinodal.**

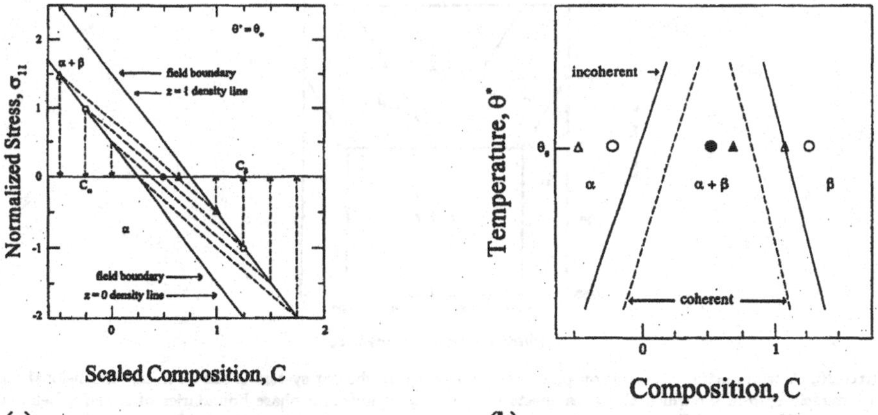

(a) (b)

Figure 25: A constant temperature, stress-composition phase diagram is shown in (a). The tie lines (dashed lines) end on the phase boundaries and give the equilibrium phase composition and stress state of the phases. The arrows show the projection of the phase compositions onto the zero-stress line. The constant stress (zero applied stress), temperature-composition phase diagram corresponding to (a) is shown in (b). The equilibrium phase compositions do not lie on the phase boundaries in this projection as the tie lines do not lie in the temperature-composition space. (Tie lines lie in the $\sigma_{11} - c$ space.)

bulk alloy composition is 0.5, denoted by the solid circle. The open triangles give the equilibrium compositions of the phases when the bulk composition is 0.62, denoted by the solid triangle.) If the bulk composition is changed at zero applied stress (for example, from 0.5 to 0.62) and the corresponding equilibrium phase compositions are plotted on the usual temperature-composition phase diagram, the compositions will shift in apparent violation[24] of the Gibbs phase rule [38, 40, 43, 47]. Two-phase coherent systems will be under a state of internal stress, even when no external stress exists, owing to misfit strains. Changing the bulk composition changes the volume fraction of the phases and the equilibrium phase compositions. Therefore constructing phase diagrams by measuring phase compositions using, for example, analytical electron microscopy may not always be correct [65].

Phase equilibria in two-phase coherent systems can be quite complex with the existence of multiple equilibrium states for a given temperature, applied stress (or pressure) and alloy composition [43].

Misfit strain can also affect equilibrium when a binary alloy can choose among three different phases. For example, consider the non-dimensional, temperature-composition phase diagram of Fig. 26 constructed in the absence of all stress, including misfit strain. For clarity, the equilibrium phase boundaries of the α and γ phases below the eutectoid temperature and in the absence of stress have been normalized to 0 and 1, respectively. Nondimensional compositions greater than one correspond to single-phase γ while nondimensional compositions less than zero lie within the single-phase α regime. The eutectoid temperature has been set to zero.

Figure 27a depicts the stress-composition phase diagram for the parallel plate geometry with traction boundary conditions corresponding to the stress-free case of Fig. 26. The temperature has been taken to be the eutectoid temperature ($\theta^* = 0$). The phase diagram, which is described in more detail below, was calculated assuming that the lattice parameters and elastic constants of the

[24] Of course, as proven in a previous subsection, a phase rule does exist for this system. The perception that the phase rule is violated arises when the thermodynamic fields and densities are not properly identified.

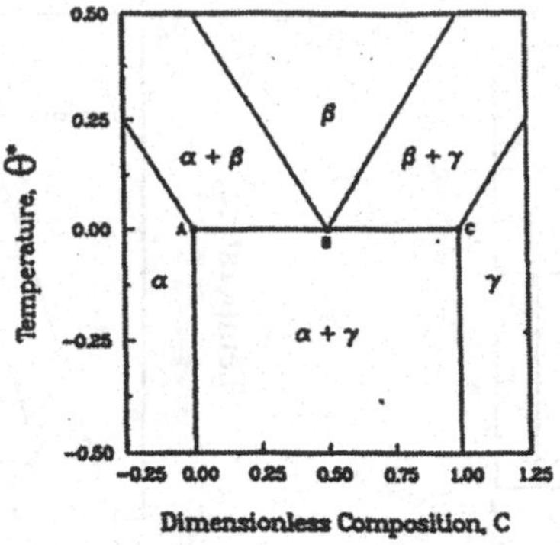

Figure 26: **A temperature-composition phase diagram for an incoherent system is shown in nondimensional units for a eutectoid system for which all stress effects vanish. The equilibrium phase boundaries of α and γ below the eutectoid have been normalized to 0 and 1, respectively, and the eutectoid temperature has been set to zero. As expected for a stress-free system, there is only one equilibrium state for a given temperature and alloy composition and that state is always stable.**

phases are independent of composition with $a_\beta > a_\gamma > a_\alpha$.

Several characteristics unique to coherent phase equilibria are found in Fig. 27a. The most apparent is the existence of several equilibrium states for a given composition and applied stress. For the material parameters employed, two regions can be identified: one for which only one equilibrium state exists and one for which three equilibrium states exist[25]. Those regions which give rise to three equilibrium states are contained within the triangle defined by vertices a, b and c. The three-phase state of $\alpha + \beta + \gamma$ is possible everywhere within the triangle abc. The equilibrium stress states and compositions of the three phases are given by the vertices a, b and c, respectively[26]. For example, if the dimensionless bulk composition and applied stress correspond to point p in Fig. 27, the equilibrium composition and stress state of the α phase is given by point a for the three-phase system. Analysis of the *stability* of the three-phase equilibrium state with respect to changes in the relative volume fractions of the phases at constant temperature, alloy composition and applied stress shows that the three-phase state is thermodynamically unstable.

Two other equilibrium states are possible at point p, both of which are thermodynamically stable [57]. The first is an $\alpha + \gamma$ two-phase system. The equilibrium compositions and stress states of the α and γ phases are given by the intersection of the tie line passing through the point p (indicated by the dashed line) and the phase boundaries delineating the α and γ single-phase fields, points t_3 and t_4, respectively. The second thermodynamically stable equilibrium state corresponding to point p is a two-phase system of $\alpha + \beta$. The equilibrium compositions and stress states are given by the intersection of the tie line with the α and β single-phase fields, points t_1 and t_2. The regions in which two linearly stable equilibrium states can be found are indicated by arrows in Fig. 27a and will now be delineated. Both an $\alpha + \gamma$ and an $\alpha + \beta$ two-phase system are stable within the triangle abe. The $\alpha + \gamma$ system is *absolutely* stable below the dotted curve that extends from point a to point c and the $\alpha + \beta$ two-phase system is linearly stable[27]. The $\alpha + \gamma$ two-phase

[25] An equilibrium state is one for which all conditions for thermodynamic equilibrium are satisfied. The state can correspond to an energy minimum, energy maximum, or a saddle point.

[26] The triangle abc is analogous to a tie-triangle observed for a three-phase region in an isothermal section of a ternary alloy.

[27] By absolutely stable, it is meant that, for all permissible variations in the phase compositions and stress states, the system has the lowest energy. Linearly stable means the system has the lowest energy in some region around the

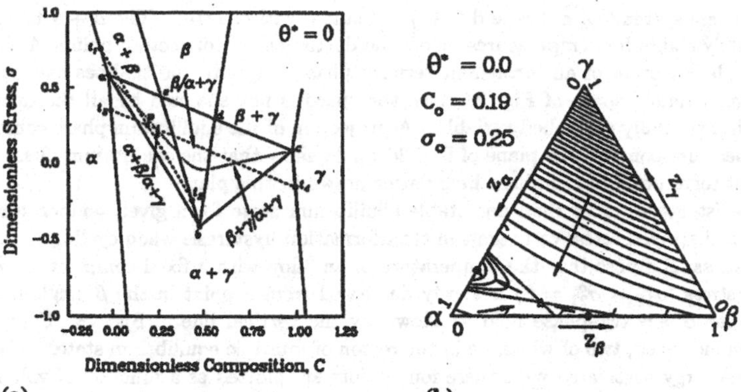

(a) (b)

Figure 27: The stress-composition phase diagram at the eutectoid temperature corresponding to the stress-free eutectoid system is shown in (a) when traction boundary conditions are imposed for the parallel plate geometry. Multiple equilibrium states are possible within triangle *abc*. (b) Lines of equal free energy are plotted as a function of the volume fractions, z_i, for the alloy composition and applied stress given by point *p*. The arrows inside the phase-fraction triangle indicate the directions of decreasing system free energy. The stable equilibrium states are indicated by filled circles whereas the unstable (saddle point) three-phase equilibrium state is denoted by the open triangle.

system is no longer an equilibrium solution for externally applied stresses that lie above the solid line *ac*. Within the triangle *bde*, both single-phase β and two-phase $\alpha + \gamma$ are thermodynamically stable while within triangle *bcd* both two-phase systems $\beta + \gamma$ and $\alpha + \gamma$ are thermodynamically stable.

Lines of constant free energy of the entire system are plotted as a function of the volume fraction of the phases for the point *p* in Fig. 27b assuming chemical, mechanical and thermal equilibrium. The arrows within the phase-fraction triangle indicate the direction of decreasing free energy. The state of three-phase equilibrium $(\alpha + \beta + \gamma)$ is indicated by the open triangle and corresponds to a saddle point in the free energy. It is unstable with respect to certain perturbations in the volume fractions and therefore would not be expected to be observed. The two stable states are given by the small, filled circles. These points are end-of-range minima in the free energy. Any physically realizable perturbation in the volume fractions about these points results in an increase in the system free energy. The stability of the two-phase equilibrium states does not arise from the presence of nucleation barriers for the formation of the equilibrium phase. In each of the thermodynamic states, formation of a non-zero volume fraction of the third phase results in an increase in the free energy.

The temperature-composition phase diagram (at constant applied stress, $\sigma = 0$) corresponding to the coherent system of Fig. 27 is shown in Fig. 28 assuming the applied stress vanishes. The phase diagram does not contain the tie lines and, hence, the equilibrium phase compositions cannot be obtained from this phase diagram. This phase diagram differs from its stress-free counterpart of Fig. 26 in that an invariant temperature does not exist. The region in which three equilibrium states can exist is given by triangle ABC. The (unstable) $\alpha + \beta + \gamma$ three-phase equilibrium state is found within triangle ABC. The two-phase systems of $\alpha + \beta$ and $\alpha + \gamma$ are stable within triangle ABE; single-phase β and two-phase $\alpha + \gamma$ are stable equilibrium states within triangle EBD; and

equilibrium state. A linearly stable state sits in an energy well.

the two-phase systems of $\alpha + \gamma$ and $\beta + \gamma$ are stable within DBC. The $\alpha + \gamma$ two-phase system is absolutely stable for temperatures below the dotted curve connecting points A and C. Similar to an isopleth section of an incoherent ternary phase diagram, the tie-lines cannot be drawn in the thermodynamic space of Fig. 28 since the space is not spanned by all the density variables among the externally controlled variables. A projection of the equilibrium phase compositions onto the temperature-composition plane of Fig. 28 would show that the equilibrium phase compositions would not correspond to the phase boundaries drawn in this plane.

The existence of more than one stable equilibrium state for a given temperature, alloy composition and applied stress can result in transformation hysteresis when cycling in temperature or applied stress. Suppose that the temperature of an alloy with a fixed composition ($c_o = 0.5$) and external stress ($\sigma_{11}^o = \sigma_{22}^o = 0$) is slowly decreased from a point in the β single-phase field to a point in the $\alpha + \gamma$ two-phase field as shown by the vertical line in Fig. 28. Four temperatures have been indicated, two of which lie in the region of multiple equilibrium states. The lines of constant free-energy associated with these four points are plotted as a function of volume fraction in Fig. 29. Figure 29a corresponds to the highest temperature and Fig. 29d to the lowest. Chemical, mechanical and thermal equilibrium has been assumed.

At the nondimensional temperature corresponding to $\theta^* = 0.1$, Fig. 29a, only one equilibrium state, identified by the solid circle and corresponding to single-phase β, exists. (The solid circle with an x appearing through it located on the $\alpha + \gamma$ two-phase line gives a minimum in the free energy of the $\alpha + \gamma$ system but is unstable with respect to the formation of β. As the temperature is reduced to $\theta^* = 0.03$, Fig. 29b, two new equilibrium states appear: the unstable three-phase system indicated by the open triangle and the stable, $\alpha + \gamma$ two-phase system. Single-phase β is still the lowest energy state (absolute minimum). As the temperature is further reduced to that given by $\theta^* = -0.05$, Fig. 29c, two two-phase equilibrium states are stable: the $\alpha + \gamma$ system and a $\beta + \gamma$ system that is mostly β phase. The absolute minimum in the free energy has jumped discontinuously from single-phase β to the $\alpha + \gamma$ two-phase system. However, if the system is cooled reversibly, it would still reside in the state of single-phase β. (The β phase remains stable with respect to formation of the γ phase until the temperature is decreased below the line BD.) Upon further cooling to $\theta^* = -0.1$, Fig. 29d, the saddle point in the free energy disappears and the $\beta + \gamma$ two-phase state becomes unstable with respect to the formation of the α phase. (The $\beta + \gamma$ two-phase state is unstable with respect to the formation of α for temperatures below the line BC.)

Now consider the case in which the same system is heated reversibly from the $\alpha + \gamma$ two-phase state at temperature $\theta^* = -0.1$ to the β single-phase state at temperature $\theta^* = 0.1$. The $\alpha + \gamma$ two-phase state remains (absolutely) stable until crossing the dotted curve and a relative minimum in the free energy until crossing the boundary AC. This indicates the system could remain in the $\alpha + \gamma$ two-phase state until it loses stability with respect to the formation of the β phase. On heating, the system would not necessarily pass through the $\beta + \gamma$ two-phase state as it did during the cooling process. It is important to note that this hysteresis is not a kinetic phenomenon due to a nucleation barrier but an equilibrium characteristic unique to a coherent system.

5 Microstructural Evolution

In this section, some general features of two-phase microstructures that are influenced by elastic stress are examined. In the first subsection, the effect of a misfit strain on the equilibrium shape

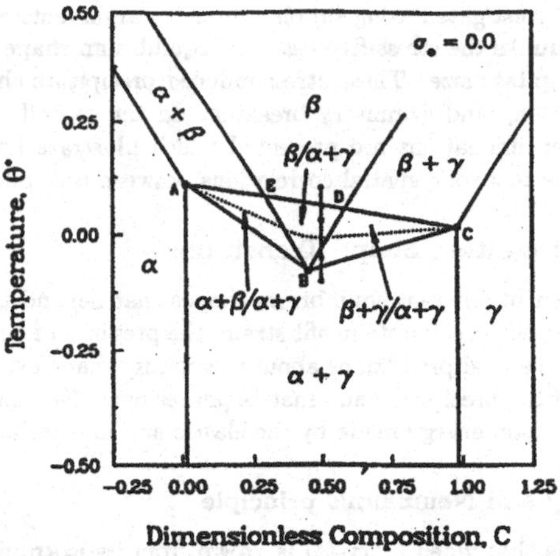

Figure 28: **The temperature-composition phase diagram for the eutectoid system with traction boundary conditions and no applied stress is shown. Multiple equilibrium states are possible within triangle *ABC*.**

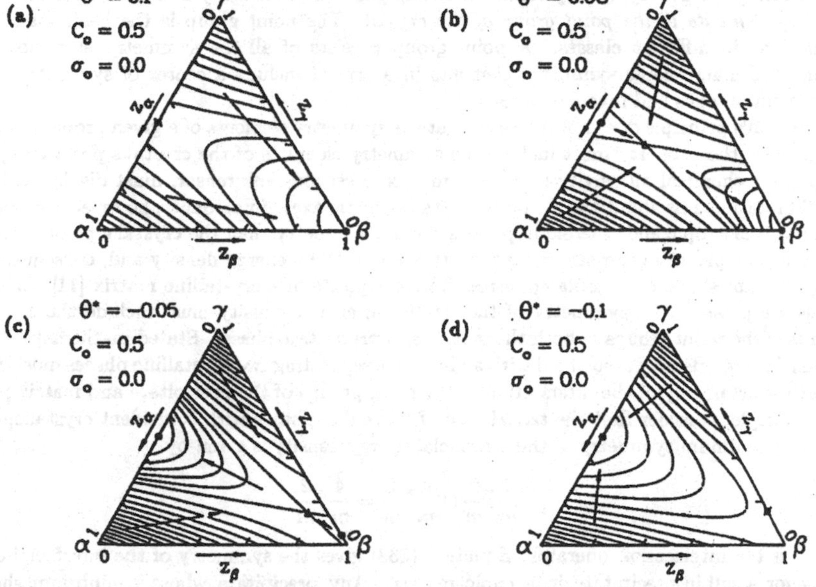

Figure 29: **Isocontours of the free energy are plotted as a function of volume fractions for four different non-dimensional temperatures at the bulk alloy composition $c_o = 0.5$ and in the absence of an applied stress. Arrows within the triangle indicate directions of decreasing free energy. The small triangle indicates the point of unstable three-phase equilibrium and corresponds to a saddle point in the free energy. The solid circles identify stable two-phase equilibrium solutions. The solid circles appearing with an X through them indicate points of two-phase equilibrium but which are unstable with respect to the formation of a third phase.**

of an isolated precipitate is investigated using simple symmetry arguments and bifurcation theory. The primary result is that, unlike the stress-free case, the equilibrium shape of the precipitate will change with increasing precipitate size. These stress-induced precipitate shape transitions are of two types, symmetry conserving and symmetry breaking. In the second subsection, the results of several different computer simulations are presented which illustrate how elastic interactions between precipitates give rise to strong spatial correlations between precipitates.

5.1 Stress-induced Precipitate Shape Transitions

Precipitates display a number of different equilibrium shapes that depend upon the anisotropy of the interfacial energy density, the precipitate misfit strain, the presence of an external field and the size of the precipitate. Some general predictions about precipitate shape evolution can be made by considering the symmetry of the precipitate and matrix phases using Neumann's principle and the relative contributions to the total energy made by the elastic and interfacial energies.

5.1.1 Crystal symmetry and Neumann's principle

One observation of nature is that when a crystal is grown from its isotropic melt, it will assume a form compatible with its crystal symmetry [11]. This observation, along with many others relating to the symmetry of various crystal properties, is contained within Neumann's principle which states [19]: *The symmetry elements of any physical property of a crystal must include the symmetry elements of the point group of the crystal.* The point group is the basis for dividing crystals into 32 different classes. A point group consists of all the symmetry elements of the crystal. The macroscopic symmetry elements in a crystal include a center of symmetry, mirror planes, rotation axes and inversion axes.

Neumann's principle does not indicate what the symmetry elements of a given property actually are, just that the property must include the symmetry elements of the crystal's point group. For example, the chemical diffusivity in a cubic matrix, a second-rank tensor, must display at least a four-fold rotation axis about the $< 100 >$ crystallographic axes. In actuality, however, the chemical diffusivity is isotropic and, therefore, possesses more symmetry than the crystal's point group.

Neumann's principle can also be applied to the interfacial energy density and, consequently, to the equilibrium shape of an isolated, stress-free precipitate in a crystalline matrix [11]. Since two crystals are present, the symmetry of the interfacial energy density must include the symmetry elements of the point groups of both the matrix and precipitate phases. Stated mathematically, the interfacial energy density associated with an interface separating two crystalline phases must display at least the symmetry of the intersection of the point groups of the precipitate and matrix phases. As an example, consider first the trivial case of two cubic phases with coincident crystallographic axes. The minimum symmetry of the interfacial energy density is given by

$$\frac{4}{m}\bar{3}\frac{2}{m} \cap \frac{4}{m}\bar{3}\frac{2}{m} = \frac{4}{m}\bar{3}\frac{2}{m} \tag{338}$$

where $\cap$ is the intersection operator. Equation (338) gives the symmetry of the interfacial energy density for a cubic precipitate in a cubic matrix. Any precipitate whose equilibrium shape is determined solely by minimization of the interfacial energy must possess a shape that displays at least the symmetry of Eq. (338). Since one symmetry can be represented by many shapes, there are a number of precipitate shapes that a coherent (unstressed) cubic precipitate in a cubic matrix can possess. For example, a sphere, a cube, an octahedron and a tetrakaidecahedron possess the symmetry given by Eq. (338) and all are possible equilibrium precipitate shapes.

Consider now the slightly more complicated case of a tetragonal precipitate nucleated in a cubic matrix. If the axis of tetragonality is assumed to lie along the x_3 axis, then the minimum symmetry of the interfacial energy density is given by:

$$\frac{4}{m}\bar{3}\frac{2}{m} \cap \frac{4}{m}mm = \frac{4}{m}mm. \tag{339}$$

At this point a distinction must be made between the symmetry of the shape of an isolated precipitate, which we have been discussing, and the symmetry of the two-phase system on a more macroscopic scale; i.e., on a scale large enough to contain many precipitates. Consider again the case of a coherent tetragonal precipitate in a cubic matrix. The symmetry of the interfacial energy density is given by Eq. (339) and the equilibrium shape of a precipitate will reflect this symmetry. However, there are three different but equivalent variants to the orientation of the precipitate; the axis of tetragonality would be expected to lie with equal frequency along the x_1, x_2 and x_3 axes. Thus, if a volume of the matrix sufficiently large to contain many precipitates of each of the three variants is considered, the cubic symmetry of the matrix phase is recovered. One would expect the physical properties of the two-phase crystal to exhibit still the cubic symmetry of the matrix phase, the same as for the matrix in its single-phase state. Of course, the values of the physical properties would change with the presence of the tetragonal precipitates[28].

From the macroscopic point of view, the matrix phase retains its symmetry when second-phase precipitates nucleate and grow coherently within it. The interfacial energy density assumes a symmetry commensurate with the intersection of the point groups of the precipitate and matrix crystals. However, the symmetry of the matrix phase is retained on a macroscopic scale owing to the different orientational variants that can be assumed by the individual precipitates. This distinction is important when discussing the symmetry-breaking precipitate shape transitions.

When an external field is present, the interfacial energy density must possess the minimum symmetry given by the intersection of the point groups of the precipitate and matrix phases and the Curie group[29] of the external field [66]. For example, when a uniaxial tensile stress is applied along a four-fold crystallographic axis, the symmetry of the interfacial energy density separating two cubic phases is at least:

$$\frac{4}{m}\bar{3}\frac{2}{m} \cap \frac{4}{m}\bar{3}\frac{2}{m} \cap \frac{\infty}{m}\frac{2}{m}\frac{2}{m} = \frac{4}{m}mm. \tag{340}$$

The presence of the external field acts to lower the symmetry of the interfacial energy density and the equilibrium precipitate shape. Likewise, all properties of the crystal will be influenced by the presence of the external field.

In the following section, we will use these symmetry arguments to define two different types of precipitate shape changes that can be induced by the presence of an elastic stress. Before doing so, it is instructive to first consider the relative contributions of the elastic and interfacial energies to the total energy of the system.

[28] For example, the yield stress of the two-phase system would still reflect the symmetry of the cubic matrix phase when the tetragonal precipitates assume all orientational variants equally. This would not be true if the axis of tetragonality of all precipitates were parallel.

[29] The Curie group contains the symmetry operations of the external field.

5.1.2 Elastic and interfacial energies

In the absence of elastic stress, the equilibrium shape of a second-phase particle (β) of fixed volume embedded in a matrix (α) is determined by minimizing the interfacial energy. If the precipitate-matrix interfacial energy density is $\sigma(\hat{n})$, where $\hat{n}$ is the outward pointing unit normal to the precipitate, the equilibrium shape is that shape which minimizes the free energy, E_t, given by [67]:

$$E_t = \int_S \sigma(\hat{n})dS, \tag{341}$$

and subject to the constraint:

$$V_\beta = \text{constant} \tag{342}$$

where V_β is the volume of the precipitate. In general, σ depends on the orientation of the interface. σ will also possess a symmetry determined by the symmetries of the precipitate and matrix phases and their relative orientation [11]. Although the interfacial energy will depend upon the particle volume, the *shape* which minimizes Eq. (341) is independent of the particle volume and is that shape given by the Wulff construction [67]. Since the shape which minimizes the interfacial energy is independent of precipitate size, no transitions in the equilibrium shape of the precipitate are expected with increasing precipitate size in the absence of elastic stress [70].

If the isolated precipitate possesses a misfit strain, ϵ_{ij}^T, the equilibrium shape is determined by minimizing the sum of the interfacial and elastic strain energies at constant particle volume, assuming there are no composition gradients [12, 13]. In this case, the total free energy is given by:

$$E_t = \int_S \sigma(\hat{n})dS - \frac{1}{2}\int_{V_\beta} \sigma_{ij}\epsilon_{ij}^T dV, \tag{343}$$

where Eq. (189) for the elastic strain energy has been used.

Before examining the dependence of Eq. (343) on precipitate shape, which usually must be done numerically, it is instructive to explore qualitatively the relative contributions of the interfacial and elastic energies to the total free energy for a precipitate of fixed shape by considering a spherical precipitate in an isotropic system. In this case, the interfacial energy density is independent of interface orientation and:

$$\int_S \sigma(\hat{n})dS = \sigma \int_S dS = 4\pi R^2 \sigma \tag{344}$$

where R is the radius of the precipitate. Using Eq. (191) for the strain energy associated with a spherical precipitate in an isotropic system, the free energy of the isolated spherical precipitate is given by:

$$E_t = 4\pi R^2 \sigma + \frac{18\mu^\alpha K^\beta (\epsilon^T)^2 V_\beta}{(3K^\beta + 4\mu^\alpha)}. \tag{345}$$

The energy per unit volume of precipitate is:

$$E_t/V_\beta = \frac{3\sigma}{R} + \frac{18\mu^\alpha K^\beta (\epsilon^T)^2}{(3K^\beta + 4\mu^\alpha)}. \tag{346}$$

For small precipitate radii, the interfacial energy clearly dominates while, for large precipitate radii, the elastic energy is the dominate contribution to the total free energy. The interfacial and elastic energy contributions are equal when:

$$R = \frac{\sigma(3K^\beta + 4\mu^\alpha)}{6\mu^\alpha K^\beta (\epsilon^T)^2}. \tag{347}$$

The critical size at which the elastic energy becomes larger than the interfacial energy depends on the elastic constants of both phases, the interfacial energy density and the square of the misfit strain.

Another way of stating the above is that the interfacial energy scales with the surface area of the particle[30] while the strain energy scales with the volume of the particle. Since the surface area is proportional to $V_\beta^{2/3}$, doubling the precipitate volume will double the elastic energy but lead to less than a doubling of the interfacial energy; the ratio of the interfacial to elastic energy thus changes with precipitate size. When the precipitate shape is not restricted to that of a sphere, precipitate shape changes can occur as the particle grows, since the interfacial and elastic energies depend on precipitate shape for a given volume of precipitate. The particle shape should tend towards that shape which minimizes the interfacial energy at small precipitate sizes, where the ratio of the interfacial energy to the elastic energy is large, and towards the shape which minimizes the elastic energy at large precipitate sizes, where the ratio of the interfacial energy to elastic energy is small [68]. The meaning of "small" and "large" in this context depends, of course, upon the material parameters.

Two general types of stress-induced precipitate shape transitions have been observed experimentally with increasing particle size in two-phase alloys possessing coherent precipitates. The first type is classified as symmetry conserving. This means that the symmetry of both the initial and final precipitate shapes is equal to or greater that the symmetry resulting from the intersection of the point groups of the precipitate and matrix phases [68, 69, 70]. An example of a symmetry-conserving shape transition was evident in Fig. 1. In this case, both the precipitate and matrix phases possess cubic symmetry and are coherent. If the equilibrium shape of an isolated precipitate is determined solely by the interfacial energy then, according to Neumann's principle, the shape of the γ' precipitate must have a symmetry given by Eq. (338). At early times the precipitate is a sphere while at later times it has a four-fold rotation axis, eventually becoming a cube. The symmetry of each of these shapes satisfy Eq. (338). This transition from a sphere to a four-fold symmetric shape is also common in other Ni-Al based alloys [71, 72]. Other documented symmetry-conserving shape transitions include the transitions between cubes, octahedra and tetrakaidecahedra for Co-rich precipitates in Cu-Co alloys [73, 74].

The second type of elastically-induced shape transition is termed symmetry-breaking [68]. In this case, the symmetry of the final precipitate shape is less than the symmetry resulting from the intersection of the point groups of the precipitate and matrix phases. The sphere-to-ellipsoid transition in Al-Zn [75] or the cube-to-cuboid transition visible in Figs. 2 and 3 are examples of symmetry-breaking shape transitions. It is important to remember that stress-induced precipitate shape transitions do not violate Neumann's principle, since different variants of precipitate orientation will exist. When a macroscopic region of the two-phase system is considered, the symmetry of the matrix crystal will be recovered.

5.1.3 Symmetry-conserving shape transitions

In addition to the anisotropy of the interfacial energy, the equilibrium precipitate shape will depend on the elastic properties of the precipitate and matrix phases as well as the misfit strain. Figure 30 illustrates the symmetry-conserving precipitate shape transition calculated numerically for a two-

[30] As used here, scaling just means that, if the surface area is doubled keeping the precipitate shape constant, the interfacial energy doubles.

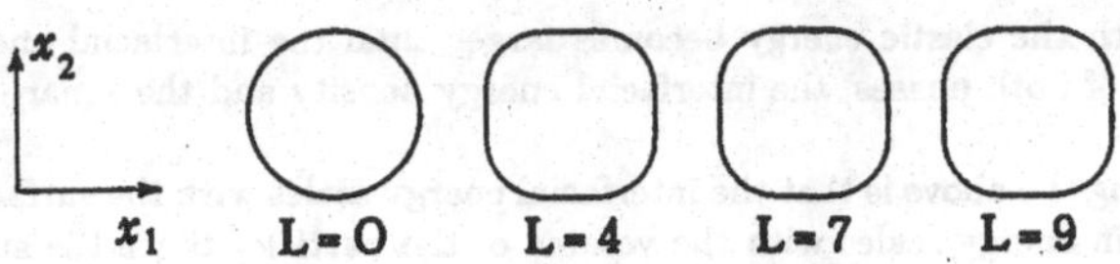

Figure 30: **Symmetry-conserving shape transitions are shown in two-dimensions for increasing precipitate size (L). Both precipitate and matrix phases possess cubic symmetry with a four-fold axis perpendicular to the plane of the paper. The particle maintains the four-fold symmetry, but the sides perpendicular to the elastically soft $< 10 >$ directions become progressively more facetted as the elastic energy becomes progressively more important at larger precipitate sizes.**

dimensional coherent precipitate embedded in a matrix with an axis of four-fold symmetry oriented perpendicular to the plane of the paper [76]. The misfit strain is dilatational and the system is elastically homogeneous. (The precipitate and matrix phases possess the same elastic constants, taken to be those of nickel.) L represents a nondimensional precipitate size defined in the same manner as for the sphere in the previous example:

$$L = (\epsilon^T)^2 C_{44} l / \sigma \tag{348}$$

where ϵ^T is the dilatational misfit strain, σ is the (isotropic) interfacial energy density and l is a measure of the particle diameter. Thus, for fixed materials parameters, larger values of L denote larger particle size. The particles become strongly facetted in the two elastically soft $< 10 >$ directions with increasing size, even though the interfacial energy is isotropic, on account of the elastic energy: A square oriented with sides perpendicular to the $< 10 >$ directions has a lower elastic energy than a circle for given material parameters. As seen here, observation of facetted precipitates does not imply anisotropy of the interfacial energy density.

The effect of anisotropy in the interfacial energy density and elastic heterogeneity on the (three-dimensional) equilibrium precipitate shape can be demonstrated by restricting the particle morphology to certain classes of geometrical shapes [74]. Figure 31 shows a tetrakaidecahedra with facets along the $< 100 >$ and $< 111 >$ directions. $A = t/a$, where t and a are defined in Fig. 31, is a shape parameter that indicates the degree of facetting. The range of A is $0 \leq A \leq 2/3$. $A = 0$ corresponds to an octahedron and $A = 2/3$ to a cube with faces perpendicular to the $< 100 >$ directions of the cubic matrix. All other values of A represent a tetrakaidecahedron. This class of shapes is observed in the Cu-Co system [73, 74].

The dependence of the elastic strain energy on the shape parameter A is shown in Fig. 32. Although A cannot be used to model a spherical precipitate, the strain energy of a tetrakaidecahedron can be compared with that of a sphere by normalizing the strain energy by the elastic energy of a spherical precipitate. The energy of the sphere thus corresponds to one. Calculations were performed using Eq. (189) and the elastic constants of Cu were used for the matrix phase. A dilatational misfit has been assumed and the elastic constants of FCC iron were used for the precipitate phase in the elastically inhomogeneous system [74]. The cube ($A = 2/3$) has a lower elastic energy than the sphere and the sphere a lower such energy than the octahedron ($A = 0$) for both the elastically homogeneous and heterogeneous systems. Changing the elastic constants of

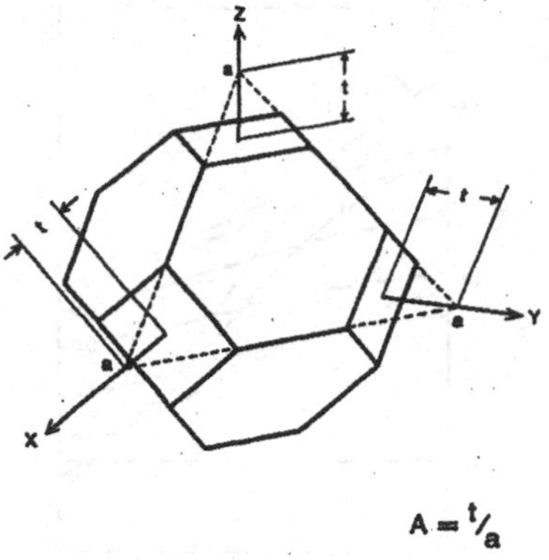

Figure 31: A tetrakaidecahedron with facets along the < 100 > and < 111 > directions of the matrix phase is shown. $A = t/a$ is a shape parameter used to describe the particle. $A = 0$ and $A = 2/3$ correspond to an octahedron and cube, respectively.

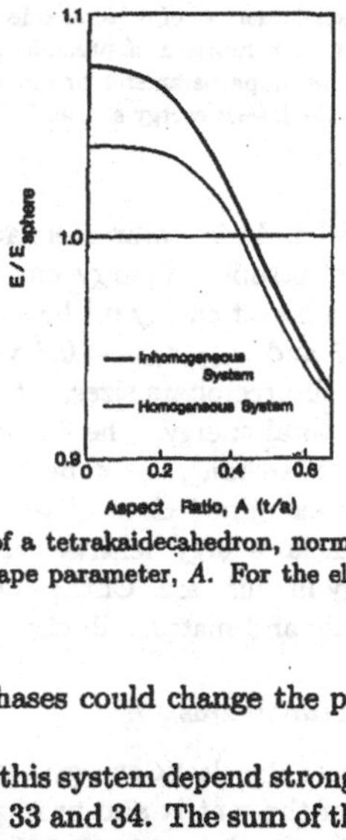

Figure 32: The elastic strain energy of a tetrakaidecahedron, normalized to the elastic energy of a homogeneous sphere, is shown as a function of the shape parameter, A. For the elastic constants employed, the cube ($A = 2/3$) has the lowest elastic energy.

either the precipitate or matrix phases could change the precipitate shape which gives the lowest elastic energy.

Predicted shape transitions for this system depend strongly upon the anisotropy of the interfacial energy density, as depicted in Figs. 33 and 34. The sum of the elastic and interfacial energies, E_{total}, normalized by the total energy of a spherical precipitate, is plotted as a function of particle volume, V_p, for several different shape parameters. When the interfacial energy is assumed to be isotropic, as in Fig. 33, the sphere possesses the lowest energy shape at small precipitate sizes while at larger sizes the cube ($A = 2/3$) has the lowest total energy. (The relative energy associated with a spherical precipitate of a given volume is represented by the horizontal line in Figs. 33 and 34.) Figure 33 is the three-dimensional equivalent of the two-dimensional situation depicted in Fig. 30 and, on the basis of the two-dimensional calculations, it should be expected that the three-dimensional shape does not necessarily change discontinuously from sphere to cube, but continuously in a manner similar to that depicted in Fig. 30. Thus, for the case of an isotropic interfacial energy density in a cubic matrix, one would not expect to observe a precipitate shaped as a tetrakaidecahedron, but rather a gradual transition from a sphere to a cube. This sequence of shape transitions is precisely that observed in the Ni-Al alloys of Figs. 1-3.

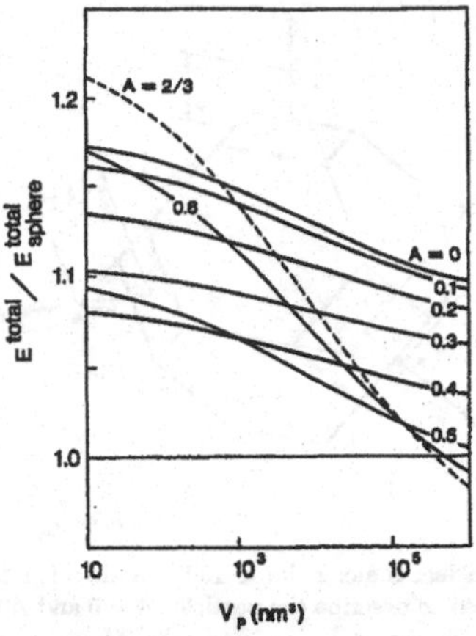

Figure 33: **The sum of the elastic and interfacial energies of a tetrakaidecahedron, normalized to the total energy of an elastically homogeneous sphere, is shown as a function of precipitate size for various shape parameters, A, assuming the interfacial energy is *isotropic*. The shape parameter or amount of facetting changes with increasing particle size. For $V_p < 2 \times 10^5 \text{nm}^3$, the sphere is the lowest energy shape. For large particles, the equilibrium shape is a cube.**

Figure 34 represents the case where all material parameters are the same as those used in Fig. 33 but the interfacial energy density is anisotropic. Energy cusps are assumed to exist along the $< 100 >$ and $< 111 >$ directions with the lowest energy in the $< 111 >$ directions. The magnitudes of the energy cusps are $\sigma_{111}/\sigma_{iso} = 0.7$ and $\sigma_{100}/\sigma_{iso} = 0.8$ where σ_{iso} is the interfacial energy density for all other orientations. At small precipitate sizes, a tetrakaidecahedron of specific shape parameter $A \approx 0.4$ possesses the lowest total energy. The energy extremizing shape changes with increasing particle size; for volumes $V_p > 10^5 \text{nm}^3$, the cube becomes the equilibrium precipitate shape. The equilibrium shape for the examined class of precipitate shapes is a cube in both Figs. 33 and 34, even though the interfacial energy densities are different. This is a result of the cube possessing the lowest elastic energy in this case. Changes in the interfacial anisotropy, misfit strain, and elastic constants of precipitate and matrix will change the equilibrium shape.

5.1.4 Symmetry-breaking shape transitions

When the precipitate shape that minimizes the elastic energy possesses a symmetry lower than that of the intersection of the point groups of the matrix and precipitate crystal lattices, a symmetry-breaking shape transition can occur during the growth of the precipitate [68]. The nature of the symmetry-breaking transitions can be understood most easily by first restricting the possible precipitate shapes to certain geometrical classes. This approach is certainly not general and the predicted precipitate shapes will not necessarily satisfy all thermodynamic equilibrium conditions, however, this approach does allow the qualitative features of the symmetry-breaking transition to become more apparent.

In order to illustrate this effect, consider an isolated precipitate in a cubic matrix whose shape is restricted to be an ellipsoid of revolution [68]. The misfit strain is dilatational, the interfacial energy density is isotropic, and the system can be elastically heterogeneous. The particle's shape and size are completely defined by its major axes or, equivalently, by its volume and a shape parameter, S ($-1 < S < 2$), given by:

$$S = 2(a_3 - a_1)/(a_3 + 2a_1) \tag{349}$$

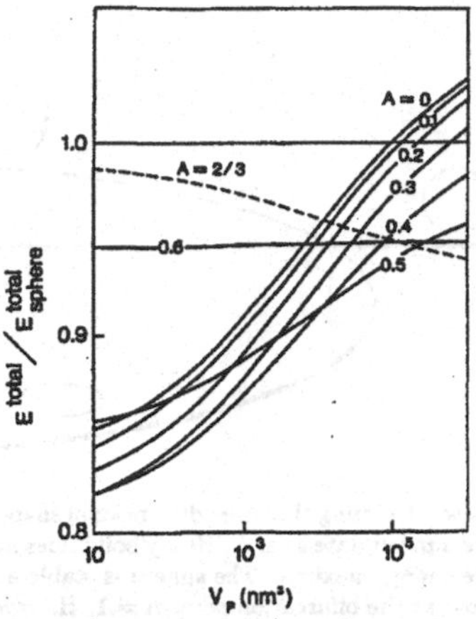

Figure 34: The sum of the elastic and interfacial energies of a tetrakaidecahedron, normalized to the total energy of an elastically homogeneous sphere, is shown as a function of precipitate size for various shape parameters. The interfacial energy is *anisotropic* with the interfacial energy density in the $<111>$ and $<100>$ orientations assumed to be 70% and 80% of that in the other directions, respectively. The shape parameter, or extent of facetting, changes with increasing particle size. For $V_p < 10^5 \text{nm}^3$, a tetrakaidecahedron has the lowest energy. For large particles, the equilibrium shape is a cube.

where the a_i are the axes of the ellipsoid (with $a_1 = a_2$). The particle is an oblate spheroid for $S < 0$ and a prolate spheroid for $S > 0$. When $S = 0$, the shape is a sphere.

The total energy, E_t, defined as the sum of the elastic strain (E_{strain}) and interfacial (E_{surf}) energies, is still determined from Eq. (343) and, for a given set of material parameters, can be expressed as a function of the particle shape parameters S and V:

$$E_t(S, V) = E_{strain}(S, V) + E_{surf}(S, V). \tag{350}$$

The energy extremizing particle shapes are those shapes which render the total energy a minimum (or maximum) for a given precipitate volume. These shapes are given by solutions to:

$$\left(\frac{\partial E_t}{\partial S}\right)_V = 0. \tag{351}$$

Formal analysis of Eq. (351) first requires solving for the elastic field of an ellipsoid of revolution and then calculating the elastic and interfacial energies associated with a precipitate in terms of S and V. The resulting expression is differentiated with respect to S at constant volume and set equal to zero. Results show that, for various combinations of materials parameters, there exists either one or three energy extremizing shapes that are ellipsoids of revolution for a given particle volume [68, 69]. These energy-extremizing solutions are plotted in the bifurcation diagrams of Fig. 35a as a function of a nondimensional precipitate size, Λ, where:

$$\Lambda = \frac{-5(\epsilon^T)^2 V^{1/3}(C_{11} + 2C_{12})}{32\pi\sigma} \left(\frac{4\pi}{3}\right)^{2/3} L_2 \tag{352}$$

where L_2 is a parameter determined numerically using the precipitate and matrix elastic constants [69]. (For the $\gamma - \gamma'$ Ni-Al system, $L_2 = -.219$).

The solid lines in Fig. 35 represent energy minima and the dashed lines energy maxima. The absolute minimum is given by the heavy solid line and the relative minima by the thinner solid

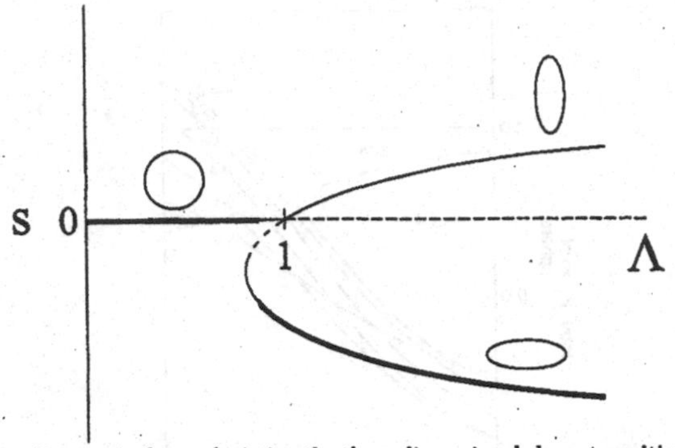

Figure 35: **A bifurcation diagram is shown depicting the three-dimensional shape transition from sphere to ellipsoid of revolution as a function of dimensionless precipitate size, Λ. Heavy solid lines are absolute minima, thin solid lines are relative minima and broken lines are energy maxima. The sphere is stable at small precipitate sizes and looses stability with respect to a prolate spheroid at the bifurcation point $\Lambda = 1$. However, the oblate spheroid ($S < 0$) has the lowest energy.**

line. The sphere ($S = 0$) is an energy extremizing shape for all precipitate sizes: It is an energy minimum for $\Lambda < 1$ and an energy maximum for $\Lambda > 1$. The sphere loses stability with respect to an ellipsoid of revolution at a critical size given by $\Lambda = 1$, when two extremizing solutions intersect at the bifurcation point [68]. (In this case, one extremizing solution is given by the sphere; i.e., the line $S = 0$ and the other by the parabola-shaped curve corresponding to an ellipsoid of revolution.) The critical precipitate radius corresponding to the bifurcation point is about 38 nm for the $\gamma - \gamma'$ Ni-Al system. For the three-dimensional problem, the shape bifurcation is termed transcritical because the solution is not symmetric about $S = 0$ and the absolute minimum in energy (heavy line) jumps discontinuously from a sphere to an oblate spheroid before the bifurcation point at $\Lambda = 1$ is reached. As all other terms on the right-hand-side of Eq. (352) are positive, a bifurcation point exists only when $L_2 < 0$; otherwise the sphere remains the shape with the minimum energy at all precipitate sizes for the restricted class of ellipsoids. In general, $L_2 < 0$. However, when the elastic anisotropy of a cubic system is close to one and the difference in elastic constants between precipitate and matrix is large, L_2 can be positive.

For an isotropic system, the precipitate volume at which the bifurcation point occurs can be calculated analytically as [68, 69]:

$$V^{1/3} = \frac{-20\pi\sigma(1-\nu)^2}{3(\epsilon^T)^2(C_{44}^p - C_{44})(1+\nu)^2}\left(\frac{3}{4\pi}\right)^{2/3}. \qquad (353)$$

A bifurcation point exists only for elastically soft precipitates; the shear modulus of the precipitate is less than that of the matrix, $C_{44}^p < C_{44}$. When the precipitate is elastically harder than the matrix, $C_{44}^p > C_{44}$, the sphere is the only equilibrium or energy-extremizing shape in an *isotropic* system for all precipitate sizes and no symmetry-breaking shape transitions are predicted. In an *anisotropic* system, symmetry-breaking shape transitions can occur for both elastically hard and soft precipitates over a wide range of materials parameters.

Symmetry-breaking precipitate shape transitions, or precipitate shape bifurcations owing to elastic stress, have also been calculated in two-dimensional anisotropic systems in which the shape of the precipitate is not restricted to a certain class [77]. An example is shown in Fig. 36 where the shape parameter a_2^R is plotted as a function of nondimensional particle size L. (a_2^R is a shape parameter that is zero for four-fold symmetric shapes and non-zero for two-fold symmetric shapes.) As in the three-dimensional case, either one or three precipitate shapes exist that extremize the sum of the interfacial and elastic energies. One solution corresponds to the four-fold symmetric

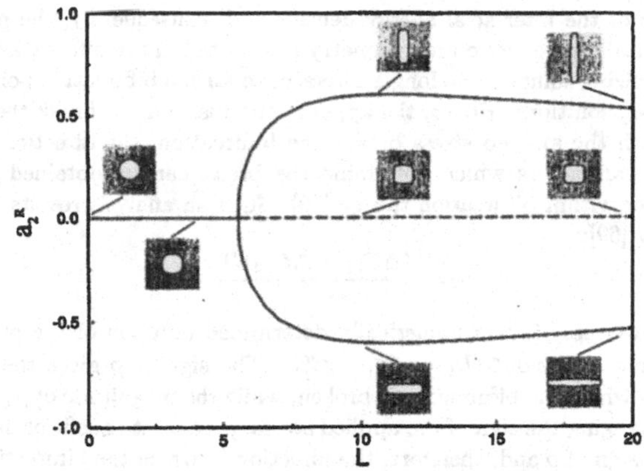

Figure 36: A bifurcation diagram is shown depicting symmetry-conserving and symmetry-breaking two-dimensional shape transitions from circle to four-fold to two-fold symmetric shapes as a function of dimensionless precipitate size, L. Solid lines are minima and broken lines are energy maxima. a_2^R is a shape parameter that is zero for four-fold symmetric shapes and nonzero for two-fold symmetric shapes. The shape bifurcation is supercritical in two-dimensions. Figure courtesy of Su and Voorhees.

solution, $a_2^R = 0$, and the other solution to two-fold symmetric shapes, $a_2^R \neq 0$. Owing to the four-fold symmetry of the matrix, the horizontally and vertically oriented shapes are identical and the second solution ($a_2^R \neq 0$) is symmetric about $a_2^R = 0$. For small sizes, the particle retains the four-fold symmetric shape. (These four-fold symmetric shapes are shown in Fig. 30.) In the vicinity of $L = 0$, the equilibrium shape is approximately a circle. However, the particle becomes progressively more square-like as the particle size increases. The four-fold shape loses stability with respect to a two-fold symmetric shape at the point where the two energy-extremizing solutions intersect; i.e., at the bifurcation point corresponding to $L = 5.6$. For $L > 5.6$, the four-fold shape is an energy maximum and is depicted with the dashed line. In the two-dimensional case, the bifurcation is termed supercritical, since the solutions are symmetric about a_2^R. In three dimensions (compare Fig. 35), the shape bifurcation is transcritical since the solution was not symmetric about $S = 0$ [78]. (In three dimensions, an oblate spheroid is different from a prolate spheroid. A plot of the sum of the interfacial and elastic energies as a function of precipitate shape for the three-dimensional case is shown in Fig. 38a.)

5.1.5 External stress field: breaking the bifurcation

The presence of an externally applied elastic load acts to "break" the bifurcation of Figs. 35 and 36 and to change the equilibrium shape of the precipitate [69]. The manner in which the bifurcation is broken can be examined using bifurcation theory in conjunction with Curie's principle. Figure 37 illustrates how the three-dimensional equilibrium precipitate shape is influenced by a uniaxial stress field applied along the x_3 axis. The shape parameter, S, that extremizes the system energy is once again plotted as a function of the nondimensional particle size Λ. The calculations were performed assuming the matrix to be nickel and the precipitate phase to be Ni_3Al (γ'). The heavy curves indicate the precipitate shapes that extremize the system energy in the presence of the uniaxial applied stress and, for comparison, the fine lines denote equilibrium shapes in the absence of an applied stress. The solid and dashed lines correspond to energy minima and maxima, respectively.

As recalled from our earlier discussion of Curie's principle, application of a uniaxial stress reduces the symmetry of the interfacial energy density and, consequently, the precipitate shapes that would be expected. This reduction in symmetry is reflected in the bifurcation plots of Fig. 37. The two energy-extremizing solutions no longer intersect to form a bifurcation point but are broken into two nonintersecting solutions. Hence, the applied stress is said to "break the bifurcation".

The extent to which the applied stress breaks the bifurcation, the direction in which it does so, and the material parameters which determine the break can be obtained by considering a perturbation parameter within bifurcation theory [79]. Such an analysis results in a perturbation parameter, p, given by [69]:

$$p = \frac{\tau(\Delta C_{11} - \Delta C_{12})H}{\epsilon^T} \tag{354}$$

where τ is the applied stress, H is a numerically determined function of the precipitate (β) and matrix (α) elastic constants and $\Delta C_{11} = C_{11}^{\beta} - C_{11}^{\alpha}$. The sign of p gives the direction on the bifurcation diagram in which the bifurcation is broken, while the magnitude of p gives the extent to which it is broken. Changing the sign of the applied stress, precipitate misfit or difference in elastic constants changes the sign of p and, therefore, the direction in which the bifurcation is broken. For example, if the material parameters are such that $p > 0$, as in Fig. 37a, an oblate spheroid ($S < 0$) is the lowest energy shape for all precipitate sizes. For $\Lambda > \Lambda_t$, another energy-extremizing solution emerges and a prolate spheroid ($S > 0$) is also an energy-minimizing precipitate shape. These solutions do not intersect when an external field is present. The prolate spheroids are metastable in that they give only a local minimum in the energy. Different behavior is predicted for negative values of the perturbation parameter, p, as shown in Fig. 37b. For small precipitate sizes, a prolate spheroid has the lowest energy. A new energy-extremizing solution appears for $\Lambda > \Lambda_t$, corresponding to an oblate spheroid. Depending upon the material parameters, the prolate spheroid can remain the lowest energy shape for all precipitate sizes or the lowest energy shape can jump discontinuously from a prolate spheroid to an oblate spheroid when the precipitate is sufficiently large.

As indicated by Eq. (354), the sign of the perturbation parameter is determined by the signs of the misfit strain (ϵ^T), the applied stress (τ) and the difference in the precipitate and matrix elastic constants. For example, if $\epsilon^T > 0$ and $\Delta C_{11} - \Delta C_{12} > 0$, then an applied tensile stress renders $p > 0$; the precipitate would be expected to grow as an oblate spheroid with its axis of revolution aligned along the direction of the applied stress according to Fig. 37a. If a compressive stress were applied to the system, $p < 0$, and, according to Fig. 37b, the growth of a prolate spheroid would be favored. Such precipitate shape changes in the presence of a uniaxial stress field have been observed experimentally [80].

The bifurcation plots predict the equilibrium precipitate shape as a function of precipitate size. The change in stability of a solution (precipitate shape) with increasing precipitate size becomes clearer when the system energy is examined. Figure 38 illustrates how the sum of the interfacial and elastic energies of the precipitate, Φ, depends upon the shape parameter S for four different nondimensional precipitate sizes (Λ). In the absence of applied stress ($p = 0$), Fig. 38a, the sphere is the lowest energy shape until just before the bifurcation point ($\Lambda = 1$) at which point an oblate spheroid has the lowest energy (compare the bifurcation diagram of Fig. 35.) When the perturbation parameter is positive ($p > 0$), as shown in Fig. 38b, the oblate spheroid has the lowest total energy for all precipitate sizes. This situation corresponds to Fig. 37a. When $p < 0$, as shown in Fig. 38c, the prolate spheroid has the lowest energy for small precipitate sizes while at larger sizes the oblate spheroid has the lowest energy. This change in stability would be reflected by a discontinuous jump

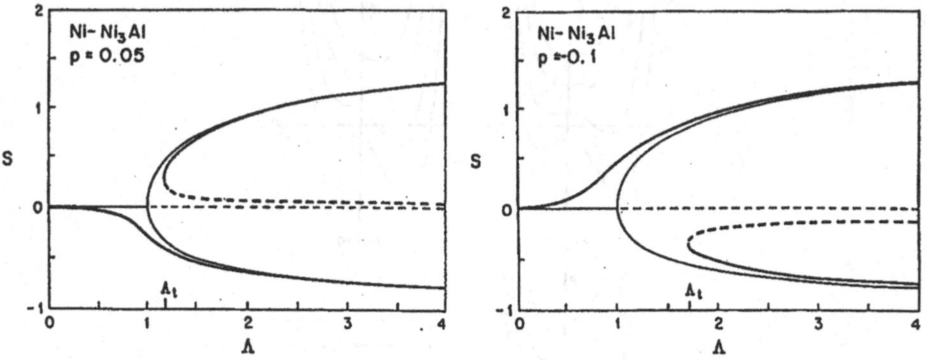

(a) (b)

Figure 37: Two bifurcation diagrams depicting the equilibrium shape parameter S as a function of dimensionless precipitate size (Λ) in the presence of an externally applied uniaxial stress is shown. Solid lines are minima and dashed lines energy maxima. The heavy lines pertain to a system with a uniaxial stress applied along the axis of revolution while the thinner lines, which show a bifurcation point at $\Lambda = 1$, depict the energy extrema in the absence of external stress. The perturbation parameter, p, is positive in (a) and negative in (b). Calculations are based on the $\gamma - \gamma'$ Ni-Al system.

of the heavy line in Fig. 37b. A prolate spheroid growing under an applied stress field might not be able to change shape to an oblate spheroid, however, as an activation energy would have to be overcome[31]. The presence of an energy barrier separating the two equilibrium shapes suggests that different microstructures might be obtainable in the presence of an external stress. If a precipitate grows in the absence of stress and an external field is applied after the precipitate has exceeded a critical size so that $p > 0$, the precipitate shape will be that of an oblate spheroid. If the external stress field is present from the initiation of precipitate growth, the precipitate shape could be a (metastable) prolate spheroid.

Restricting the class of possible shapes which a growing precipitate can assume in the presence of an applied stress to ellipsoids provides useful information on how the sign and magnitude of the applied stress and misfit strain and elastic heterogeneity affect precipitate shape. However, the actual shapes assumed by a precipitate [81, 82] will certainly be more complex than an ellipsoid of revolution and will also depend upon the degree of supersaturation in the matrix phase [17].

5.1.6 Precipitate splitting

Differences in the elastic constants between precipitate and matrix, the misfit strain, and elastic anisotropy of the matrix phase all influence the morphological evolution of a second-phase particle. Figure 39, which shows the two-dimensional evolution of an initially rectangular precipitate in an isotropic matrix using the discrete atom method [83, 84], illustrates the complexity of microstructural evolution. The shear modulus of the particle is twice that of the matrix and the equilibrium shape of the particle is a circle for all precipitate sizes. In reaching this shape, the particle first splits into two particles which then coarsen leaving just one circular particle.

Experimental confirmation of precipitate splitting has been reported by Miyazaki, Imamura and Kozakai [15] and others [16, 17]. Several types of splitting transitions have been observed experimentally in cubic materials including the splitting of an isolated cuboid into either two

[31] It is also possible that the precipitate could assume shapes other than a spheroid to circumvent the energy barrier.

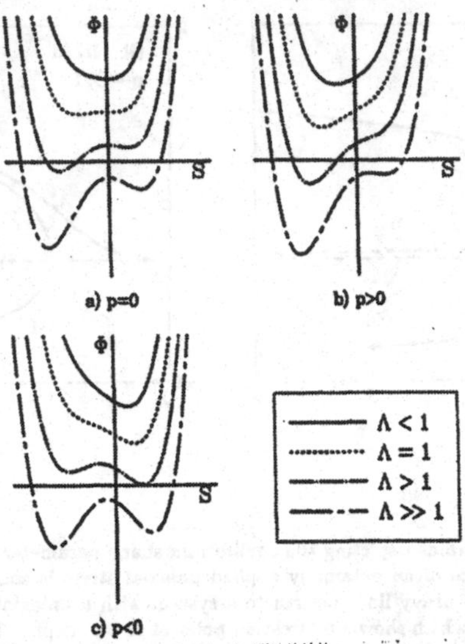

Figure 38: **The sum of the interfacial and elastic energies, Φ, is plotted as a function of the shape parameter S for four different precipitate sizes, Λ. In (a) there is no applied field, in (b) the perturbation parameter p is positive and in (c) the perturbation parameter is negative. A sphere ($S = 0$) is never an energy extremizing shape when an external field is applied to the system in (b) and (c).**

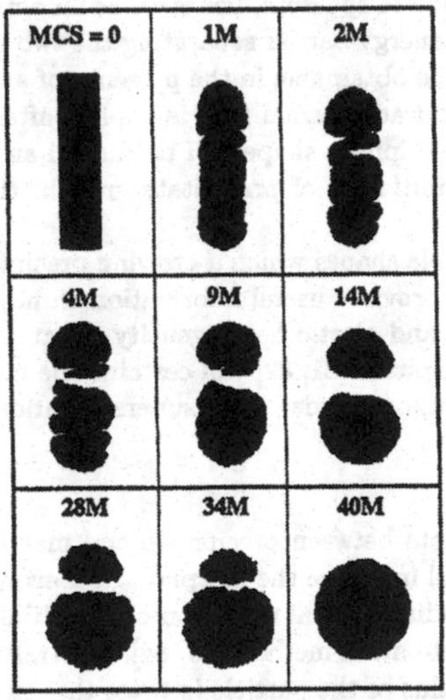

Figure 39: **The two-dimensional morphological evolution of an initially rectangular and misfitting precipitate in an isotropic matrix. The shear modulus of the precipitate is twice that of the matrix. The particle splits in two before eventually evolving into the equilibrium shape of a circle. Figure courtesy of J. K. Lee.**

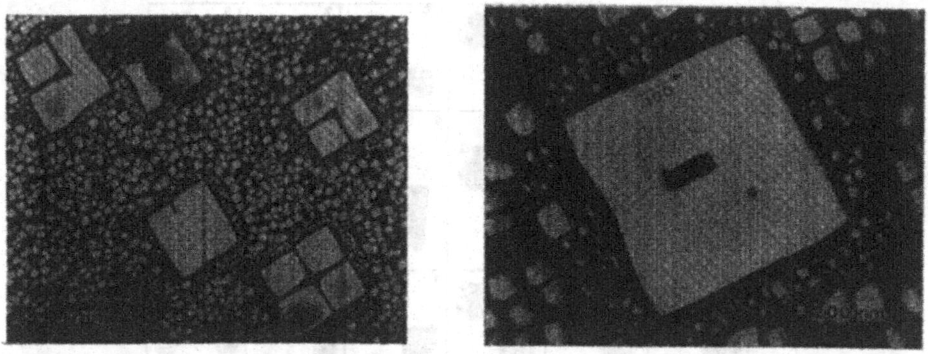

(a) (b)

Figure 40: Dark-field TEM micrographs of the $\gamma - \gamma'$ Ni-Al system showing (Ni_3Al) particle splitting. Micrographs courtesy of M. J. Kaufman.

parallel plate-like particles (doublet) or eight smaller cuboids (octet) arranged in a cubic geometry with their faces parallel to the < 100 > cube directions [72]. Calculations of the interfacial energy, elastic self-energy and elastic interaction energy indicate that the splitting process in some of the cubic alloys occurs because the total energy of the system can be reduced. The increase in interfacial energy is more than offset by the decrease in elastic energy [85-88]. An example of a precipitate in the process of splitting is given in Fig. 40.

5.2 Elastic Interactions during Coarsening

In this subsection, the results of several different computer simulations accounting for the elastic and diffusional interaction between precipitates are presented [89-106]. The simulations, which are both two- and three-dimensional in nature, are intended to demonstrate the influence of misfit strain, applied stress and elastic heterogeneity on microstructural evolution in two-phase systems. In particular, they serve to help explain the stress-assisted morphological evolution depicted in Figs. 2 and 3.

5.2.1 Influence of precipitate misfit strain

Elastic interactions between precipitates affect both the shape of the individual precipitates and the evolution of the microstructure. These effects are usually most noticeable during the coarsening regime when matrix supersaturation and chemical driving force are small and the driving force for further microstructural evolution can be approximated by considering just the interfacial and elastic energies.

Figure 41 illustrates the effect of precipitate elastic constants and misfit strain on the evolution of a simple system consisting initially of two circular precipitates using the discrete atom method of J. K. Lee [84]. The elastic constants of the matrix phase possess a four-fold symmetry with elastically soft directions in the horizontal [10] and vertical [01] directions. The size, shape, and relative position of the precipitates at the start of the simulation are identical in each of the four simulations, as shown in the first column of Fig. 41. On account of the identical initial conditions,

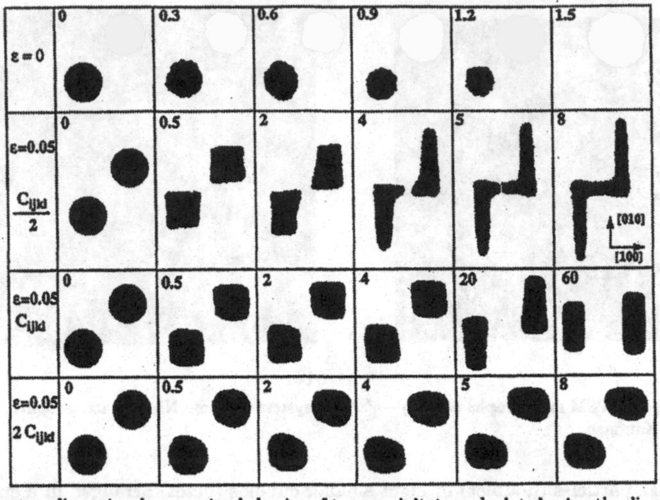

Figure 41: The two-dimensional coarsening behavior of two precipitates calculated using the discrete atom method is shown. (Numbers in the upper left corner are measures of the elapsed time.) The matrix is cubic with the cube directions lying in the horizontal and vertical directions. The first row shows classical, capillarity-driven coarsening (no misfit strain) while the next three rows show the effect on microstructural evolution of changing the precipitate elastic constants with a misfit strain $\epsilon^T = 0.05$. The precipitates are elastically softer than the matrix in the second row, have the same elastic constants as the matrix in the third row and are elastically harder than the matrix in the bottom row. The initial configuration of each of the four simulations, shown in the first column, are identical. Figure courtesy of J. K. Lee.

differences in microstructural evolution between the different systems, as displayed in each of the rows of Fig. 41, indicate the effects of different misfit strains and precipitate elastic constants on the coarsening behavior. In the top row, the misfit strain vanishes and the particles do not interact elastically; coarsening proceeds classically, with the larger particle growing at the expense of the smaller particle [89, 90]. The second row depicts the coarsening process when a misfit strain is present and the precipitates are elastically softer than the matrix phase. The individual particles initially undergo a symmetry-conserving shape transition from circle to square, slightly modified by the elastic interaction between the precipitates. The elastic interaction also induces a translation of the center of mass of the precipitates and a change in shape, eventually leading to precipitate coalescence. The elastic interaction energy between the two precipitates is positive for all indicated values of the precipitate elastic constants when the precipitates are aligned along the diagonals ($< 11 >$ directions) and negative when they are aligned along one of the elastically soft directions ($< 10 >$). The elastic interaction induces atom-by-atom precipitate migration, increasing the alignment along the elastically soft direction, before coalescence can occur.

When the precipitate and matrix elastic constants are equal, as in the third row of Fig. 41, the repulsive interaction between precipitates aligned along the [11] direction at first stabilizes the precipitates with respect to size and shape. The interaction also induces precipitate migration which tends to align the precipitates more closely along the [10] direction. This migration lowers the (initially positive) elastic interaction energy and eventually leads to an attraction between precipitates. At longer times, these precipitates coalesce into one precipitate with a rectangular shape, as shown in Fig. 42 [84].

The elastic interaction between precipitates induces a strong preferential coarsening. Conversely, significant precipitate migration is not observed in stress-free systems [91, 92, 93]. Figure 43

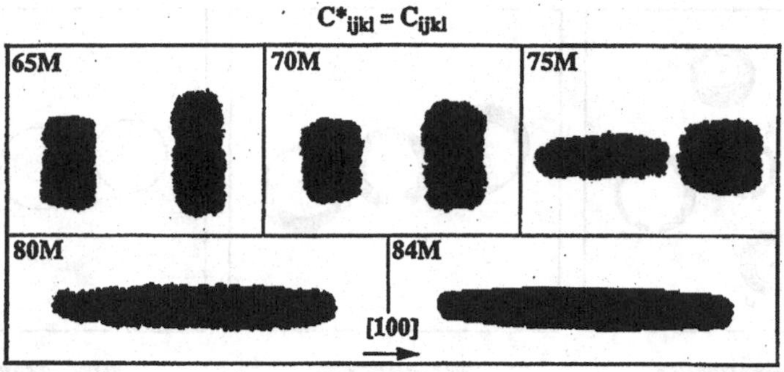

Figure 42: The long-term, two-dimensional coarsening behavior of the two elastically homogeneous precipitates depicted in the third row of Fig. 39 is shown. Eventually, the particles fuse to form a rectangular shape. Figure courtesy of J. K. Lee.

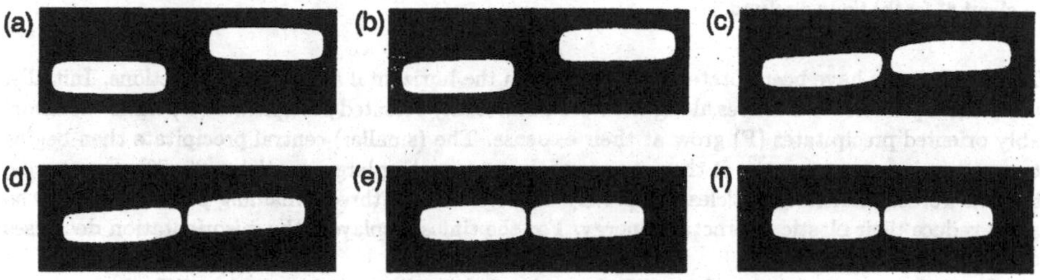

Figure 43: The precipitate migration and alignment of two rectangular shaped precipitates is shown. The precipitates and matrix possess the same cubic elastic constants with the elastically soft directions along the $< 10 >$ (horizontal and vertical) directions. The misfit strain is dilatational. Figure courtesy of P. W. Voorhees.

depicts the two-dimensional simulation of particle migration during coarsening using a boundary integral method [94]. Aging time increases from Fig. 43a to Fig. 43f. The two rectangular shaped precipitates migrate so as to align themselves along the elastically soft [10] (horizontal) direction while decreasing their intercenter distance of separation. This migration decreases the total elastic energy of the system as the interaction energy becomes increasingly negative as the precipitates become progressively more aligned along the $< 10 >$ direction.

Figure 44 illustrates how both elastically-induced particle migrations and preferential coarsening can contribute to the development of strong spatial correlations among precipitates. In this example, the simulations are performed in three dimensions by solving the coupled equations for elasticity and diffusion. The precipitates are constrained to remain spherical and are arrayed with their centers located on the $z-x$ plane [95, 96]. The elastic constants of the precipitate and matrix phases are the same and are taken to be isotropic. The misfit strain is tetragonal with the axis of tetragonality parallel to the z (vertical) axis. Particles aligned along their axis of tetragonality possess a positive elastic interaction energy while those aligned perpendicular to the axis of tetragonality are favorably oriented and have a negative interaction energy [85]. The radius of the central precipitate is 10% smaller than the radii of the four outer precipitates in the starting configuration.

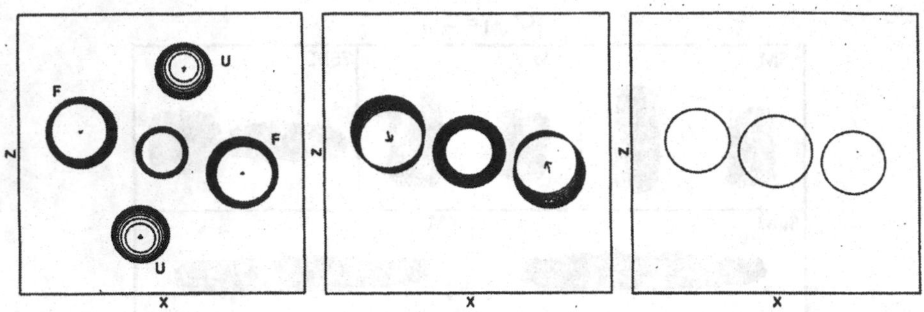

(a) Preferential Coarsening (b) Particle Migrations (c) Final Configuration

Figure 44: **Preferential coarsening, inverse coarsening, and precipitate migration and alignment of three-dimensional spherical precipitates possessing a tetragonal misfit is shown. (Time increase from (a) to (c).) The precipitates and matrix possess the same isotropic elastic constants. The central precipitate begins to dissolve then grows (inverse coarsens) at the expense of the surrounding precipitates. Precipitate migration reduces the misorientation from 18° to about 8° for the times displayed.**

The precipitates have been rotated about 18° from the horizontal and veritcal directions. Initially, the central precipitate dissolves along with the unfavorably oriented precipitates (U) while the favorably oriented precipitates (F) grow at their expense. The (smaller) central precipitate then begins to grow, or inverse coarsen[32], at the expense of the surrounding larger particles [85, 86]. Eventually, the unfavorably oriented particles completely dissolve and the three remaining particles migrate so as to reduce their elastic interaction energy. For the times displayed, the misorientation decreases from 18° to about 8°.

Although stress-induced preferential coarsening and precipitate migration are difficult to observe directly in experiment, the manifestation of these processes on microstructural evolution is often clear. Figure 45 shows the simulated time evolution of two, two-dimensional microstructures [97]. In the top row, (a)-(d), microstructural evolution in the absence of precipitate misfit strain is shown. In Figs. 45e-45h, the same system is allowed to evolve when a precipitate misfit strain is present. In the stress-free case, of the precipitates labelled 1-5, only precipitates 3 and 5 remain after the simulation. When a misfit strain is present, however, the elastically favorable orientation of precipitates 1-5 acts to stabilize precipitate 1 with respect to the other precipitates such that, after aging, precipitate 1 is the only precipitate remaining. This is another example of inverse coarsening.

5.2.2 Influence of applied stress

The effect of external stress on microstructural evolution is depicted in Figs. 46-48 [98, 99]. Figure 46 displays the starting configuration used in the simulations of Figs. 47 and 48. The precipitates are spherical with their centers placed on a common plane for ease of viewing. Thus the diameter

[32]Inverse coarsening refers to the growth of smaller precipitates at the expense of larger precipitates. Although interfacial energy increases during this process, the system free energy decreases owing to a decrease in the elastic strain energy.

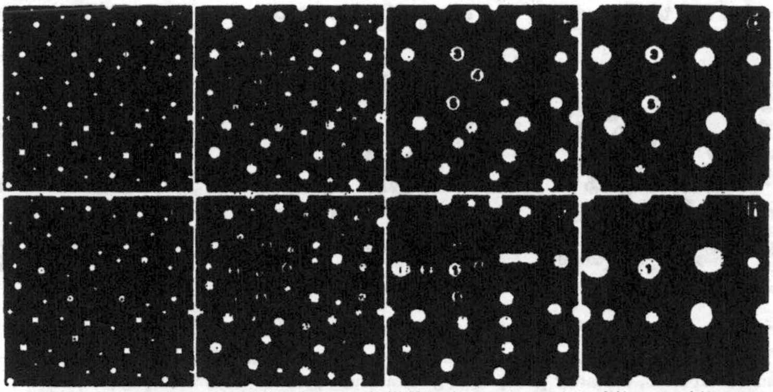

Figure 45: **The two-dimensional temporal evolution of a system of stress-free precipitates is simulated in the top row, figures (a) through (d), while an identical system of misfitting precipitates with the same starting configuration is simulated in the bottom row, figures (e) through (h). The system is cubic and elastically homogeneous and periodic boundary conditions have been employed. Note the difference in the coarsening behavior of precipitates labelled 1-5 in each case. Figure courtesy of Y. Wang, L.-Q. Chen and A. G. Khachaturyan.**

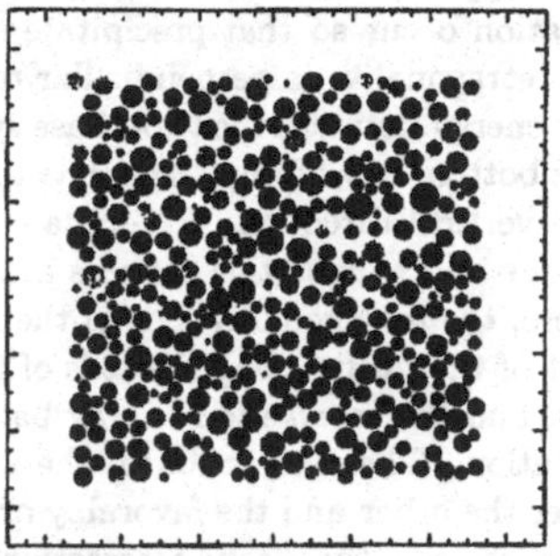

Figure 46: **The starting configuration for the simulations of Figs. 47 and 48 containing 400 spherical precipitates possessing the same elastic constants as the isotropic matrix. Simulations are three-dimensional with the center of the spheres placed on a plane for clarity. Black and grey spheres represent two different variants of the tetragonal misfit strain.**

of a circle in a figure corresponds to the actual diameter of the sphere. In the simulations with misfit strains and applied stress, the precipitates possess a tetragonal misfit: the two shades of gray indicate two different variants of the misfit strain. The axis of tetragonality lies in the direction of the line (black-horizontal, gray-vertical); the misfit is slightly larger in the direction of the axis of tetragonality. There are 400 precipitates in the initial configuration and the precipitates and matrix are assumed to have the same isotropic elastic constants. The microstructures are obtained by solving simultaneously the elastic and diffusional field equations [95].

Figure 47 depicts the microstructural evolution in the absence of all stress effects. Precipitate coarsening is driven solely by the reduction in interfacial energy (and area in this isotropic system) with the larger precipitates growing at the expense of the smaller precipitates [89, 90]. Coarsening proceeds until only one precipitate remains in the system.

Figure 48 illustrates how the microstructure is altered by the presence of a tetragonal misfit strain and applied stress field. In the first row of Fig. 48, the coarsening behavior of the tetragonally misfitting precipitates is shown in the absence of an external stress field. Precipitates coarsen

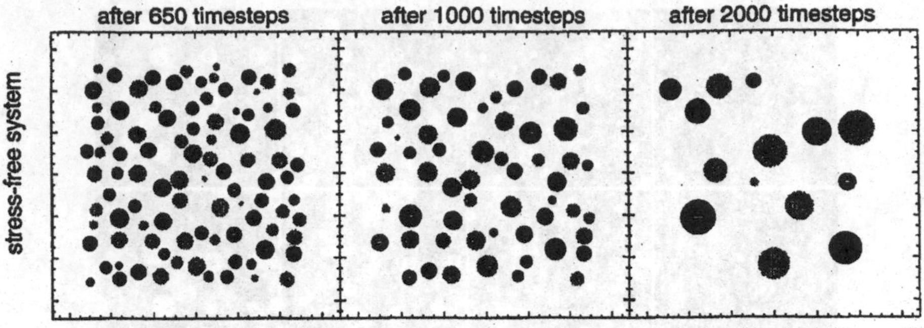

Figure 47: Simulation of the coarsening behavior of the initial distribution shown in Fig. 46 when all stress effects are neglected. As expected, the larger precipitates grow at the expense of the smaller precipitates.

preferentially so as to form stringers with their axis of tetragonality parallel to one another. This configuration lowers the elastic interaction energy between precipitates, and stringers with both variants of the misfit strain are observed to form. In the second row of Fig. 48, a uniaxial compressive stress in the vertical direction is applied to the system of misfitting precipitates. In this case, coarsening and precipitate migration occur so that precipitate stringers develop in the direction of the applied field. The axis of tetragonality is perpendicular to the applied compressive field as this lowers the elastic interaction energy between the two-phase system and the loading mechanism applying the external stress. The bottom row of Fig. 48 depicts the microstructural evolution when a tensile stress is applied in the vertical direction. Precipitate stringers still develop with their axis of tetragonality parallel to one another, as this induces a negative elastic interaction energy between precipitates. Furthermore, the tetragonality axis of the precipitate lies in the direction of the applied field. This is a result of the misfit along the axis of tetragonality being slightly larger than in the other two directions, and the interaction energy between the precipitate and applied stress is decreased in this orientation. The interaction of the applied field and the misfit strain favors one precipitate variant over the other and the favorably oriented precipitates coarsen at the expense of the other variant. In each case, the elastic interaction between precipitates favors their alignment so that their axes of tetragonality remain parallel.

6 Acknowledgements

I am indebted to H. I. Aaronson, A. J. Ardell, L.-Q. Chen, M. Fährmann, W. H. Hort, M. Kaufman, J. K. Lee, P. H. Leo and P. W. Voorhees for valuable discussions and for the use of various materials. I am also grateful to the National Science Foundation under Grant DMR-9496133 for the support of this work.

7 Appendix: Thermodynamic Perturbations at a Coherent Interface

Referring to Fig. 18, consider a point on the planar, coherent interface separating the α and β phases. Let $\delta\bar{y}$ be a perturbation in the position of the interface as measured in the actual system. Physically, this perturbation can accrue from two different sources: an elastic deformation of each crystal without undergoing a phase transformation or a phase transformation of one phase into the

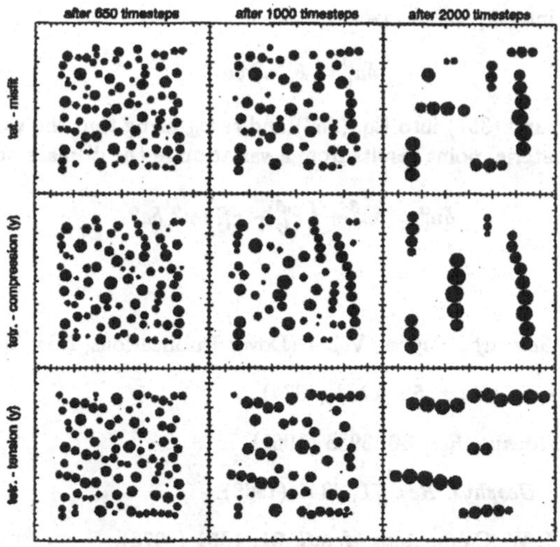

Figure 48: **Microstructural evolution of tetragonally misfitting precipitates for the initial distribution shown in Fig. 46. First row: absence of an applied stress; second row: compressive stress applied along the vertical axis; third row: tensile stress applied along the vertical axis. The evolved microstructure is strongly influenced by the applied stress field.**

other without elastic deformation [6, 10]. (Of course, a combination of the two is also possible.) Since the interface is constrained to be coherent, the α and β phases must remain in contact at the interface after the perturbation. This requires that, for every point on the interface:

$$\delta\vec{y}^{\alpha} = \delta\vec{x}^{\alpha} + \delta\vec{a}^{\alpha} = \delta\vec{x}^{\beta} + \delta\vec{a}^{\beta} = \delta\vec{y}^{\beta} \tag{355}$$

or, in indicial notation:

$$\delta x_i^{\alpha} + \delta a_i^{\alpha} = \delta x_i^{\beta} + \delta a_i^{\beta} \tag{356}$$

where $\delta\vec{x}$ is the change in interfacial position of the indicated phase owing to a variation in the elastic field and $\delta\vec{a}$ is the change in interfacial position owing to a virtual phase transformation, measured in the direction of the outward pointing normal of the indicated phase. $\delta\vec{a}$ is an accretion term.

The accretion term can be referred back to the reference state configuration as shown in Fig. 18. It is simplest to assume that matter is accreted along the outward pointing normal direction in the reference state; $\hat{n}^{\alpha'}$ for the α phase and $\hat{n}^{\beta'}$ for the β phase.

The accretion vectors in the actual and reference states can then be related using the deformation gradient tensor as in Eq. (45):

$$\delta a_i^{\alpha} = F_{ij}^{\alpha}\delta a_j^{\alpha'} = F_{ij}^{\alpha} n_j^{\alpha'} \delta a^{\alpha'} \tag{357}$$

and

$$\delta a_i^{\beta} = F_{ij}^{\beta}\delta a_j^{\beta'} = F_{ij}^{\beta} n_j^{\beta'} \delta a^{\beta'}. \tag{358}$$

Since the interface must remain coherent:

$$\delta a^{\alpha'} \hat{n}^{\alpha'} = \delta a^{\beta'} \hat{n}^{\beta'} \quad \text{or} \quad \delta a^{\alpha'} n_j^{\alpha'} = \delta a^{\beta'} n_j^{\beta'} \tag{359}$$

As $\hat{n}^{\alpha'} = -\hat{n}^{\beta'}$, then:

$$\delta a^{\alpha'} = -\delta a^{\beta'} \tag{360}$$

Substituting Eq. (359) into Eq. (358) gives:

$$\delta a_i^\beta = F_{ij}^\beta n_j^{\alpha'} \delta a^{\alpha'} \tag{361}$$

Substituting Eqs. (361) and (357) into Eq. (356) and recognizing that the variation in the position coordinate of a given material point results from a variation in the displacement field; i.e., $\delta\vec{x} = \delta\vec{u}$, gives:

$$\delta u_i^\alpha = \delta u_i^\beta + \left(F_{ij}^\beta - F_{ij}^\alpha\right) n_j^{\alpha'} \delta a^{\alpha'}. \tag{362}$$

References

[1] J. W. Gibbs, *The Scientific Papers*, Vol. 1 (Dover Publications, New York, 1961).

[2] P.-Y. F. Robin, *Amer. Miner.* **59**, 1286 (1974).

[3] W. B. Kamb, *J. Geophys. Res.* **66**, 3985 (1961).

[4] A. G. McLellan, *J. Geophys. Res.* **71**, 4341 (1966).

[5] F. C. Larché and J. W. Cahn, *Acta Metall.* **21**, 1051 (1973).

[6] F. C. Larché and J. W. Cahn, *Acta Metall.* **26**, 1579 (1978).

[7] W. C. Johnson and J. I. D. Alexander, *J. Appl. Phys.* **59**, 2735 (1986).

[8] W. W. Mullins and R. F. Sekerka, *J. Chem. Phys.* **82**, 5192 (1986).

[9] P. H. Leo and R. F. Sekerka, *Acta metall.* **37**, 3119 (1989).

[10] W. C. Johnson, *J. Amer. Cer. Soc.* **77**, 1581 (1994).

[11] J. W. Cahn and G. Kalonji, *Int. Conf. Solid-solid Phase Transformations* (ed. by H. I. Aaronson, D. E. Laughlin, R. F. Sekerka and C. M. Wayman) p. 3, TMS-AIME, Warrendale, PA (1982).

[12] J. K. Lee, D. M. Barnett and H. I. Aaronson, *Metall. Trans.* **8A**, 973 (1977).

[13] A. G. Khachaturyan, *Theory of Structural Transformations in Solids* (Wiley, Berlin, 1983).

[14] T. Mura, *Micromechanics of Defects in Solids* Second Edition, (Martinus Nijhoff Pub., Dordrecht 1987).

[15] T. Miyazaki, H. Imamura and T. Kozakai, *Matls. Sci. Engr.* **54**, 9 (1982).

[16] M. J. Kaufman, P. W. Voorhees, W. C. Johnson and F. S. Biancaniello, *Metall. Trans.* **20A**, 2171 (1989).

[17] S. J. Yeon, D. Y. Yoon and M. F. Henry, *Metall. Trans.* **A24**, 1975 (1993).

[18] M. Fährmann, P. Fratzl, O. Paris, E. Fährmann and W. C. Johnson, *Acta metall. mater.* **43**, 1007 (1995).

[19] J. F. Nye, *Physical Properties of Materials* (Clarendon Press, Oxford, 1985).

[20] G. Arfken, *Mathematical Methods for Physicists* (Academic Press, Inc., Orlando, FL, 1985) ch. 3.

[21] L. E. Malvern, *Introduction to the Mechanics of a Continuous Medium* (Prentice-Hall, Englewood Cliffs, NJ 1969).

[22] Y. C. Fung *A First Course in Continuum Mechanics* (Prentice Hall, Inc., New Jersey, 1969).

[23] R. Aris *Vectors, Tensors, and the Basic Equations of Fluid Mechanics,* (Prentice Hall, 1962).

[24] J. D. Eshelby, *Proc. Roy. Soc.* **A241**, 376 (1957).

[25] J. D. Eshelby, *Proc. Roy. Soc.* **A252**, 561 (1959).

[26] J. D. Eshelby, *Prog. Sol. Mech.* **2**, 89 (1961).

[27] N. F. Mott and F. R. N. Nabarro, *Proc. Phys. Soc.* **52**, 86 (1940).

[28] W. C. Johnson and P. W. Voorhees, *Metall. Trans.* **16A**, 337 (1985).

[29] A. H. Cottrell and M. A. Jawson, *Proc. Royal Soc.* **A199**, 104 (1949).

[30] D. M. Barnett, G. Wang and W. D. Nix, *Acta metall.* **30**, 2035 (1982).

[31] J. W. Christian, *The Theory of Transformations in Metals and Alloys Part I: Equilbirum and General Kinetic Theory* (Pergamon Press, Oxford, 1975).

[32] J. P. Hirth and J. Lothe, *Theory of Dislocations* (John Wiley and Sons, New York, 1982).

[33] J. Piller, M. K. Miller and S. S. Brenner in *Proc. 29th Int. Field Emmisions Symp.*, H. O. Andren and H. Norden, eds., (Goteborg, Sweden, 1982) p. 473.

[34] M. E. Gurtin and P. W. Voorhees, *Proc. Roy. Soc. London* **A440**, 323 (1993).

[35] M. E. Gurtin and P. W. Voorhees, *Acta Mater.* **44**, 235 (1996).

[36] W. C. Johnson, *Metall. Mater. Trans.*, (1997).

[37] J. W. Cahn, *Acta metall.* **10**, 907 (1962).

[38] R. O. Williams, *Metall. Trans.* **11A**, 247 (1980).

[39] R. O. Williams, *CALPHAD* **8**, 1 (1984).

[40] J. W. Cahn and F. C. Larché, *Acta metall.* **32**, 1915 (1984).

[41] A. L. Roitburd, *Sov. Phys. Sol. St.* **26**, 1229 (1984).

[42] A. L. Roitburd, *Sov. Phys. Sol. St.* **27**, 598 (1984).

[43] W. C. Johnson and P. W. Voorhees, *Metall. Trans.* **18A**, 1213 (1987).

[44] W. C. Johnson, *Metall. Trans.* **18A**, 1093 (1987).

[45] W. C. Johnson and C.-S. Chiang, *J. Appl. Phys.* **64**, 1155 (1988).

[46] C.-S. Chiang and W. C. Johnson, *J. Mater. Res.* **4**, 678 (1989).

[47] Z.-K. Liu and J. Ågren, *Acta metall. mater.* **38**, 561 (1990).

[48] M. S. Paterson, *Rev. Geophys. Space Phys.* **11**, 355 (1973).

[49] A. A. Mbaye, D. M. Wood and A. Zunger, *Phys. Rev. B* **37**, 3008 (1988).

[50] I. Müller, *Cont. Mech. Thermodyn.* **1**, 1 (1989).

[51] W. C. Johnson and W. H. Müller, *Acta metall. mater.* **39**, 89 (1991).

[52] R. D. James, *J. Mech. Phys. Sol.* **34**, 359 (1986).

[53] F. Falk, *Acta metall.* **28**, 1773 (1980).

[54] M. J. Pfeiffer and P. W. Voorhees, *Acta metall. mater.* **39**, 2001 (1991).

[55] W. C. Johnson, *Mater. Res. Soc. Symp. Proc.* (ed. by T. W. Barbee, F. Spaepen and L. Greer) Vol. 103, p. 61 (1987).

[56] J.-Y. Huh and W. C. Johnson, *Acta metall. mater.* **43**, 1631 (1995).

[57] J.-Y. Huh and W. C. Johnson *Mater. Res. Soc. Symp. Proc.*, Vol. 311, Eds. M. Atzmon, J. M. E. Harper, A. L. Greer and M. R. Libera, p. 119 (1993).

[58] R. B. Griffiths and J. W. Wheeler, *Phys. Rev. A* **2**, 1047 (1970).

[59] M. Hillert, *Int. Metal. Rev.* **30**, 45 (1985).

[60] M. Hillert, this volume.

[61] J. E. Hilliard in *Phase Transformations*, ed. by H. I. Aaronson (ASM, Metals Park, OH 1970), chap. 1.

[62] G. B. Stringfellow, *J. Electon. Mater.* **11**, 903 (1982).

[63] F. Glas, *J. Appl. Phys.* **62**, 15 (1987).

[64] M. Neuberger, *Handbook of Electronic Materials* (IFI/Plenum, New York, 1971) vol. 2, p. 66.

[65] J.-Y. Huh, J. M. Howe and W. C. Johnson, *Acta metall. mater.* **41**, 2577 (1993).

[66] Y. I. Sirotin and M. P. Shaskolskaya, *Fundamentals of Crystal Physics*, Mir, Moscow (1982).

[67] G. Wulff, *Zeit. f. Kristallog.* **34**, 449 (1901).

[68] W. C. Johnson and J. W. Cahn, *Acta metall.* **32**, 1925 (1984).

[69] W. C. Johnson, M. B. Berkenpas and D. E. Laughlin, *Acta metall.* **36**, 3149 (1988).

[70] W. C. Johnson and P. W. Voorhees, *Sol. State Phen.* **23**, 87 (1992).

[71] A. J. Ardell and R. B. Nicholson, *Acta metall.* **14**, 1295 (1966).

[72] M. Doi, T. Miyazaki and T. Wakatsuki, *Matls. Sci. Engr.* **74**, 139 (1985).

[73] V. A. Phillips, *Acta Metall.* **14**, 271 (1966).

[74] S. Satoh and W. C. Johnson, *Metal. Trans.* **23A**, 2761 (1992).

[75] G. Kostorz, *Physica* **120B**, 387 (1983).

[76] P. W. Voorhees, G. B. McFadden and W. C. Johnson, *Acta metall. Mater.* **40**, 2979 (1992).

[77] C. H. Su and P. W. Voorhees, *Acta metall. mater.* **44**, 1987 (1996).

[78] M. B. Berkenpas, W. C. Johnson and D. E. Laughlin, *J. Mater. Res.* **1**, 635 (1986).

[79] G. Iooss and D. D. Joseph, *Elementary Stability and Bifurcation Theory*, (Springer-Verlag, New York, 1980).

[80] J. K. Tien and S. M. Copley, *Metall. Trans.* **2**, 215 (1971).

[81] H.-J. Jou, P. H. Leo and J. S. Lowengrub, *J. Comp. Phys.* **131**, 109 (1997).

[82] H.-J. Jou, P. H. Leo and J. S. Lowengrub, *Int. Conf. Solid-solid Phase Transformations* (Ed. W. C. Johnson, J. M. Howe, D. E. Laughlin, and W. A. Soffa) p. 635, TMS-AIME, Warrendale, PA (1994).

[83] J. K. Lee, *Scripta metall. mater.* **32**, 559 (1995).

[84] J. K. Lee, *Metall. Mater. Trans.* **27A**, 1449 (1996).

[85] A. G. Khachaturyan and G. A. Shatalov, *Sov. Phys. Solid State* **11**, 118 (1969).

[86] W. C. Johnson, *Acta metall.* **32**, 465 (1984).

[87] A. G. Khachaturyan, S. V. Semenovskaya and J. W. Morris Jr., *Acta metall.* **36**, 1563 (1988).

[88] T. Miyazaki *Int. Conf. Solid-solid Phase Transformations* (Ed. W. C. Johnson, J. M. Howe, D. E. Laughlin, and W. A. Soffa) p. 573, TMS-AIME, Warrendale, PA (1994).

[89] I. M. Lifshitz and V. V. Slyozov, *J. Phys. Chem. Solids* **19**, 35 (1961).

[90] C. Wagner, *Z. Elektrochem.* **65**, 581 (1961).

[91] P. W. Voorhees and W. C. Johnson, *Phys. Rev. Lett.* **61**, 2225 (1988).

[92] W. C. Johnson, P. W. Voorhees and D. E. Zupon, *Metall. Trans.* **20A**, 1175 (1989).

[93] W. C. Johnson, T. A. Abinandanan and P. W. Voorhees, *Acta metall. mater.* **38**, 1349 (1990).

[94] C. H. Su and P. W. Voorhees, *Acta metall. mater.* **44**, 2001 (1996).

[95] T. A. Abinandanan and W. C. Johnson, *Acta metall. mater.* **41**, 17 (1993).

[96] T. A. Abinandanan and W. C. Johnson, *Acta metall. mater.* **41**, 27 (1993).

[97] Y. Wang, L.-Q. Chen and A. G. Khachaturyan, *Acta metall. mater.* **41**, 279 (1993).

[98] W. H. Hort and W. C. Johnson, *Int. Conf. Solid-solid Phase Transformations* (Ed. W. C. Johnson, J. M. Howe, D. E. Laughlin, and W. A. Soffa) p. 629, TMS-AIME, Warrendale, PA (1994).

[99] W. H. Hort and W. C. Johnson, *Metal. Mater. Trans.* **27A**, 1461 (1996).

[100] K. Kawasaki and Y. Enomoto, *Physica A* **50**, 463 (1988).

[101] Y. Wang, L.-Q. Chen and A. G. Khachaturyan, *Scripta metall. mater.* **25**, 1387 (1991).

[102] H. Nishimori and A. Onuki, *Phys. Rev. B* **42**, 980 (1990).

[103] A. Onuki and H. Nishimori, *J. Phys. Soc. Jap.* **60**, 1 (1991).

[104] H. Nishimori and A. Onuki, *J. Phys. Soc. Jap.* **60**, 1208 (1991).

[105] J. Gayda and D. J. Srolovitz, *Acta metall.* **37**, 641 (1989).

[106] P. H. Leo and H.-J. Jou, *Acta metall. mater.* **41**, 2271 (1993).

THEORY OF CAPILLARITY

Rohit Trivedi
Ames Laboratory US-DOE and Department of Materials Science and Engineering
Iowa State University, Ames, IA 50011, USA

In thermodynamics, one considers the bulk properties of materials. However, when the system is of a finite size, the atoms at the boundaries see a different environment so that the average properties of atoms at the boundary is different from those in the bulk. This change in energy due to the presence of a boundary becomes very important when the number of atoms at the boundary are comparable to the number of atoms in the bulk, i.e. when the size of the system is small. These changes are generally represented by interface properties. Because interface effects in solids are complex, the basic ideas on interfaces are generally developed for incompressible fluids. We shall therefore initially consider interface effects in incompressible fluids, and then briefly discuss the concepts which must be modified when solid systems are considered.

Whenever two homogeneous phases are brought in contact, most properties of each phase are slightly disturbed in the vicinity of the surface of contact. These surface effects, however, become much smaller than volume effects as the bulk phases become large since the surface area increases slower than the volume, and thus they are usually ignored when dealing with the bulk properties of the body. However, surface effects become quite important when the surface to volume ratio is appreciable. Such cases occur during phase transformation when an interface has large curvature or when the dimensions of crystals or precipitates are small, typically in the micron range or below. Some of the examples in which surface effects play a prominent role include nucleation, growth of dendrites and Widmanstätten precipitates, eutectics and eutectoids, precipitate coarsening, sintering and grain growth.

Capillarity theory in general deals with the origin of surface forces and the kinetics of changes occurring due to these forces. We shall first examine the concept of an interface, outline the thermodynamic description of a system which contains an interface, and then discuss the equilibrium conditions in the presence of a curved interface. Subsequently, we shall emphasize the driving forces or potentials associated with a surface or an interface, and briefly indicate how these driving forces lead to morphological changes.

1. DEFINITION OF AN INTERFACE

In the thermodynamics of heterogeneous masses in contact, the effects at the interfaces of the masses are not explicitly considered. The density, entropy, energy, etc., of the masses are assumed to be uniform right up to the plane of contact. However, as pointed out by Gibbs (1), this cannot be true because, for example, if the densities of the masses were uniform up to the surface of contact, the energy would not be continuous because of the finite range of atomic interaction. Consider a system of two fluid phases in equilibrium, as shown in Fig. 1. Because of the equilibrium conditions in each phase and a short range atomic interaction, each phase can be considered to be homogeneous throughout its interior except in the vicinity of the other phase where a transition from one to the other takes place within a thin layer. The thickness of the transition layer is not arbitrary, but is uniquely determined from the equilibrium constraint that the chemical potential of each species present be constant throughout the system.

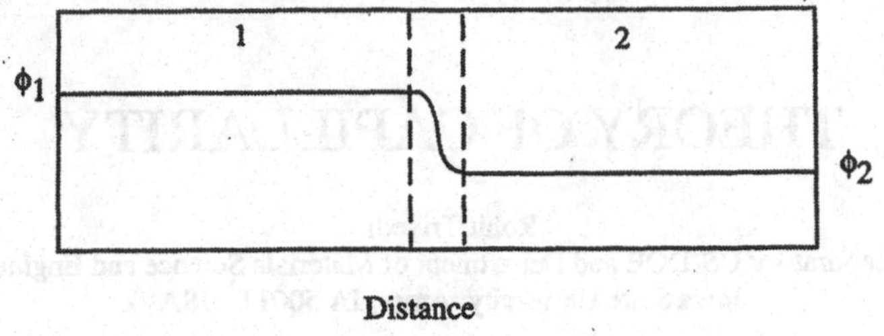

Fig. 1. The value of the thermodynamic function f in two phases separated by an interface.

A detailed thermodynamic study of the above system can be made from the properties of each homogeneous phase and the detailed properties of the equilibrium transition layer. The free energy of the system is generally defined for uniform composition. When a gradient in composition exists, such as that in the transition region, then the free energy depends not only on composition but also on the gradient of composition. We shall first describe the simple treatment of interface proposed by Gibbs (1), and then briefly present a description of the interface in terms of inhomogeneous thermodynamics developed by Cahn and Hilliard (2,3).

1.1 Sharp Interface

The description of the nonhomogeneous interface region gives us an insight into the nature of the interface between two phases. However, it is quite complex and Gibbs has pointed out that a simple phenomenological treatment of the problem can be given if the transition layer is replaced by a hypothetical geometrical interface which he called a *dividing surface*. Each phase is assumed to be homogeneous right up to this geometrical interface so that thermodynamics of homogeneous phases can be applied to each phase. In this case, any extensive thermodynamic property ϕ can be considered as the sum of volume and interface contributions. The volume contributions are the sum of ϕ_1 and ϕ_2 which the two phases would have if they remained homogeneous right to the dividing surface, and the surface contribution, ϕ_s, gives the necessary correction arising from the actual presence of a transition layer. Thus, we may write

$$\phi = \phi_1 + \phi_2 + \phi_s \qquad [1a]$$

or

$$\phi_s = \phi - (\phi_1 + \phi_2) \qquad [1b]$$

The above division, though artificial, enables us to define various thermodynamic properties for an interface since ϕ, ϕ_1 and ϕ_2 are well defined for equilibrium between the two phases. The value of ϕ can be calculated by using the thermodynamics of inhomogeneous systems as will be described in the next section.

The question now arises as to the choice of such a dividing surface. Since the dividing surface is artificial, the properties of the system should not depend on how such an interface is selected. Fig. 2 shows the transition region and its vicinity in both phases. The dividing surface is constructed to lie within the transition layer and is chosen to pass through points in that layer such that each point chosen has nearly the same surroundings as the neighboring points on the interface. Such a geometrical interface uniquely determines only the normal to the interface. Since any other surface

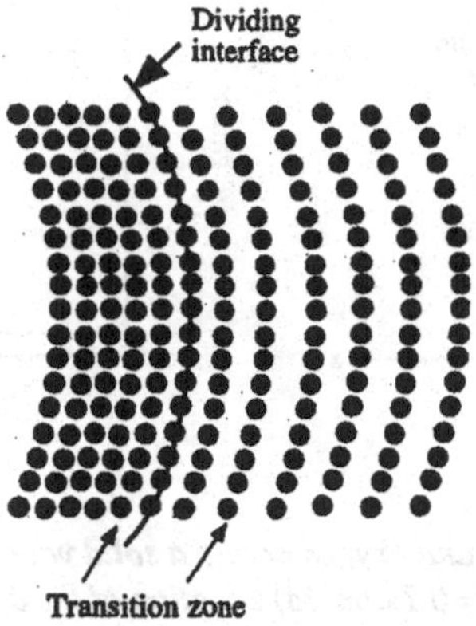

Fig. 2. Transition zone and the dividing surface.

in the transition layer which is parallel to this interface will also satisfy the definition of the dividing surface, some other convention is needed to define uniquely the location of the geometrical interface. The actual position of the interface is not critical if the interface is flat (curvature equal to zero) since the interface properties are characterized by area only, and any normal displacement will not change the area. However, for a curved interface, knowledge of the precise position of the interface is essential since parallel curved interfaces will have different areas, and thus different interface properties. The problem would be particularly serious if the radius of curvature of the interface is comparable to the thickness of the transition layer. However, when the radii of curvature are much larger than the thickness of the interface, the error introduced by the arbitrary location of the dividing surface will be negligible, so that we can choose any convenient convention in locating the interface. Note that when the thickness of the transition layer is comparable to the radii of curvature or to the dimension of the crystal, as in nucleation and in spinodal decomposition, the problem becomes quite complex and there is no advantage in using the dividing surface concept. A more rigorous statistical treatment, or a treatment which takes into account the thermodynamics of inhomogeneous systems as given by Cahn and Hilliard (3), should be used in these cases.

We shall now consider the case in which the thickness of the transition layer is much smaller than the radii of curvature so that surface energy variation with curvature and all other effects of similar magnitude can be neglected. Under this condition the properties of the interface are a function of area only. All thermodynamic extensive properties can now be expressed as sum of volume and surface contributions. The volumes V_1 and V_2 of the two phases are still not uniquely defined since they depend upon the exact location of the dividing surface. Gibbs suggested a unique division by locating the dividing surface such that the surface density of atoms is zero in a one component system. Such an interface location is given by the condition $N_1 + N_2 = N$, or $N_s = 0$, where N is the total number of particles in the system and N_1 and N_2 are the number of particles in each phase. Once the volume of each phase is uniquely defined by the above convention, we can characterize all thermodynamic properties of the system by volume and surface contributions.

For a multicomponent system, the dividing surface is located such that the surface density of any one component (usually the principal component) is zero. The surface densities of solvent and solute atoms are defined in terms of surface excess $\Gamma_i = (N_s)_i/A$. Figure 3 shows the data from Buttner et al {6} in the silver-oxygen system for adsorption of oxygen on a planar silver surface (7). Fig. 3a shows the definition of the interface with $\Gamma_1 = 0$ for silver. Fig. 3b then shows that the surface excess of oxygen, Γ_2, on the surface will not be zero.

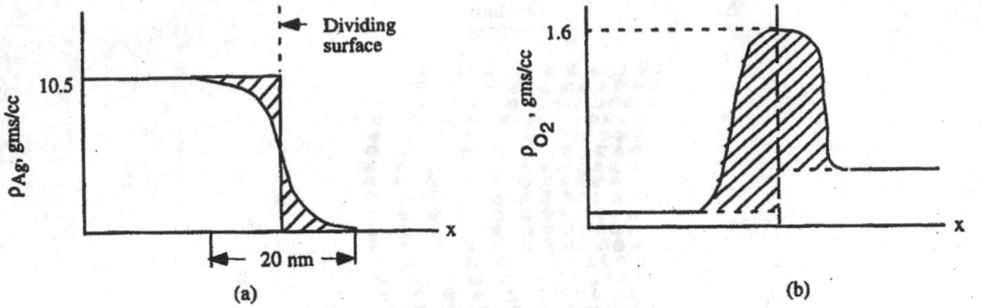

Fig. 3. (a) Density profile of silver and oxygen across a solid vapor intereface at 1185 K and partial pressure of oxygen = 0.1 atm. (a) Location of the dividing surface by the convention $\Gamma_1 = 0$. (b) The surface excess, Γ_2, of oygen is proportional to the hatched area. [7].

We shall now summarize the major points of this section. 1) Actual interfaces are finite in thickness although all thermodynamic properties are referred to a sharp, two-dimensional mathematical interface. Surface properties do not correspond to the properties of the transition layer, but to the difference in properties between the transition layer and a hypothetical case in which the transition layer is replaced by two homogeneous phases separated by a dividing surface, which lies in the transition zone. 2) Surface energy has a unique value only under equilibrium conditions. 3) The capillarity model (the dividing surface concept) is useful only when the thickness of the transition zone is smaller than the radii of curvature of the interface. 4) The dividing surface is chosen such that the surface density of the principal constituent is zero.

1.2 Diffuse interface

In order to evaluate the interface thermodynamic quantity, ϕ_s, it is necessary to obtain the value of ϕ for a nonhomogeneous system. In addition, it is often necessary to consider a diffuse interface when the size of the new phase is comparable to the transition layer. Important examples of such cases include spinodal decomposition and fine microstructural evolution that is modeled by the phase-field approach (8). In this case it is necessary to obtain a free energy function for a nonhomogeneous system, which was first developed by Cahn and Hilliard (2).

Consider a system in which the composition is uniform. For this uniform composition, c, let the Helmholtz free energy per atom be given by f(c), as shown by point P in Fig. 4. Now if we redistribute the solute so as to create a concentration gradient, then we have to do some isothermal reversible work to transport some solute atoms from one part of the system to the other, and this work to first order is proportional to the square of the concentration gradient. This additional work will increase the free energy of the system. Cahn and Hilliard (2) described the Helmholtz free energy per unit volume of an inhomogeneous system as

$$F = n_v \int_{-\infty}^{\infty} \left[f(c) + k\left(\frac{dc}{dx}\right)^2 \right] A\, dx \qquad [2]$$

where the first term in the integrand, f(c), corresponds to the Helmholtz free energy per molecule of a homogeneous system of average composition c, and the second term, called the gradient energy term, gives the contribution due to the presence of composition gradient. n_v is the number of atoms

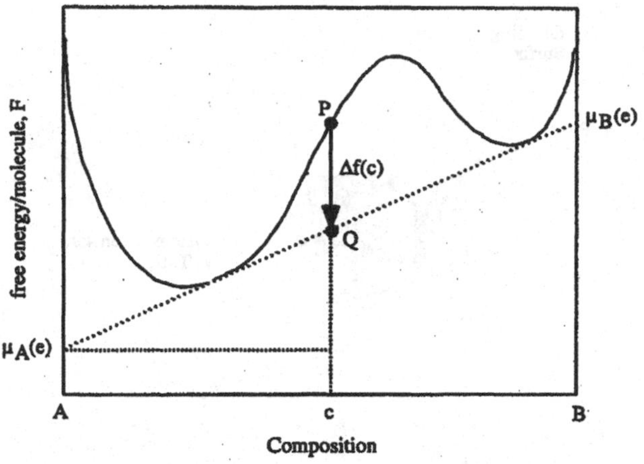

Fig. 4. Molar Helmholtz free energy - composition diagram showing the free energy of a system of average composition c which is separated into equilibrium compositions. The free energy of the equilibrium mixture of two phases is given by point Q.

per unit volume, A is the cross-sectional area and k, the gradient energy coefficient, is approximately constant for small gradients. The actual value of the free energy of the system depends on the concentration profile and thus on the thickness of the transition zone, and has a unique value only under equilibrium conditions when the transition zone is fixed by the requirement that the free energy, F, given by equation [2], be a minimum.

The surface free energy can now be calculated by subtracting from the actual free energy of the system, equation [2], the energy which it would have if the properties of each phase were continuous throughout, i.e., if no interface effect was present. The latter contribution is given by the change in free energy of the system when a uniform composition, c, transforms to two compositions which are in equilibrium, as shown in Fig. 4. The free energy, $F_1 + F_2$, of the equilibrium mixture of the two phases is given by point Q, whose value can be obtained from the geometrical construction shown in Fig. 4, as

$$F_1 + F_2 = \mu_A(e) + [\mu_B(e) - \mu_A(e)]\,c$$

$$= [c\mu_B(e) - (1-c)\mu_A(e)]. \qquad [3]$$

where $\mu_A(e)$ and $\mu_B(e)$ are the equilibrium chemical potentials of species A and B in the two phases. If we consider unit area of the interface, then the interface energy is given by $\gamma = [F - (F_1+F_2)]/A$ Substituting the values of F and F_1+F_2 gives

$$\gamma = n_v \int_{-\infty}^{\infty} \left[\Delta f(c) + k\left(\frac{dc}{dx}\right)^2 \right] dx \qquad [4]$$

in which

$$\Delta f(c) = f(c) - [c\mu_B(e) + (1-c)\mu_A(e)] \qquad [5]$$

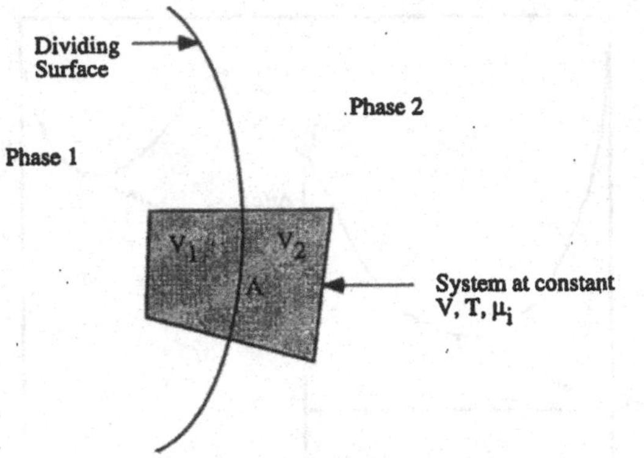

Fig. 5. Conical subsystem which contains an interface as well as parts of two homogeneous phases. The sides are chosen to be conical since the variation in thermodynamic properties in the direction normal to the conical sides will be zero from the definition of the dividing surface.

is the free energy change when a homogeneous composition transforms to an equilibrium mixture of the two phases. The value of $\Delta f(c)$ can be obtained from the free energy-composition diagram, Fig. 4, by drawing a vertical line PQ at any composition c of interest such that the point P lies on the free energy curve and the point Q lies on the common tangent which represents the equilibrium condition.

2. INTERFACIAL ENERGY

Having established the basic criteria for the thermodynamic description of a system containing an interface, we shall now define interfacial energy and derive basic relationships which show the dependence of surface energy on system variables.

2.1 Thermodynamic definition

When interface effects are important, it is not advantageous to use pressure to characterize the state of the system since, as we shall shortly see, pressure is not the same in two phases separated by a curved interface (4, 5). Thus, besides volume, we use two quantities which have the same value in the two phases, i.e. temperature (T) and chemical potential (μ), to describe the state of the system. *Interfacial energy is then defined as the reversible work required to create a unit area of the interface at constant temperature, volume and chemical potential.*

Consider a small system in the vicinity of the dividing interface which is described by a doubly truncated cone whose sides are normal to the interface and which contains an area A of the interface, as shown in Fig. 5. Let the cone extend into the homogeneous parts of each phase. If dW is the reversible work at constant T, V, and μ_i required to increase the area by dA, then surface energy is defined as

$$\gamma = \left(\frac{dW}{dA}\right)_{T,V,\mu_i} \quad [6]$$

We shall now examine the thermodynamic function that gives the reversible work under the constraints of constant T, V and μ. For the creation of an interface, the condition of constant chemical potential is required since the interface is uniquely defined only under equilibrium conditions. Different species can thus accumulate or deplete at the interface to maintain constant chemical potential in the system. Consider a system, as shown in Fig. 5, in which temperature and volume are constant. If dW is the external work done on the system at constant V, then from the first and second law of thermodynamics (4):

$$dE = T\,dS + \Sigma\mu_i\,dN_i + dW \qquad [7a]$$

or

$$dW = dE - T\,dS - \Sigma\mu_i\,dN_i \qquad [7b]$$

The external work required to create an interface at constant T, V and μ_i can thus be obtained from the above equation as

$$dW = d\,(E - TS - \Sigma\mu_i\,N_i) = d\Omega \qquad [8]$$

where

$$\Omega = E - TS - \Sigma\mu_i N_i. \qquad [9]$$

The above equation shows that the thermodynamic function Ω gives the reversible work done on the system under constant temperature, volume and chemical potentials. Thus, the equilibrium criterion for systems under constant T, V and μ_i is that Ω has a minimum value. The surface energy can now be defined as

$$\gamma = \left(\frac{dW}{dA}\right)_{T,V,\mu_i} = \left(\frac{d\Omega_s}{dA}\right)_{T,V,\mu_i} \qquad [10]$$

Note that the free energy function Ω can be expressed as: $\Omega = F - G$. Since for a homogeneous bulk phase, $G = \Sigma\mu_i N_i = F + pV$, one obtains $\Omega = -pV$. For an interface with isotropic interface energy, $d\Omega_s = \gamma\,dA$, or $\Omega_s = \gamma\,A$. Thus, interface energy and interface area for interfaces are analogous to (negative) pressure and (positive) volume for bulk fluid phases. If the value of γ varies with orientation of the element of the interface, the total surface work is given by the general expression

$$\Omega_s = \int_s \gamma\,dA. \qquad [11]$$

where the integral is evaluated over the entire interface. Some typical values of different types of interfaces are listed in the Appendix.

2.2 Surface Excess

We note that the creation of the interface under the constraint of constant chemical potentials requires flow of the ith species, dN_i, to the interface from the large bulk phases outside of our conical system. The quantity dN_i/dA is the surface excess, Γ_i, of the ith component, and it can have a positive or a negative value depending upon whether at equilibrium the ith species segregates or depletes at

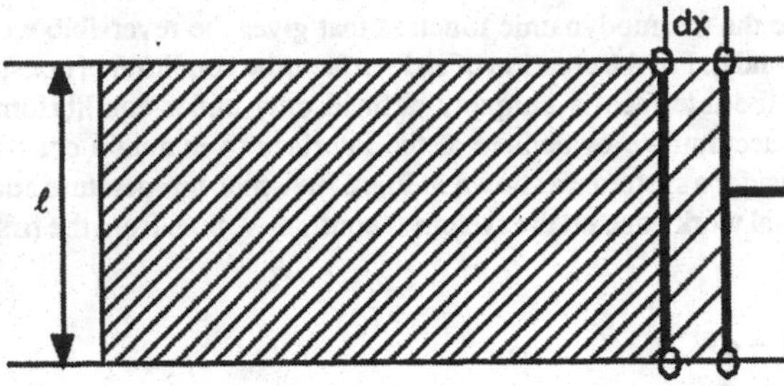

Fig. 6. A film of liquid in a U-shaped wire frame is stretched by a distance dx. The width of the film is ℓ.

the interface. Combining equation [6] with the definition of Ω_s, equation [9], we obtain for constant T and μ_i,

$$d\Omega_x = \gamma dA = dF_s - \sum_i \mu_i dN_i$$

or

$$\gamma = \frac{dF_s}{dA} - \sum_i \mu_i \left(\frac{dN_i}{dA}\right) \qquad [12]$$

or

$$\gamma = f_x - \sum_i \mu_i \Gamma_i, \qquad [13]$$

where f_s is the Helmholtz free energy per unit area of the interface. Since the dividing surface is chosen such that $\Gamma_i = 0$, surface energy is equal to the Helmholtz free energy per unit area of the surface in one component systems only.

2.3 Surface tension

In order to examine the equilibrium conditions or the changes in surface configuration due to surface energy effects, it is advantageous to consider the forces exerted by the surface. For this, it is simpler to consider the case of a fluid-fluid or fluid-vapor interface for which the surface energy is equal to the surface tension. The term surface tension, σ, represents the force per unit length in contrast to surface energy, γ, which is the reversible work required to create a unit area of the surface (at constant T, V and μ_i). In order to visualize the surface tension force, consider a one component fluid film stretched over a rectangular wire frame in which one side of the frame of length ℓ is movable (5), as shown in Fig. 6. Then work done in reversibly stretching the film by extending the movable side by distance dx is

$$\text{Work} = 2\sigma_{xx} \ell dx \qquad [14]$$

where surface stress, σ_{xx}, acts normal to the edge of the frame so as to tend to contract the film. The surface stress at any point on the surface is defined as the force acting across any line on the surface which passes through this point in the limit dS, the length of the line, goes to zero. The factor of two in equation [14] comes from the two surfaces of the film. Before the film was stretched, an equilibrium surface configuration with surface energy γ was present. Once the film is stretched, the surface will tend to maintain this equilibrium configuration which requires interior atoms from the bulk of the film to move to the surface so that the increased surface area has the same surface density of atoms as the unstretched film. This is possible for fluid phases where atomic mobility is large enough so that atoms can be freely supplied to the surface. Also, the thickness of the film can adjust freely to prevent any volume strains in the liquid. Since the extended surface of the film has precisely the same surface configuration as the unextended film, and thus the same surface energy, we have in effect done work to create an extra surface of area $2\ell dx$ and energy γ. Thus, from the definition of surface energy, work done in extending the film can also be written as

$$\text{work} = 2\,\gamma\,\ell\,dx \qquad [15]$$

Equating Eqs. [14] and [15] gives $\sigma_{xx} = \gamma$. If the film were stretched in the y direction under the constraint that the length in x direction remains constant, then the surface stress in the y direction, σ_{yy} would also be equal to γ. Surface tension, σ, is generally defined as the average of the surface stresses in two mutually perpendicular directions, and is equal to $\sigma = (\sigma_{xx}+\sigma_{yy})/2=\gamma$. Thus, surface tension and surface energy are identical for a fluid-fluid interface. This equality between surface energy and surface tension for fluid-fluid interfaces allows one to examine the equilibrium condition at the junction of several boundaries.

Note that the definition of surface tension in two dimensions is analogous to the definition of hydrostatic pressure in three dimensions. The atomic displacement of surface atoms is such that the stress perpendicular to the surface is zero so that the liquid surface is in a state of plane stress. Although the atomic arrangement on a fluid surface is isotropic, the surface atomic configuration is different than that in the bulk since surface atoms do not have neighbors above the surface plane. The equilibrium surface configuration exhibits an increased separation of atoms in directions parallel to the surface compared to that of the bulk atoms. This effect of a stretched surface is identified as the surface tension, and it is equivalent to the negative pressure parallel to the surface.

Examples:

(a) Equilibrium Condition at a Triple Point

Consider the intersection of three surfaces which are perpendicular to the plane of the paper. At the triple point, each surface exerts a force to reduce its area. This can be represented as a vector whose magnitude is given by surface tension, which is equal to the surface energy for the fluid-fluid interface. Since the surface tension force tends to shrink the interface, the direction of the vector is along the interface, away from the triple point, as shown in Fig. 7a. The equilibrium condition, for isotropic interface energies, is given by the balance of surface tension (or surface energy when $\sigma = \gamma$) at the triple point, as

$$\sum_{i=1}^{3} \gamma_i \bar{t}_i = 0 \qquad [16]$$

where $\bar{t}_i$ is a unit vector in the ith interface perpendicular to and directed away from the line of intersection. If θ_1, θ_2, and θ_3 are the angles between the vectors $(\vec{t}_2,\vec{t}_3)$, $(\vec{t}_3,\vec{t}_1)$ and $(\vec{t}_1,\vec{t}_2)$, respectively, then the equilibrium condition is obtained as

$$\frac{\gamma_1}{\sin\theta_1} = \frac{\gamma_2}{\sin\theta_2} = \frac{\gamma_1}{\sin\theta_3}$$

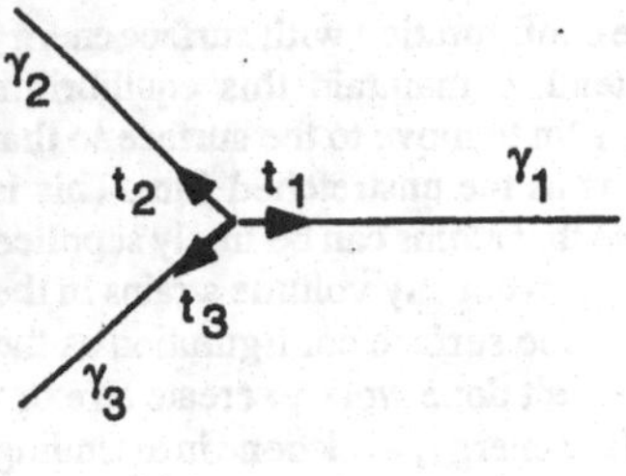

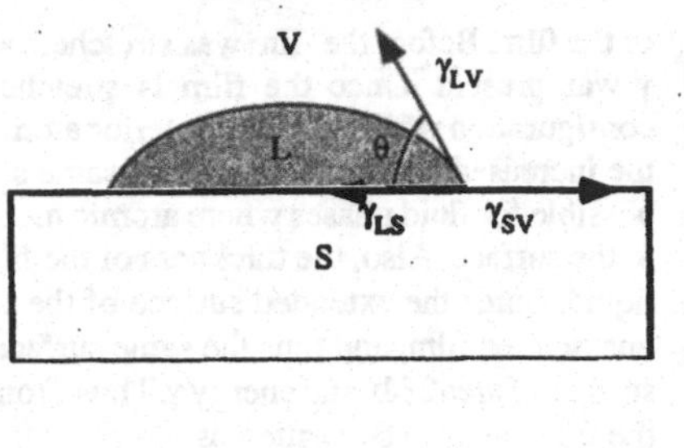

Fig. 7. Equilibrium conditions at the triple point junction. (a) Intersection of three boundaries. (b) Drop of liquid on a substrate.

Thus, if the surface energies are equal, then the angles will also be equal, and each angle will be 120°.

(b) Equilibrium of a liquid drop on a substrate

For a liquid drop on a solid surface, Fig. 7b, the balance of forces in the horizontal direction leads to the result

$$\gamma_{SV} = \gamma_{LS} + \gamma_{LV} \cos\theta \qquad [17]$$

where appropriate interface energies and the contact angle θ are defined in Fig. 7b. This equilibrium condition is known as the Young equation

2.4 The effect of temperature and composition on interfacial energy

Interface energy is a function of temperature and composition (or chemical potential), and this variation often gives rise to forces that cause motion of liquid or cause solute segregation (or depletion) at the boundaries. We shall thus examine the important thermodynamic relationships that predict the effect of temperature and chemical potential on interface energy.

Using the definition of Ω, Eq. [9], and the expression for the change in Helmholtz free energy $dF = -SdT - PdV + \sum \mu_i dN_i$ gives, for bulk phases,

$$d\Omega = -S\,dT - P\,dV - \sum N_i\,d\mu_i \qquad [18]$$

Subtracting $d\Omega = -d(PV)$ from the above equation gives the Gibbs-Duhem equation for the bulk phases

$$-S\,dT + V\,dP - \sum N_i\,d\mu_i = 0 \qquad [19]$$

Similar relationships for the interface will be

$$d\Omega = -S\,dT + \gamma\,dA - \sum N_i\,d\mu_i$$

and

$$-S\,dT - A\,d\gamma - \sum N_i\,d\mu_i = 0$$

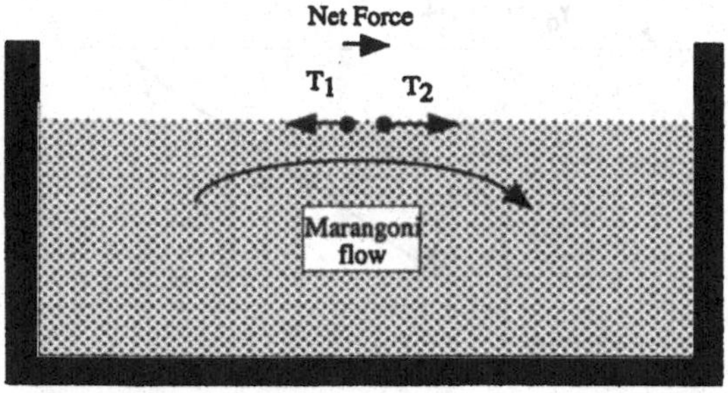

Fig. 8. A difference in surface temperature of the liquid showing different surface tension effects leading to Marangoni flow.

or

$$d\gamma = -s\,dT - \sum \Gamma_i\, d\mu_i \qquad [20]$$

where s is the surface entropy per unit area of the interface, $\Gamma_i = N_i/A$ is the surface excess of the ith component, and the term $d\gamma$ corresponds to the change in surface energy due to change in temperature and chemical potentials only.

Examples:

We now consider a one component system (with $\Gamma_1 = 0$), and examine the effect of temperature on the surface energy. The variation in surface energy with temperature is given by

$$d\gamma/dT = -s \qquad [21]$$

Since entropy is positive, surface energy decreases as the temperature is increased.

We now consider a surface of a liquid which is subjected to a temperature gradient. Thus a gradient of surface energy exists along the surface. We consider two elements of this surface at different temperatures, as shown in Fig. 8. If $T_1 > T_2$, then $\gamma_1 < \gamma_2$. Since surface energy in liquids is equal to surface tension, which acts to minimize the surface area, the different surface tensions at these two elements will result in a net force along the free surface in the direction of lower temperature, as shown by the arrow in Fig. 8. Since the atoms in liquids are mobile, this net force resulting from the gradient in surface tension will cause flow of liquid at a free surface. This capillary driven flow is known as the Marangoni convection.

We next consider the effect of composition on surface energy at constant temperature in a binary system. In this case

$$d\gamma/d\mu_2 = -\Gamma_2 \qquad [22]$$

where the dividing surface is selected such that $\Gamma_1 = 0$. In a dilute solution for which Henry's law is satisfied, $d\mu_2 = kT d\ln x$, so that

$$d\gamma/d\ln x = -\Gamma_2 kT$$

Thus, for $\Gamma_2 > 0$, interface energy decreases as the solute concentration is increased. Note that if

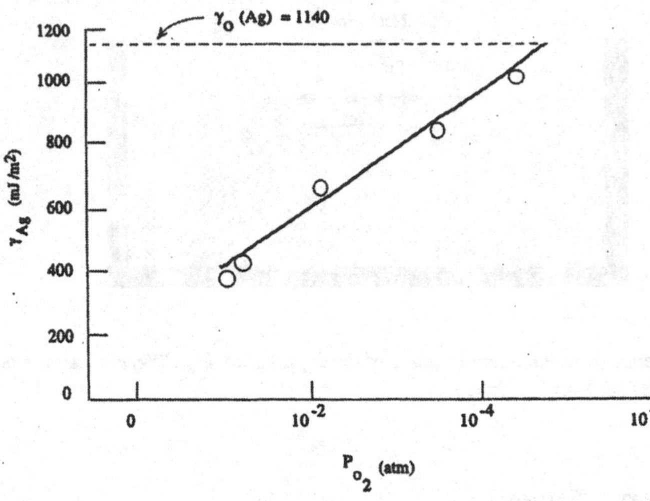

Fig. 9. Experimental data on the variation in the surface energy of silver as a function of oxygen pressure at T= 1205 K. The slope in the linear region gives the value of (-Γ_2/RT) (6).

composition gradients exist at a free liquid surface, then there will be a gradient of surface energy which will also cause solute driven Marangoni convection.

We shall now examine the effect of the surface energy of silver in the presence of oxygen whose pressure is p, so that $d\mu_2 = kT \, d\ln p$. Then

$$d\gamma/d\ln p = -(\Gamma_2 kT) \quad [23a]$$

Thus, as the pressure is increased, the surface energy decreases. Figure 9 shows experimental results on the variation in surface energy of silver with oxygen pressure, which confirm the above relationship (7,8). We may rewrite the above result as

$$\gamma = \gamma_0 - \int_{p_0}^{p} \Gamma_2 \left[\frac{kT}{p}\right] dp \quad [23b]$$

where γ_0 is the surface energy of pure material in absence of oxygen. The above result consists of two parts: the first term gives the reversible work required to create a unit area of surface of pure material, and the second term gives the work released due to the adsorption of oxygen atoms on the surface.

3. EQUILIBRIUM CONDITIONS FOR A CURVED INTERFACE

We shall first examine the effect of curvature of the interface on mechanical equilibrium. We shall then show how this pressure difference influences the chemical potential of atoms at a curved interface. Finally, we shall discuss the chemical equilibrium between the two phases separated by a curved interface.

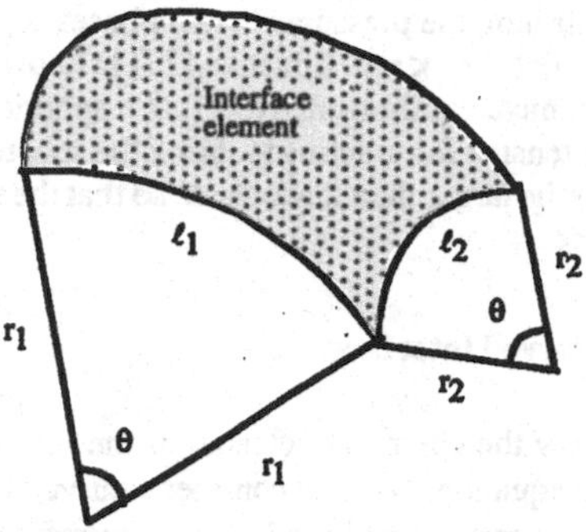

Fig. 10. An element of a curved surface with principal radii r_1 and r_2

3.1 Laplace equation

In order to obtain equilibrium conditions between two incompressible fluids separated by a curved interface, we consider a subsystem of fixed volume V, as shown in Fig 5, in which T and μ_i are equal in the two phases. At equilibrium the free energy Ω is stationary for any infinitesimal change in surface configuration so that

$$d\Omega = 0 = d\Omega_1 + d\Omega_2 + d\Omega_s$$

$$= -P_1 dV_1 - P_2 dV_2 + \gamma dA \qquad [24]$$

However $dV = 0 = dV_1 + dV_2$ which gives

$$(P_1 - P_2) = \gamma\,(dA/dV_1) \qquad [25]$$

Now, for a smoothly curved interface, $dA/dV_1 = \kappa$, the mean curvature of the interface. To visualize this relationship, we consider an element of the interface of arc lengths ℓ_1 and ℓ_2 in the principal directions, as shown in Fig. 10. If r_1 and r_2 be the radii of curvature, then we choose the element such that $\ell_1 = r_1\theta$ and $\ell_2 = r_2\theta$. If this element is displaced by a distance dr, then

$$dA = (r_1 + dr)(r_2 + dr)\theta^2 - r_1 r_2 \theta^2 \cong (r_1 + r_2)\theta^2 dr$$

and

$$dV = (r_1\theta)(r_2\theta)dr = r_1 r_2 \theta^2 dr.$$

Dividing these results, we obtain

$$\frac{dA}{dV} = \frac{1}{r_1} + \frac{1}{r_2} = \kappa \qquad [26]$$

Substituting these values in the above equation gives

$$P_1 - P_2 = \gamma\kappa \qquad [27]$$

This result shows that at equilibrium, the pressures in two phases separated by an interface are not the same unless the interface is flat, i.e. $\kappa = 0$. Equation [27] is known as the Laplace equation and it gives the condition for mechanical equilibrium. Note that a spherical particle will tend to reduce its area so that there is a surface tension force acting to shrink the sphere. At equilibrium, the pressure inside the sphere must therefore be larger than that outside so that the net outward pressure balances the inward surface tension force.

3.2 Chemical potential at a curved interface

We shall first consider how the chemical potential at the interface changes as a function of curvature by using the Laplace equation. We first consider a curved interface which is at a pressure $P_1 = P_0 + \gamma k$, where P_0 refers to the pressure in phase 1 for a flat interface, i.e. $\kappa = 0$. Thus, the chemical potential at a curved interface is given by

$$\mu_1(T,P_l) = \mu_1(T,P_0 + \gamma\kappa) \qquad [28]$$

If $\gamma\kappa << P_0$, then we can expand the right hand side of the above equation in a Taylor expansion to give

$$\mu_1(T,P_0 + \gamma\kappa) = \mu_1(T,P_0) + (\partial\mu_1/\partial p)\,\gamma\kappa, \qquad [29]$$

$$= \mu_1(T,P_0) + \gamma v_1 \kappa, \qquad [30]$$

where we have substituted, $(\partial\mu_1/\partial P)_T = v_1$, the molar volume of phase 1. Substituting the value of $\mu_1(T,P_1)$ from equation. [30] into equation [28], we obtain

$$\mu_1(T,P_1) = \mu_1(T,P_0) + \gamma v_1 \kappa \qquad [31]$$

Note that $\mu_1(T,P_0)$ is the chemic al potential at a flat interface, i.e. when $\kappa = 0$. Thus, the above equation shows how the chemical potential in the phase 1 changes as a function of curvature, and it is the most useful equation for examining the changes in morphology that occur due to surface energy effects.

Example:

As an example of morphological changes that occur due to surface energy effects, we consider a cylindrical particle that can undergo instability and breakup into small spheres, i.e. go through the process of spherodization. To examine the stability, we perturb the shape of the cylinder in the z direction in the form, $r = \delta \sin 2\pi/\lambda$, as shown in Fig. 11, where r is the radial distance from the axis of the cylinder, λ is the wavelength of perturbation, and δ is the amplitude of perturbation which is infinitesimal. Whether the cylinder is stable or not depends on whether the atoms move from B to A or from A to B. In which directions the atoms move will be dictated by the chemical potentials at A and B which will be different due to different curvature values at points A and B. By examining the chemical potentials at points A and B in terms of total curvature at these points, it can be shown that the cylinder is stable if $\lambda < 2\pi R$ and unstable if $\lambda > 2\pi R$.

3.3 Equilibrium conditions between the two phases in one-component system

The equilibrium between two phases separated by a curved interface, in a one-component

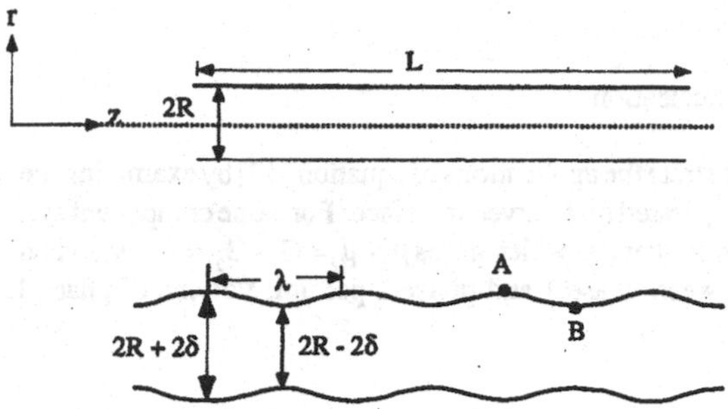

Fig. 11. A cylinder which is perturbed in the z-direction to examine the stability of the shape with respect to spherodization.

system, can now be obtained by equating the chemical potentials in the two phases, taking into account that the pressures in the two phases are not the same. Let the pressure in the phase 2 (i.e. the matrix phase) be P_0 which is equal to the external pressure in the system. Thus, the equilibrium condition can be written as:

$$\mu_1(T,P_1) = \mu_2(T,P_0) \qquad [32]$$

By substituting the value of $\mu_1(T,P_1)$ from equation [31], we obtain

$$\mu_1(T,P_0) + \gamma v_1\kappa = \mu_2(T,P_0)$$

or

$$\mu_1(T,P_0) - \mu_2(T,P_0) + \gamma v_1\kappa = 0 \qquad [33]$$

The above equation is known as the Gibbs-Thomson equation, and it gives the equilibrium condition for a curved interface whose interface energy is isotropic. It can also be derived by considering equilibrium condition in which dF = 0 when dn moles are transferred from *bulk phase* 2 to *bulk phase* 1 in a system at constant V and T. This gives

$$(\partial F_1/dn_i)_{T,V,nj}\, dn_i - (\partial F_2/dn_i)_{T,V,nj}\, dn_i + \gamma\, dA = 0$$

which is equivalent to equation [33] since $dA/dn_i = v_1\kappa$.

Equation [33] is the most useful equilibrium relationship because the chemical potentials in both phases are considered at the same pressure. These chemical potentials are referred to as bulk chemical potentials, i.e. the chemical potentials the phases would have in the absence of an interface ($\kappa = 0$). The three terms can be considered as contributions from phase 1, from phase 2, and from the interface. The equilibrium condition [33] is more useful than equation [32] for numerical calculations since chemical potentials or partial molar free energies are usually tabulated for 1 atm. pressure. Also, equation [33] contains volume and surface contributions in the same way as the other thermodynamic properties previously discussed. Such a division is useful when solid phases are considered since capillary pressure may be fictitious in the case of solids as we shall discuss in subsequent sections.

Example:

(a) Critical radius of nucleation

We shall now illustrate the applications of equation [33] by examining the equilibrium condition between two phases separated by a curved interface. For a one component system chemical potential is equal to the molar free energy, which gives $\mu_1 - \mu_2 = G_1 - G_2 = \Delta G_v v_1$, where ΔG_v is the Gibbs free energy difference between phase 1 and phase 2 per unit volume of phase 1. Thus equation [33] becomes

$$\Delta G_v + \gamma\kappa = 0 \tag{34}$$

If the disperse phase 1 is spherical, then substituting $\kappa = 2/r$ gives the equilibrium radius of curvature (critical radius of nucleation) as a function of ΔG_v as $r = -2\gamma/\Delta G_v$. Thus, the critical nucleation condition is the same as the equilibrium condition for a curved interface.

(b) Melting point depression

A curved interface during solidification reduces the freezing point of the liquid. Substituting in equation [34] the value $\Delta G_v = -\Delta S_v(T - T_M)$, where ΔS_v is the entropy change upon melting per unit volume of solid (which is positive) and T_M the freezing point for a flat interface, we obtain for the equilibrium temperature the expression

$$T = T_M - \gamma\,\kappa/\Delta S_v.$$

or

$$T = T_M - \Gamma\kappa. \tag{35}$$

where $\Gamma = \gamma/\Delta S_v$ is a capillary constant. For pure nickel, $T_M = 1728K$, $\gamma = 225 \times 10^{-3}\ J/m^2$ and $\Delta S_v = 1.594 \times 10^6\ J/deg\ m^3$, which gives $T = 1728 - 0.00000014\ \kappa$, where the curvature is in m^{-1}. Thus, interface effects become significant only when $\kappa > 10^6\ m^{-1}$ or $r < 1\ \mu m$.

The reason for freezing-point lowering at a curved interface can be seen graphically from Fig. 12. The free energy of the solid is increased due to the higher pressure in the solid, and this increase is equal to V dP or $\gamma V_l \kappa$. The equilibrium condition is thus shifted to a lower temperature. Alternately, if the freezing point at a curved interface were T_M, then the solid will dissolve since it can eliminate its surface area, and thus lower the free energy of the system. Thus, for equilibrium, the free energy of the liquid must be higher than that of the solid by the amount necessary to balance any decrease in free energy due to surface area reduction, i.e., $G_L - G_S = \gamma v_l k$. Note that as curvature increases, $G_L - G_S$ must also increase which causes the equilibrium temperature to decrease as seen in Fig. 12.

3.4 Equilibrium conditions between the two phases in an n-component system

So far we have considered one component systems only. These results can be readily extended to m-component systems in which a virtual displacement of the interface in Fig. 5 is achieved by transferring $dn_1, dn_2, \ldots dn_i$ moles of the ith species from phase 2 to phase 1 at constant temperature and volume of the system. Then the equilibrium condition [32] can be written in a general form as

$$\sum_{i=1}^{m} [\mu_{i1}(P_1) - \mu_{i2}(P_2)]\, dn_i = 0 \tag{36}$$

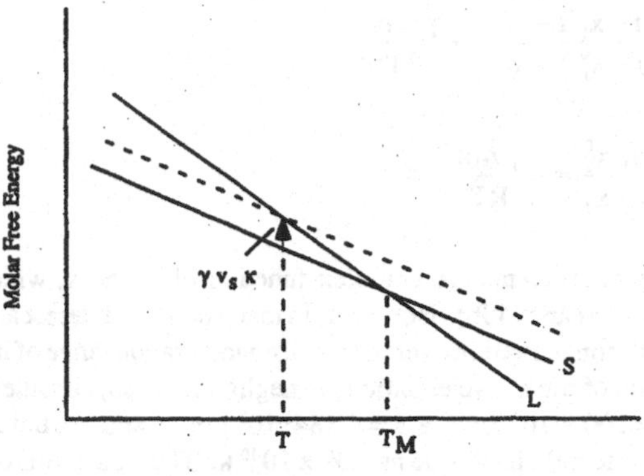

Fig. 12. Free energy variation with temperature of a solid and a liquid phase.

where the chemical potentials of the component i in phases 1 and 2 are referred to at pressures P_1 and P_2, respectively. If both the phases are solid solutions, the dn_j are independent, and we obtain the equilibrium condition for each ith species as $\mu_{i1}(T, P_1) = \mu_{i2}(T, P_2)$, which is analogous to equation. [32].

One can now write the result of equation [36] in terms of bulk chemical potentials which are evaluated at the same pressure, i.e. the external pressure. In this case, the equilibrium condition, equation [36], becomes

$$\mu_{i1} - \mu_{i2} + \gamma \bar{v}_{i1} \kappa = 0 \qquad [37]$$

where v_{i1} is the partial molar volume of the ith component in phase 1. We shall now obtain the equilibrium partitioning of solute in two phases separated by a curved interface for a two component system, and note that this treatment can be extended to an m-component system in general. For a two-component system, let subscripts 'A' and 'a' refer to solvent and solute atoms, respectively. At equilibrium

$$\mu_{A1} - \mu_{A2} + \gamma \bar{v}_{A1} \kappa = 0 \qquad [38a]$$

$$\mu_{a1} - \mu_{a2} + \gamma \bar{v}_{a1} \kappa = 0 \qquad [38b]$$

If x_1 and x_2 are the equilibrium atom fractions of solute in phases 1 and 2, respectively, then the simultaneous solution of equation [38] will enable us to calculate the variation of x_1 and x_2 as a function of κ. Letting superscript zero refer to the condition in which $\kappa = 0$, or $\mu_{A1}{}^0 = \mu_{A2}{}^0$ and $\mu_{a1}{}^0 = \mu_{a2}{}^0$, we may rewrite equations [38] as

$$(\mu_{A1} - \mu^0_{A1}) - (\mu_{A2} - \mu^0_{A2}) + \gamma \bar{v}_{A1} \kappa = 0 \qquad [39a]$$

and

$$(\mu_{a1} - \mu^0_{a1}) - (\mu_{a2} - \mu^0_{a2}) + \gamma \bar{v}_{a1} \kappa = 0 \qquad [39b]$$

We now use the relationship $\mu - \mu_o = RT \ln a/a_o$, where a is the activity which can be written in terms of the corresponding activity coefficient (v) and concentration of solute (x) to obtain

$$\ln\frac{v_{A1}}{v^0_{A1}}\frac{v^0_{A2}}{v_{A2}}+\ln\frac{1-x_1}{1-x^0_1}\frac{1-x^0_2}{1-x_2}=-\frac{\gamma\bar{v}_{A1}\kappa}{RT} \qquad [40a]$$

$$\ln\frac{v_{a1}}{v^0_{a1}}\frac{v^0_{a2}}{v_{a2}}+\ln\frac{x_1}{x^0_1}\frac{x^0_2}{x_2}=-\frac{\gamma\bar{v}_{a1}\kappa}{RT} \qquad [40b]$$

In general, activity coefficients are complex function of x_1 and x_2 which require numerical evaluation of equations [40a] and [40b]. However, in many cases of interest activity coefficients do not vary appreciably with concentration over the small concentration range of interest so that the first term on the left hand side of the above equations is negligible. Also, considering typical values of the parameters, viz. $\gamma = 200 \times 10^{-3}$ J/m^2, RT = 4.184×10^3 J/mole and partial molal volume = 5.0×10^{-6} m^3/mole, we obtain the right hand side as 2.39×10^{-10} κ. Thus, the terms on the right hand sides of the above equations are much smaller than unity even for large interface curvature, e.g. $\kappa \sim 10^8$ m^{-1}, so that equations [40] can be simplified as follows:

$$\frac{1-x_2}{1-x_1}\frac{1-x^0_1}{1-x^0_2}=\exp\left(\frac{\gamma\bar{v}_{A1}\kappa}{RT}\right)\simeq 1+\frac{\gamma\bar{v}_{A1}\kappa}{RT} \qquad [41a]$$

and

$$\frac{x_2}{x_1}\frac{x^0_1}{x^0_2}=\exp\left(\frac{\gamma\bar{v}_{a1}K}{RT}\right)\simeq 1+\frac{\gamma\bar{v}_{A1}K}{RT} \qquad [41b]$$

Multiplying equation [41a] by $(1-x_1)$ and equation [41b] by x_1, and adding, we obtain after some simple algebraic manipulation, the result

$$x_2 = x^0_2\left[1 - \frac{1-x^0_2}{x^0_2-x^0_1}\frac{\gamma v_1\kappa}{RT}\right] \qquad [42]$$

where v_1 is the molar volume of phase1 defined as $v_1 = (1 - x_1)\,\bar{v}_{A1} + x_1\,\bar{v}_{a1}$. Substituting equation [42] in equation [41b] we obtain for the concentration x_1 the result

$$x_1 = x^0_1\left[1 - \frac{1-x^0_1}{x^0_2-x^0_1}\frac{\gamma v_1\kappa}{RT}\right] \qquad [43]$$

Equations [42] and [43] show that equilibrium concentrations in both phases vary linearly with curvature. It can be shown that a linear relationship also holds true even when the activity coefficients are not constant but vary linearly with composition (9). For this case the activity coefficient terms in equations [40a] and [40b] are retained, and their solution yields

$$x_1 = x^0_1\left[1 - \Gamma_{c1}\kappa\right], \qquad [44a]$$

$$x_2 = x^0_2\left[1 - \Gamma_{c2}\kappa\right], \qquad [44b]$$

where Γ_{c1} and Γ_{c2} are capillarity constants defined as

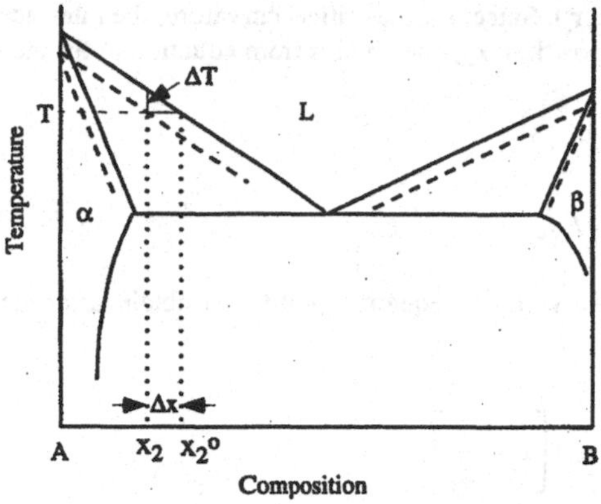

Fig. 13. Phase diagram showing the efefect of curvature on theequilibrium phase boundaries. The dotted line represent the equilibrium between a curved interface for a given value of the curvature.

$$\Gamma_{c1} = \frac{1-x_1^0}{x_2^0-x_1^0}\frac{v_1}{RT}\frac{\gamma}{\varepsilon_1} \quad \text{and} \quad \Gamma_{c2} = \frac{1-x_2^0}{x_2^0-x_1^0}\frac{v_1}{RT}\frac{\gamma}{\varepsilon_2}, \qquad [45]$$

in which parameters ε_1 and ε_2 in phases 1 and 2, respectively, are defined by the expression

$$\varepsilon_i = 1 + \left[\frac{\partial \ln v_{ai}}{\partial \ln x_i}\right]_{x_i^0} \qquad [46]$$

Note that parameter ε_i is the same for a solute or a solvent species in a given phase. The parameters ε_1 and ε_2 take into account the deviation from ideality and have constant values when activity coefficients vary linearly with composition. It is interesting to note that the sign of the capillarity constants, Γ_{c1} and Γ_{c2} depends on the relative values of x_1^0 and x_2^0. For negative, or solute-poor precipitates in which the solute is rejected, i.e., $x_1^0 < x_2^0$, the capillarity constants are positive so that the equilibrium solute concentrations decrease with increasing curvature. In contract, for a positive or solute-rich precipitate, i.e. $x_1^0 > x_2^0$, the capillarity constants are negative so that the solute concentrations in both phases increase with increasing curvature. How the effect of interface energy modifies equilibrium conditions for these two cases is shown schematically on a phase diagram in Fig. 13. The reason for this difference in behavior between the positive and negative precipitates can also be seen from the free energy-composition diagrams by including the effect of pressure in the precipitate phase, as discussed by Hillert in this book. The change in solute composition in phase 2, i.e. Δx, due to curvature at constant temperature can also be expressed as a change in equilibrium temperature at constant composition by using the slope, m, of the equilibrium phase boundary, i.e. $\Delta T = m\,\Delta x$, as shown in Fig. 13.

During phase transformations, equilibrium conditions between phases 1 and 2 are not always attained. However, local equilibrium conditions are often established at the curved interface between these phases. In such cases, equation [44b] is often rewritten in terms of the critical curvature κ_c (or critical radius of curvature, r_c). Since, for the critical curvature, the interface composition in phase 2 is equal to the bulk composition, x_∞, one obtains from equation [44b] the relationship:

$$x_\infty = x_2^0 - x_2^0 \Gamma_{c2} \kappa_c$$

or

$$x_2^0 \Gamma_{c2} = \left(x_2^0 - x_\infty\right) / \kappa_c \qquad [47]$$

Substituting the above result in equation [44b], we obtain after some rearrangement the relationship:

$$x_2 - x_\infty = (x_2^0 - x_\infty)\left[1 - \frac{\kappa}{\kappa_c}\right] \qquad [48a]$$

In terms of critical radius of curvature, the above equation becomes:

$$x_2 - x_\infty = (x_2^0 - x_\infty)\left[1 - \frac{r_c}{r}\right] \qquad [48b]$$

Equations [44a] and [44b] show the variation of equilibrium concentrations with curvature in two phases when both the phases are solid solutions. In many cases of practical interest the precipitate phase is a compound, viz. Fe_3C, Cu_2Al, etc. Calculations in which phase 1 is a compound of fixed composition, whereas phase 2 is a solid solution, show that equation [44a] is still valid in this case and the values of parameters x_1^0 at the interface and v_1, for the compound $A_\alpha a_\beta$, are given as $x_1^0 = \beta / (\alpha + \beta)$ and $v_1 = v_c / (\alpha + \beta)$, where V_c is the molar volume of the compound.

4. CAPILLARITY EFFECTS IN SOLIDS

So far we have discussed surface energy concepts only for incompressible fluids. When one or both phases are solid further complexities arise, and the capillarity theory developed so far must be properly corrected to include these additional effects. Two primary concepts which are important in solids are: a) surface stress (or surface tension) is not equal to surface energy as is the case with fluids, and b) surface energy is anisotropic in solids. We shall now discuss these concepts in some detail, and inquire into their effects on the equilibrium between two phases separated by a curved interface.

4.1 Surface Energy and Surface Stress

The discussion leading to equations [14] and [15] shows that the area of the liquid surface can be increased only by the addition of atoms to the surface. In contrast to this a solid surface can increase its area either by addition of atoms to the surface or by merely stretching the bonds of existing surface atoms. Since solids can sustain stresses, the work required to stretch a surface (surface stress) is quite different from the work required to create additional surface having the same configuration as the original surface. The existence of surface stress in solids can be visualized by considering the equilibrium configuration at absolute zero of a two dimensional surface plane only (9). The equilibrium spacing of atoms in this two dimensional lattice will be different from that in the bulk of a crystal since the number of neighbors will be different in the two cases. If this two dimensional

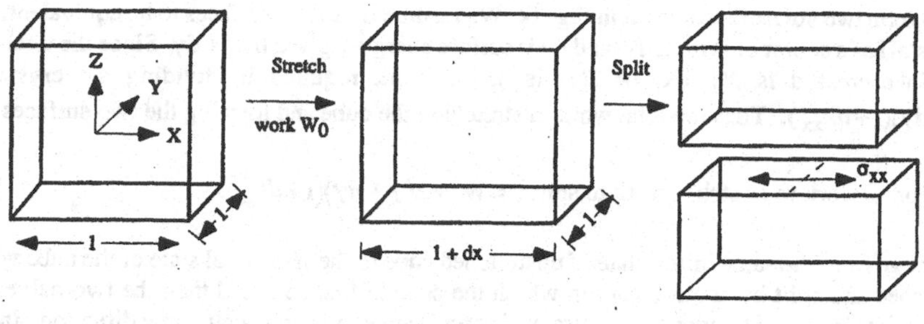

Fig. 14. Figure illustrating the difference between surface stress and surface energy.

plane is to become the surface plane, then some force must be applied to the edges of this plane so that there is a matching of the lattice spacings. These surface forces can be tensile or compressive depending upon whether the atomic spacing in the two dimensional lattice is smaller or larger than that in the equivalent plane in the bulk of the crystal. The applied force necessary to produce matching is reduced slightly by the adjustments in atom spacings in the second and succeeding layers below the surface. Besides the surface plane, some small tangential forces will also need to be applied to the successive planes below the surface to keep these planes in equilibrium. The sum of all these forces per unit length of edge gives the surface tension of the solid surface. The fact that this surface stress is not usually the same as surface energy can be clearly seen by considering an ideal case of the nearest neighbor pairwise interaction model in which the nearest neighbor distances are fixed solely by the lowest energy configuration of a two-atom model. In such a case the surface configuration is a precise extension of the bulk lattice. If such a crystal is cut reversibly to obtain two surfaces, the work required will correspond to the sum of the surface energies of the two newly formed surfaces. However, since the surface configuration is precisely identical to the configuration these atoms had in the bulk, no surface force is needed and the surface stresses are identically equal to zero. In real crystals, however, some surface stresses will always be present, though a part of which could be relaxed if defects such as dislocations are present.

In addition to the components σ_{xx} and σ_{yy}, surface stress will also have a shear component σ_{xy}. The equilibrium condition that any element of surface experience no resultant couple to rotate it requires that $\sigma_{yx} = \sigma_{xy}$. Thus, surface stresses present on solid surfaces are σ_{xx}, σ_{yy} and σ_{xy}, i.e., a solid surface is in the condition of plane stress only. It can be easily shown that these surface stresses are components of a second-order tensor, a result which further emphasizes the difference between surface stress and surface energy in solids since the latter is a scalar quantity.

In crystalline solids surface stress must satisfy the symmetry properties of the surface. Thus, across a line of mirror symmetry the shear stresses are zero so that the direction of this line and a direction perpendicular to it are the principal directions. Also, if a crystal surface has a rotational symmetry of threefold or greater then the normal stress components across <111> lines in the face are equal and all shear stresses are zero. Thus, for a cubic crystal, surface shear stresses are zero and normal stresses are equal for {111} and {100} surfaces. Surface stress in these cases can be characterized by one variable only which can be represented by $\sigma = (\sigma_{xx} + \sigma_{yy})^{1/2}$.

We shall now obtain a quantitative relationship between surface stress and surface energy by following the procedures given by Shuttleworth (10) and Mullins (11). Consider a unit cube of one-component solid which is stretched reversibly in the x direction by a distance dx under the constraint of constant length in the y direction. Let this work be W_0. The atomic displacements during stretching give rise to a strain increment $d\varepsilon_{xx} = dx/1$ in the lattice. We now reversibly divide the stretched cube to form two surfaces as shown in Fig. 14. We assume these two surfaces to be equivalent, i.e., the crystal has a center of symmetry, and their surface energy is given by $\gamma + d\gamma$. Since the total area of surface created is 2(1+dx) or $2(1+d\varepsilon_{xx})$, the work required in dividing the crystal is $2(\gamma + d\gamma)(1 + d\varepsilon_{xx})$. Thus the total work in stretching the cube and forming the two surfaces is

$$\text{Work} = \text{work to stretch} + \text{work to split} \quad = W_0 + 2(\gamma + d\gamma)(1 + d\varepsilon_{xx}). \qquad [49]$$

We now go from the same initial state of unstretched cube to the same final state of the cube which is stretched and split by another path in which the cube is first split and then the two halves are stretched by distance dx under the constraint that no change in length occurs in y direction. In this case, the total work done will be

$$\text{Work} = \text{work to split} + \text{work to stretch} = 2\gamma + W_1, \qquad [50]$$

where W_1, the work required to stretch the two halves will be different from the work W_o due to the presence of surface stresses in the second operation. Their difference will be related to the surface stress and surface strain by the relationship $W_1 - W_o = 2\sigma_{xx} d\varepsilon_{xx}$. Equating equations [49] and [50], and neglecting the $d(d\varepsilon_{xx})$ term, gives

$$\sigma_{xx} = \gamma + \frac{d\gamma}{d\varepsilon_{xx}}. \qquad [51]$$

If the above procedure is carried out by stretching the cube in the y direction, keeping the length in the x direction constant, we would obtain

$$\sigma_{yy} = \gamma + \frac{d\gamma}{d\varepsilon_{yy}} \qquad [52]$$

If instead of stretching the cube we apply a shear strain $d\varepsilon_{xy}$ the work done in shearing and splitting the cube will be $W_0' + 2\gamma + 2d\gamma$ only since the area remains unchanged during a shear. The work done in first splitting the cube and then shearing the two halves will be $2\gamma + W_1'$. Equating these work increments, and substituting $W_1' - W_0' = 2\sigma_{xy} d\varepsilon_{xy}$ we obtain

$$\sigma_{xy} = \frac{d\gamma}{d\varepsilon_{xy}} \qquad [53]$$

Equations [51] - [53] can be written in a general form as

$$\sigma_{ij} = \delta_{ij}\gamma + \frac{\partial\gamma}{\partial\varepsilon_{ij}}, \qquad i, j = 1, 2 \qquad [54]$$

where δ_{ij} is Kronecker delta whose value is unity if $i = j$ and zero if $i \neq j$. The orientation of the surface is assumed to be constant in calculating the $\partial\gamma / \partial\varepsilon_{ij}$ term.

The above result is general and is valid for solid as well as liquid surfaces. It shows that surface stress is equal to surface energy only if γ does not change with the stretching process. Since surface energy depends on the configuration of atoms at the surface, the key point that determines whether surface stress is equal to surface energy is whether the atomic configuration on the surface is sensibly

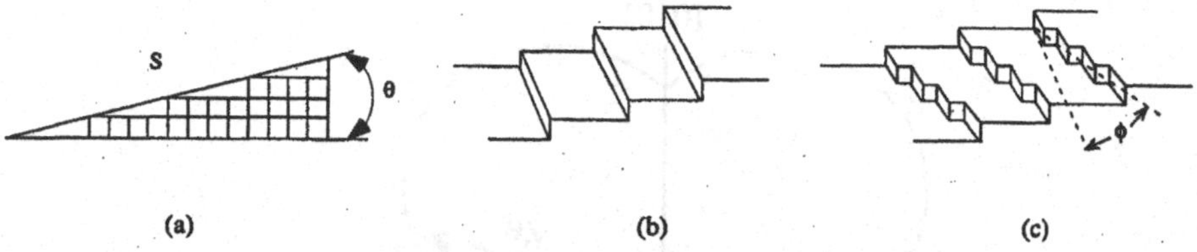

Fig. 15. Terrace-ledge-kink model of a surface. (a) A 2-dimensional representation. (b) Presence of ledges on the surface. (c) Presence of ledges and kinks on the surface.

altered by the stretching process (10). In the case of solids, where long-range order in the position of atoms is present, a distorted configuration of atoms can exist when boundary atoms are subject to small constrained displacements. In case of liquids no long-range order is present and the constrained surface configuration can achieve its undistorted configuration by the motion of atoms from the interior to the surface. Thus the equality of surface stress and surface energy depends on the mobility of atoms and thus on the relaxation time required for surface atoms to assume their undistorted configuration by the atomic migration process. The relaxation time for liquids is usually less than the duration of stretch so that surface energy and surface stress are identical. Also, in solids where long-range periodicity is interrupted by the presence of incoherent boundaries, the relaxation time is finite so that such boundaries can be easily created or destroyed by the atomic diffusion process if the rate of extension is less than the reciprocal of the relaxation time. In such cases surface energy and surface stress are identical and such boundaries behave as liquid films. This condition was used by Udin, Shaler and Wulff (12) in devising a zero creep experiment to measure the surface tension of a vertically suspended polycrystalline wire (13).

4.2 Anisotropy of Surface Energy

For crystalline solids surface energy will vary with the crystallographic orientation. To illustrate the nature of the orientation dependence of γ we consider a solid-vacuum interface for a simple cubic crystal at absolute zero. Let surface S make a small angle θ with the (100) surface, as shown in Fig. 15a in which the surface is perpendicular to the page and of unit length in that direction. Such a surface will then consist of a number of steps or ledges on an otherwise flat, terrace-like (100) orientation, as shown in Fig. 15b. If the surface plane is not perpendicular to the page then the ledges will not be normal to the page and will make an angle ϕ with it. Such an orientation can be visualized by considering the presence of kinks in a ledge as shown in Fig. 15c. Thus, any crystal orientation within the basic stereographic triangle can be developed in terms of a specified density of ledges and kinks on an appropriate terrace orientation.

To illustrate the variation of surface energy with orientation, consider the terrace-ledge-kink model of a simple cubic system at zero K. For simplicity we assume ϕ to be zero so that the surfaces consist of straight ledges separated by terraces. As the orientation of such a surface changes, the new surface will look just like the original surface except that the density of ledges will be different. The properties of these surfaces then can be separated into contributions from the ledges and from the terraces. Let γ_t and γ_ℓ be the surface energies per unit area of the terrace and the ledge, respectively. If the area of the terrace is taken to be unity, then the surface energy of any orientation, θ, can be written as the sum of the terrace energy, γ_t, and the ledge energy, $\gamma_\ell \tan\theta$ divided by the area of the inclined surface, $1/\cos\theta$. Thus

$$\gamma(\theta) = (\gamma_t + \gamma_\ell \tan\theta)\cos\theta = \gamma_t \cos\theta + \gamma_\ell \sin\theta \qquad [55]$$

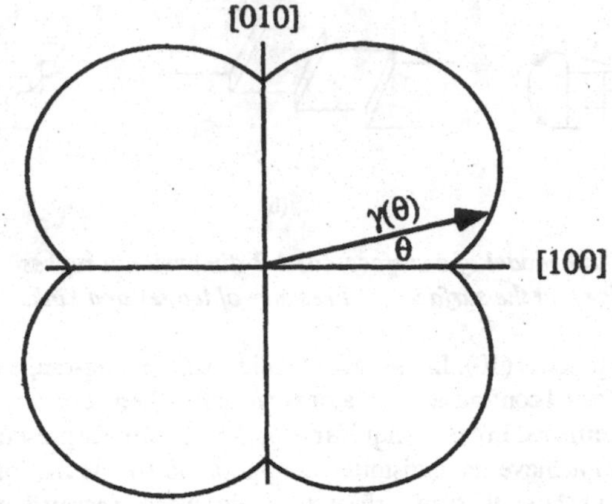

Fig. 16. Polar plot or g-plot illustrating surface energy variation with surface orientation.

This is the equation of a circle passing through the origin and having a diameter of length $\left(\gamma_t^2 + \gamma_\ell^2\right)^{1/2}$ which extends from the origin along a line inclined at tan $\left(\gamma_\ell / \gamma_t\right)$. If $\gamma_t = \gamma_\ell = \gamma_o$, as is the case with the nearest neighbor bond model, the variation of surface energy with orientation for $0 < \theta < \pi/2$ is a portion of a sphere passing through origin with a diameter $2\gamma_o$ along a line inclined at 45°. For a (100) plane with four fold rotational symmetry, the surface energy variation as a function of orientation θ consists of portions of four circles as shown in Fig. 16. Such a polar diagram is known as a γ-plot or Wulff plot. The surface energy at any orientation, θ, is given by the length of the vector drawn from the origin at an angle θ which ends on the γ-plot.

Two important conclusions can be drawn from the above result. 1) Energy cusps (discontinuous $d\gamma/d\theta$) exist at some orientations. In the above calculations we have assumed that the energy of a ledge is independent of the spacing between the ledges. In reality some ledge-ledge interaction will be present so that γ_ℓ will be a function of spacing between the ledges and thus of the surface orientation. If such an interaction is taken into account, one would observe cusps at all other rational orientations. 2) As the orientation is changed slightly, the surface energy also changes. Thus, a surface has a tendency to rotate toward a direction of lower surface energy configuration. Hence the surface energy γ tends to contract the surface and $\partial\gamma / \partial\theta$ called the torque term, tends to rotate the surface. These torque terms are quite important in solids and must be taken into account when equilibrium between different phases is considered.

4.3 Equilibrium Between Two Phases

The equilibrium conditions derived for fluid phases which are separated by a curved interface should be modified to include the effects of surface energy anisotropy and surface stresses when at least one of the phases is a solid. The equilibrium condition can be derived by considering the virtual displacement of the interface as was done with the fluid-fluid system.

For solid phases, the surface stresses are balanced by the presence of volume stresses in the bulk phases which are inhomogeneous in nature. During virtual displacement of the interface these volume stresses should therefore be taken into consideration. However, these volume stresses are of the same order of magnitude as the variation of surface energy with curvature so that they are negligible if we assume that the thickness of the inhomogeneous layer is small compared to the radius of curvature. However, for a curved interface element, surface stresses on the element will have a net component along the normal to the center of the element and in the direction of the center of curvature of the element. These forces are balanced by the volume stresses in the bulk phases which are quite appreciable and which, in case of fluids, give rise to a pressure difference between the two phases. This concept of pressure difference was found to be quite valuable in studying equilibrium properties of fluid systems. In case of solids, however, surface stresses may be absent or may be different than γ, so that the pressure concept is not as useful. We shall therefore obtain equilibrium conditions by considering the thermodynamic properties of the system to be comprised of bulk and surface contributions, the bulk volumes being fixed by the location of the interface at $\Gamma_1 = 0$.

Since bulk chemical potentials in two phases are not the same when a curved interface is present, it will be advantageous to use the Helmholtz free energy function to calculate the work done during the virtual displacement of the interface at constant T and V of the system. Let $F_1(T,N_i)$ and $F_2(T,N_i)$ be the bulk Helmholtz free energies of phases 1 and 2, respectively. The equilibrium condition can be obtained by requiring the change in Helmholtz free energy of the system, dF, to be zero when dN_i moles of the ith component are transferred from the bulk of phase 2 to the bulk of phase 1. dF will consist of three parts, the change in free energies of phase 1 and phase 2, and the change in free energy of the interface which is given by d(γA). This gives

$$dF = \left(\frac{\partial F_1}{\partial N_1}\right)_{T,N_j} dN_i - \left(\frac{\partial F_2}{\partial N_i}\right)_{T,N_j} dN_i + \gamma dA + A d\gamma = 0$$

or

$$\mu_{i1} dN_i - \mu_{i2} dN_i + \gamma dA + A d\gamma = 0 \qquad [56]$$

The surface work accomplished during virtual displacement is calculated as follows. The interface initially had an area A and energy γ. After virtual displacement let the area be A + dA and energy be $\gamma + d\gamma$, so that the surface work done is $(\gamma + d\gamma)(A + dA) - \gamma A \approx \gamma dA + A d\gamma$. The change in area, dA, occurs due to the change in the number of surface atoms as well as to the stretching apart of surface atoms caused by the change in surface strains. The term $d\gamma$ arises due to the change in orientation of the interface as well as the change in surface strains. The contribution from the change in γ due to a change in composition is zero by the Gibbs-Duhem equation. We may thus write the surface work as $\gamma dA + A d\gamma(\theta) + A d\gamma(\varepsilon)$, where $d\gamma(\theta)$ and $d\gamma(\varepsilon)$ are the effects of anisotropy and surface strains respectively, on γ. Substituting this relationship in equation [30], we obtain

$$\mu_{i1} dN_i - \mu_{i2} dN_i + \gamma dA + A d\gamma(\theta) + A d\gamma(\varepsilon) = 0 \qquad [57]$$

We now consider a simple case of an interface, such as an incoherent interface, in which $\sigma_{xx} = \sigma_{yy} = \gamma$ and $\sigma_{xy} = 0$, or the term $A d\gamma(\varepsilon) = 0$. The surface work in this case, given by $\gamma \delta A + A \delta\gamma(\theta)$, was derived by Herring (14) to give the following result:

$$\gamma dA + A d\gamma = \left[\gamma(\kappa_1 + \kappa_2) + \left(\gamma_x'' \kappa_1 + \gamma_y'' \kappa_2\right)\right] dV \qquad [58]$$

where κ_1 and κ_2 are principle curvatures, and γ_x'' and γ_y'' are the respective second derivatives of surface energy with orientation in the directions of principle curvatures κ_1 and κ_2. Substituting equation [58] in equation [56], we obtain the equilibrium condition

$$\mu_{i1} - \mu_{i2} + \left(\gamma + \gamma_x''\right) v_{i1} \kappa_1 + \left(\gamma + \gamma_y''\right) v_{i1} \kappa_2 = 0. \qquad [59]$$

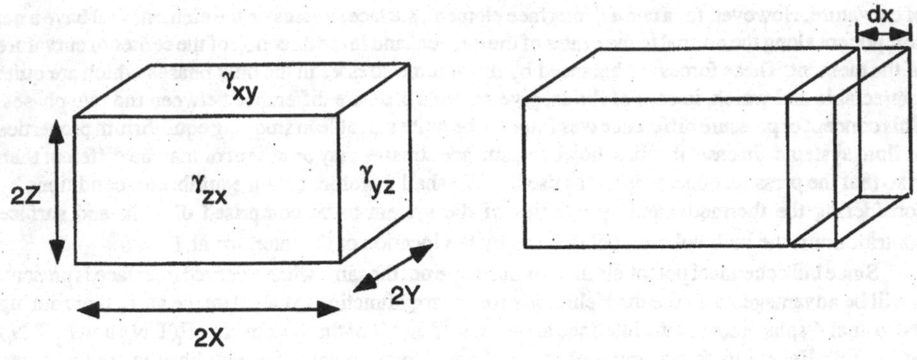

Fig. 17. (a) A rectangular parallelopiped crystal. (b) The change in the configuration of the crystal when dn moles are transfewrred across the yz interface.

When surface energy is isotropic, the γ'' terms are zero and the above equation coincides with equation [33] derived for the fluid-fluid system. Equation [59] is a general form of the Gibbs-Thompson equation which takes into account the anisotropy of interface energy.

Equation [59] assumes that second derivative of γ exists and is finite, i.e., the surface does not have a cusp orientation. For a cusped surface, surface work can be calculated by a procedure similar to that used earlier except that the value of the term dA/dV depends upon the shape of the crystal, as shown by Herring (14).

Examples

Equilibrium Condition at a Triple Point

When the interfacial energies are anisotropic, the torque terms also need to be taken into account, so that the equilibrium conditions obtained for fluid interfaces, equation [16], must be modified to include the torque terms. This gives a more general equilibrium condition as:

$$\sum_{i=1}^{3}\left[\gamma_i \bar{t}_i + \left(\frac{\partial\gamma}{\partial\theta}\right)_i \bar{n}_i\right] = 0 \qquad [60]$$

where $\bar{t}_i$ is a unit vector in the ith interface perpendicular to and directed away from the line of intersection, and $\bar{n}_i$ is a unit vector perpendicular to both $\bar{t}_i$ and the line of intersection.

4.4 Chemical potential at a faceted interface

For isotropic interfaces, we have considered the effect of interface curvature on chemical potentials. However, chemical potentials depend on interface energy, and they will be different even for a flat interface if the size of the crystal is small. In order to examine the chemical potential at a flat or a faceted interface as a function of size, consider a rectangular solid of a pure material, shown in Fig. 17a, whose lengths are 2X, 2Y and 2Z, respectively in the x, y and z directions, and whose interface energies are given by γ_{xy}, γ_{yz} and γ_{zx}, for the planes xy, yz and zx, respectively.

We first consider the chemical potential at the interface plane yz by considering the change in free energy when dn moles are transferred across this interface, as shown in Fig. 17b. The crystal grows by a distance dx, with the sides y and z remaining constant. For *local equilibrium*, the free energy change must be zero, which gives

$$\mu_{yz} - \mu_0 + (2Y\gamma_{xy} + 2Z\gamma_{zx})(dx/dn) = 0 \qquad [61]$$

Since $dx/dn = (dx/dV)(dV/dn) = V_m/(4YZ)$, we obtain

$$\mu_{yz} - \mu_0 + [(\gamma_{xy}/2Z) + (\gamma_{zx}/2Y)]\,V_m = 0 \qquad [62a]$$

Thus chemical potential will be altered significantly if the dimension of the crystal is small in the y or z direction. One can obtain similar relation ships for the planes zx and xy, as

$$\mu_{zx} - \mu_0 + [(\gamma_{xy}/2Z) + (\gamma_{yz}/2X)]\,V_m = 0 \qquad [62b]$$

and

$$\mu_{xy} - \mu_0 + [(\gamma_{zx}/2Y) + (\gamma_{yz}/2X)]\,V_m = 0 \qquad [62c]$$

A similar approach can be used to calculate the chemical potential at a flat interface of any faceted interface.

4.5 Equilibrium shape of a crystal

For an isotropic interface energy, a small particle will assume a spherical shape at equilibrium since a sphere has the smallest area to volume ratio. However, when the surface energy is anisotropic, the shape of the crystal depends on the polar γ-plot, and the condition for the equilibrium shape of a given phase, equation [11], needs to be modified as follows to take into account the variation in surface energy with orientation.

$$\Omega_s = \int_s \delta(\gamma dA) = \text{minimum} \qquad [63]$$

In order to examine the application of this equation, we first consider the simple shape of a rectangular parallelepiped. The relative areas of different interface energy planes that are present in the equilibrium shape, described earlier, are shown in Fig. 17. We examine the equilibrium shape by requiring that the chemical potential on all sides must be the same, i.e. $\mu_{xy} = \mu_{yz} = \mu_{zx}$. From equation [62]

$$\frac{\gamma_{xy}}{Z} = \frac{\gamma_{yz}}{X} = \frac{\gamma_{zx}}{Y}. \qquad [64]$$

or

$$X:Y:Z = \gamma_{yz} : \gamma_{zx} : \gamma_{xy} \qquad [65]$$

The above relationship shows that the distance from the center of the crystal in a given direction is proportional to the surface energy of the plane that is perpendicular to that direction. Thus, an interface with a smaller surface energy will be closer to the center and thus would have a larger area.

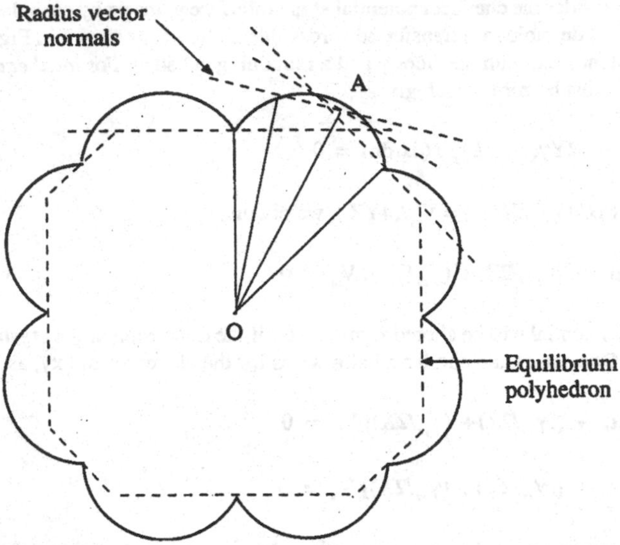

Fig. 18. A schematic polar plot of interface energy versus orientation, and the Wulff construction to determine the equilibrium shape (14).

A generalization of this result for determining the equilibrium shape of the crystal from a γ-plot is obtained by the Wulff construction. In this case one erects vectors from the origin to the γ-plot for all surface orientations. Each vector has a length proportional to interfacial energy at that boundary orientation. One then draws planes *perpendicular* to end points of the vectors. The inner envelope of all these planes gives the equilibrium shape of the crystal, as shown in Fig. 18 . Note that the distance from the origin (or point of symmetry in the crystal) to the surface of the crystal is proportional to the interface energy of that orientation so that the equilibrium shape of the crystal will have the smallest total interface energy. Once the γ-plot is known, the equilibrium shape is unique. The inverse is not true in that it is not always possible to draw a unique γ-plot from the equilibrium shape of the crystal unless some specific assumptions are made about the form of the γ-plot or all orientations are present in the equilibrium shape of the crystal. The relative values of surface energies can however be determined experimentally from the equilibrium shape of the crystal.

The equilibrium shape of a small crystal is often used to obtain the relative surface energy values of different orientations. If the anisotropy is small such that all orientations are exhibited in the equilibrium shape, a complete γ-plot can be constructed. Huang and Glicksman (15) have measured the shape of a liquid droplet in pure solid in pure succinonitrile, and thereby determined a cross-section of the γ-plot. Different techniques have been developed to measure solid:vapor, liquid-vapor, solid-liquid and solid-solid interface energies, and these techniques are described in references (13), (16) and (17).

In the absence of strain energy, the equilibrium shape of the crystal is also the shape of the critical nucleus. Thus, the knowledge of equilibrium shape sllows one to determine the work required for its formation and the nucleation rate can then be computed. The effect of anisotropic interface energy on nucleation in solid-solid transformations is discussed in the next chapter by Aaronson and Lee.

ACKNOWLEDGMENTS

The author would like to acknowledge Professor H. I. Aaronson for his valuable comments on the manuscript. This work was supported by the NASA Microgravity Science and Application Division under Grant No. NCC898. This work was carried out at Ames Laboratory, which is operated for the U. S. Department of Energy by Iowa State University, under contract no. W-7405-Eng-82, supported by the director of Energy Research, Office of Basis Energy Sciences.

References

1. J. W. Gibbs, "On the Equilibrium of Heterogeneous Substances", The Scientific Papers of J. Willard Gibbs, vol. 1, Dover Publications, Inc., New York, 1961.
2. J. W. Cahn and J. E. Hilliard, J. Chem. Phys., 1959, vol. 28, p. 258.
3. J. W. Cahn and J. E. Hilliard, J. Chem. Phys., 1958, vol. 31, p. 688.
4. L. D. Landau and E. H. Lifshitz, *Statistical Physics*, Pergamon Press Ltd., London, 1958.
5. A. W. Adamson, *Physical Chemistry of Surfaces*, Interscience Publishers, New York,
6. F. H. Buttner, E. R. Funk and H. Udin, J. Phys. Chem. 1952, vol. 56, p. 657.
7. J. P. Hirth, *Energetics in Metallurgical Phenomena, vol. II*, Ed. by W. M. Mueller, Gordon and Breach, 1965, p. 7.
8. J. A. Warren and W. J. Boettinger, Acta Metall. Mater. 1995, vol. 43, p. 689.
9. G. R. Purdy, Metal Sci. J., 1971, vol. 5, p. 81.
10. R. Shuttleworth, Proc. Phys. Soc., 1950, vol. A63, p. 444.
11. W. W. Mullins, "Metal Surfaces: Structure, Energetics and Kinetics," Ed. by W. D. Robertson and N. A. Gjostein, American Society for Metals, Metals Park, Ohio, 1963, p.17.
12. H. Udin, A. J. Shaler, and J. Wulff, Trans. AIME, 1952, vol. 185, p. 186.
13. R. Trivedi and J. D. Hunt, in: The Mechanics of Solder Alloys, Ed. by F. G. Yost, F. M. Hosking and D. R. Frear, Van Nostrand Rheinhold, N.Y. 1993, pp. 191-226.
14. C. Herring, *Physics of Powder Metallurgy*, Ed. by W. E. Kingston, McGraw-Hill Book Co., New York, 1951, p. 143.
15. S.C. Huang and M.E. Glicksman, Acta Metall., 1981, vol. 701, p. 29.
16. B. J. Keene, *Surface Tension of Pure Liquids*, National Physical Laboratory, Teddington, Middlesex, Crown Publishers. 1991.
17. R. G. Linford, Solid State Surface Science, Vol. 2, Mercer Dekker, New York, 1973, p. 85.
18. D. Turnbull, J. Appl. Phys., 1950, Vol. 21, p. 1022.
19. J. P. Hirth and J. Lothe, Theory of Dislocations, McGraw-Hill, New York, 1968, p. 764.
20. J. H. Perepezko, K. H. Rasmussen, I. E. Anderson and C. R. Loper Jr., in: Solidification and Casting of Metals, Metals Soc., London, 1979, p. 169.

Appendix

Selected values of surface energy ($\gamma_{S\text{-}V}$) measured in ultra high vacuum or helium atmosphere (17), solid-liquid interfacial energy ,$\gamma_{S\text{-}L}$,(18,20), disordered grain boundary energy. γ_{GB}, (19), twin boundary energy, γ_{TB},(19), and stacking fault energy, γ_{SF}, (19). All interface energies are in mJ/m^2.

Metal	$\gamma_{S\text{-}V}$	$\gamma_{S\text{-}L}$	γ_{GB}	γ_{TB}	$\gamma_{\Sigma\Phi}$
Gold	1370	132	364	10	55
Silver	1140	126	790	-	17
Platinum	1310	240	1000	196	~95
Nickel	1860	255	690	-	~400
Aluminum	1140	93	625	120	~200
Copper	1750	177	646	44	73
Iron	1950	204	780	190	-
Tin	680	70.6	160	-	-

THE KINETIC EQUATIONS OF SOLID→SOLID NUCLEATION THEORY AND COMPARISONS WITH EXPERIMENTAL OBSERVATIONS

H. I. Aaronson* and J. K. Lee**

*Department of Materials Science and Engineering
Carnegie Mellon University
Pittsburgh, Pennsylvania 15213
**Department of Metallurgical and Materials Engineering
Michigan Technological University
Houghton, Michigan 49931

Introduction

A large proportion of all known phase transformations in solid metals takes place by diffusional nucleation and growth. Understanding the mechanisms of these processes is thus helpful in the manipulation of a wide range of metallurgical microstructures. This article will present a fairly detailed development of the theory for the nucleation component of diffusional transformations.

Nucleation theory was founded by Gibbs (1), and has since been extensively developed by many physical chemists, physicists and metallurgists (2). In most of these studies the theory was designed for application to vapor→liquid and vapor→ solid transformations. Much of the theory, however, is applicable to solid→solid transformations, and adaptation of the remainder, though not yet complete, is continuing.

Before developing the equations of solid→solid nucleation theory, it will be helpful to provide an essentially qualitative sketch of the fundamental aspects of the theory. Nucleation may be defined as the fluctuational process through which the smallest metastable aggregate of a more stable phase develops from the matrix phase. This process is the only one through which changes in both composition and crystal structure can be simultaneously accomplished. However, diffusional nucleation can also produce only a change in crystal structure (as during a massive transformation) or only a change in composition (when the matrix and precipitate have the same crystal structure). Martensitic nucleation, on the other hand, is able to provide only a change in crystal structure; spinodal decomposition (which does not involve a nucleation process) can effect only a change in composition (3).

Nucleation nearly always occurs through a series of biatomic reactions. The process begins with single atoms (which, following the physical chemists, we shall term monomers) joining together to produce atom pairs, or dimers, of the new phase; dimers are in turn "promoted" to trimers through single atom additions, and so forth. Particularly in a solid, the probability of a triatomic reaction in which two monomers simultaneously undergo diffusional jumps which join them to another monomer to form a trimer, is sufficiently low so that it can be dismissed. n-mers which are just large enough so that addition of the next monomer yields a decrease in the standard Helmholtz free energy change, ΔF_n^o, associated with the formation of this n-mer, are known as "critical nuclei", containing n* atoms. Smaller size n-mers are termed "embryos"; in the context of larger n-mers which may still revert through statistical fluctuations to monomers it will be convenient to refer to them also as embryos.

The free energy change associated with the nucleation process is often described in terms of ΔG, the change in Gibbs free energy. However, Cahn and Hilliard (4) have pointed out that the use of ΔG is incorrect, as it already incorporates the change in pressure associated with the nucleus:matrix interfacial energy, thereby eliminating the principal barrier to the nucleation process. This occurs through the $\Delta H = \Delta E + V \cdot \Delta P$ component of ΔG, where ΔH, ΔE and ΔP are respectively the enthalpy, energy and pressure changes associated with nucleation, with $\Delta P = 2\gamma V/r$ for a spherical embryo of radius r and molar volume V. The correct version of the free energy change in this situation is ΔF, the Helmholtz free energy change (4). Recall that ΔF is expressed in terms of ΔE rather than of ΔH and that ΔF applies at constant temperature and volume, the conditions operative during isothermal nucleation.

The <u>driving force</u> for nucleation is the volume free energy change attending the transformation, ΔF_V , minus the volume strain energy (if any) arising from the size and/or shape misfit between the cluster and the matrix, ΔF_ε. The <u>barrier</u> to nucleation is the interfacial free energy of the cluster:matrix interface, ΔF_S. As sketched in Figure 1, ΔF_n^o for the formation of an embryo consisting of n atoms can be considered to have two components: ΔF_S, which is positive,

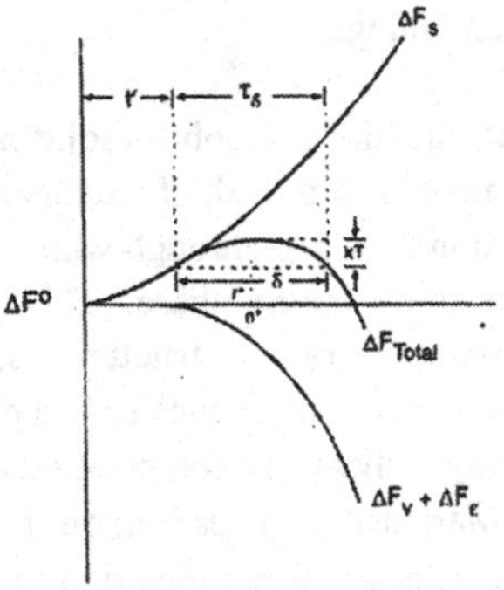

Figure 1: Variation of ΔF^o for embryo formation with radius, r and n, the number of atoms in the embryo.

and the sum of ΔF_V and ΔF_ε, which is negative. ΔF_S is proportional to the second power of the embryo radius, r, whereas $\Delta F_V + \Delta F_\varepsilon$ is proportional to r^3. Hence ΔF_S increases more rapidly than $\Delta F_V + \Delta F_\varepsilon$ decreases when r is very small and the reverse situation obtains at larger values of r. When the contribution of r is first able to prevent ΔF^o from further increasing, as just noted the radius at which this occurs is termed r*, the critical radius; the corresponding ΔF_n^o is termed ΔF^*, the free energy of formation for the critical nucleus. However, even though a further increase in r is attended by a decrease in ΔF_n^o, not until r corresponds to $\Delta F^* - kT$ (where k = Boltzmann's constant and T = absolute temperature) is the nucleus safe against loss of enough atoms through thermal fluctuations to make r < r*, and thus virtually ensure, as next discussed, dissolution into component monomers.

As illustrated in Fig. 2 the addition of another monomer requires that an activation barrier equal to the sum of the additional ΔF_n^o and the free energy of activation for diffusion ΔF_D, be surmounted. On the other hand, loss of a monomer from an n-mer requires only passage over the activation barrier for diffusion. Hence the probability of losing an atom when r < r* is always greater than that of adding an atom. Assuming that the probabilities of individual monomer additions are independent, recalling that such probabilities are multiplicative and noting that a critical nucleus may contain as many as 50-100 atoms, it can be readily understood why the probability of successful fluctuations to critical nucleus size is so very small.

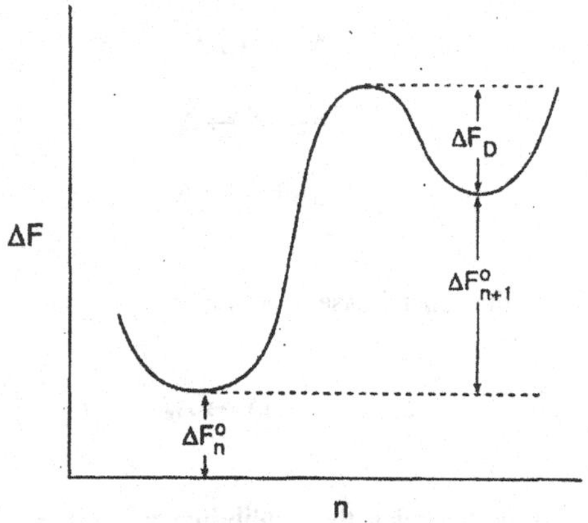

Figure 2: "Fine structure" for the addition of a single atom, in the ΔF_{TOTAL} curve in Fig. 1, incorporating ΔF_D, the free energy of activation for volume diffusion.

The general, time-dependent equation for the rate of nucleation, J*, is (5):

$$J^* = Z\beta^* N \cdot \exp(-\Delta F^*/kT) \cdot \exp(-\tau/t) \quad (1)$$

where J* may also be termed the flux of embryos through ΔF_n^o vs. n-space at r* or n*, β* = the frequency factor, the rate at which single atoms join the critical nucleus, N = number of atomic nucleation sites per unit volume of the matrix phase, τ = incubation time and t = isothermal reaction time. In the next three sections, development of this equation will be undertaken as follows. Calculation of the metastable equilibrium concentration of embryos in the first section will provide the starting point for the derivation of the steady state nucleation rate in the second. A modification of this derivation will lead to the development of Eq. (1), the time-dependent nucleation rate, in the third. Subsequent sections will examine calculation of the individual terms in Eq. (1) in more detail. In the concluding section, the predictions of the theory will be compared with the results of experiment.

Development of the Nucleation Rate Equations

Derivation of the Metastable Equilibrium Cluster Concentration

Let A = a single atom, and A_n = a cluster of n atoms. Formation of the critical nucleus through biatomic reactions is described in these terms by the following sequence of equations:

$$A + A \leftrightarrow A_2$$

$$A_2 + A \leftrightarrow A_3$$

$$A_{(n-l)} + A \leftrightarrow A_n \quad (1)$$

Summing these equations for the general case of an n-mer:

$$nA \leftrightarrow A_n \quad (2)$$

From standard chemical thermodynamics, the equilibrium constant, K, for this reaction is:

$$K = \frac{x_n/x_n^o}{(x_1/x_1^o)^n} \quad (3)$$

where x_n and x_n^o = mol fractions of n-mers in the actual and standard states, and x_1 and x_1^o = actual and standard state mol fractions of monomer, respectively. From the van't Hoff isotherm,

$$\Delta F_n^o = -kT \ln K \tag{4}$$

where ΔF_n^o = free energy of forming an n-mer in standard state x_n^o from monomer in standard state x_1^o. In the limit of a dilute solution of embryos of all sizes, most of the atoms are present as monomers. Accordingly, substitution of $x_1 = x_1^o$ and $x_n^o = 1$ into Eq. (3), utilizing Eq. (4) and defining $C_n = x_n N$, where C_n = number of n-mers per unit volume and N = total number of atoms per unit volume:

$$C_n = N \exp\left(-\Delta F_n^o/kT\right) \tag{5}$$

A plot of C_n vs. n, shown in Figure 3, is seen to pass through a minimum at n*, as should be expected from Figure 1 and this equation. At $n > n^* - \delta/2$, the curve is drawn dashed to indicate the instability of such large embryos and nuclei.

<u>Derivation of the Steady State Nucleation Rate</u>

This relationship was mainly developed by Farkas (6), Becker and Doering (7), and Zeldovich (8).

Retaining the designation C_n from Eq. (5) for the equilibrium or metastable equilibrium concentration of n-mers, C_n' will be employed to represent the concentration actually present. Thus the net rate at which embryos of size n become embryos of size n+1 is:

$$J_n = \beta_n C_n' - \alpha_{n+1} C_{n+1}' \tag{6}$$

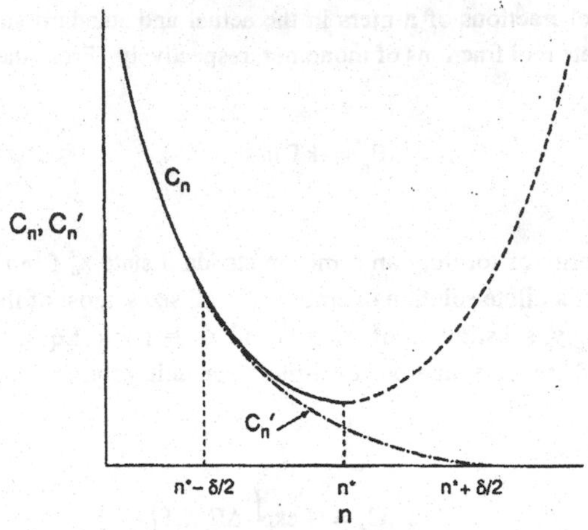

Figure 3: Dependence of the equilibrium or metastable equilibrium concentration of embryos, C_n, and the number of embryos actually present, C_n', upon n.

where β_n= rate at which single atoms impinge upon an n-mer cluster and α_{n+l} = rate at which single atoms leave a cluster of size n+l. The dependence of α upon n arises because the driving force for departure of an atom from a cluster is capillarity, and is thus size-dependent. Since α is more difficult to evaluate than β, the former will be eliminated through use of the principle of time reversal (8). This principle states that <u>at equilibrium every process and its reverse taking place by statistical fluctuations occur at exactly the same rate</u>. Toward this end, Eq. (6) is re-written for equilibrium conditions:

$$J_n = 0 = \beta_n C_n - \alpha_{n+1} C_{n+1} \tag{7}$$

Rearranging,

$$\alpha_{n+1} = \frac{\beta_n C_n}{C_{n+1}} \tag{8}$$

Substituting Eq. (8) into Eq. (6) and rearranging,

$$J_n = \beta_n C_n \left[\frac{C_n'}{C_n} - \frac{C_{n+1}'}{C_{n+1}} \right] \tag{9}$$

C_n'/C_n is nearly constant, except in the region δ of Figure 1, over which it declines gradually at first and then with increasing rapidity from nearly 1 to 0, as indicated in Figure 3. (A physical analogy to this situation is the diffusion of gas through a membrane. Upstream and downstream densities are nearly constant and the concentration drops linearly across the thickness of the membrane.) The expression in brackets can thus be re-written, with little loss of accuracy, as a differential:

$$\left[\frac{C_n'}{C_n} - \frac{C_{n+1}'}{C_{n+1}} \right] \cong \frac{\partial (C_n'/C_n)}{\partial n} \cdot [n - (n+1)] = -\frac{\partial (C_n'/C_n)}{\partial n} \tag{10}$$

Substituting Eq. (10) into Eq. (9)

$$J_n = -\beta_n C_n \cdot \frac{\partial (C_n'/C_n)}{\partial n} \tag{11}$$

Given an ample supply of monomers, a quasi-steady state situation must eventually develop in which all C_n' are independent of time, and hence all J's including J* are independent of both n and time. With J_n constant, integration of equation (11) is now possible. Suitable boundary conditions are:

$$C_n'/C_n \rightarrow 1 \text{ as } n \rightarrow 1 \tag{12}$$

$$C_n'/C_n \rightarrow 0 \text{ as } n \rightarrow \infty \tag{13}$$

Both of these conditions follow from the circumstance that, even at steady state, $\bar{C}_n$ falls below C_n because some embryos ultimately achieve promotion into the growth stage. Eqs. (12) and (13) are equivalent to holding the monomer concentration constant, and to removing all very large embryos from the system, respectively. This reasonably well approximates the common conditions of little monomer consumption and few large embryos. Since β is in general a weak function of n (as compared to the exponential dependence of C_n upon n), it may be replaced by β^* and equation (11) becomes:

$$J^* \int_1^{\infty} \frac{dn}{C_n} = -\beta^* \int_1^{0} d\left(\frac{C_n'}{C_n}\right) = -\beta^* \left[\frac{C_n'}{C_n}\right]_1^0 = \beta^* \tag{14}$$

Rearranging,

$$J^* = \frac{\beta^*}{\int_1^{\infty} \frac{dn}{C_n}} \tag{15}$$

Substituting Eq. (15) for C_n, and noting that $C_1 \approx N$ during homogeneous nucleation,

$$J^* = \frac{\beta^* N}{\int_1^{\infty} e^{\Delta F_n^0/kT} dn} \tag{16}$$

To integrate the denominator, ΔF_n^0 is expanded in a Taylor's Series about the maximum at n = n* and terms to the second power in the exponent are retained (the so-called "quadratic approximation"). This is a good approximation here as most of the value of the integral comes from the region δ (Fig. 1) near n*. One form of this series is:

$$f\{z\} = f\{z_0\} + (z - z_0) f'\{z_0\} + \frac{(z - z_0)^2}{2} f''\{z_0\} + \ldots \tag{17}$$

where $f\{z\}$ = function of z and $f'\{Z_0\}$ and $f''\{Z_0\}$ are the first and second derivatives, respectively, of the function evaluated at z_0. Thus ΔF_n^0 may be expanded about n* to give:

$$\Delta F_n^o = \Delta F^* + \frac{(n - n^*)^2}{2} \cdot \left.\frac{\partial^2(\Delta F^0)}{\partial n^2}\right|_{n^*} \tag{18}$$

This relationship is then substituted into Eq. (16). Since ΔF^* is constant it may be moved outside the integrand. The resulting integral is of the form:

$$\int_{1-n^*}^{\infty} e^{-aU^2} dU$$

where $a = -(1/2)kT\left(\partial^2 \Delta F_n^0/\partial n^2\right)_{n^*}$ and $U = n - n^*$. The integrand is virtually zero when $U < (1-n^*)$, so to a good approximation the lower limit may be extended to $-\infty$. This integrand, which is of standard form, integrates to $\sqrt{2}/\pi$. Hence the steady state nucleation rate, J_s^*, is:

$$J_s^* = Z\beta^* N e^{-\Delta F^*/kT} \tag{19}$$

where:

$$Z = \left[-\frac{1}{2\pi kT} \cdot \left.\frac{\partial^2(\Delta F_n^0)}{\partial n^2}\right|_{n^*} \right]^{1/2} \tag{20}$$

is the Zeldovich non-equilibrium factor. The dot-dashed line in Figure 3 represents C_n', which begins to fall noticeably below C_n at $n^* - \delta/2$ and becomes zero at $n^* + \delta/2$.

Time-Dependent Nucleation Rate and an Introduction to Time-Reversal

We would really like to know the variation of nucleation rate with time, beginning from the state when only monomers are present. Whereas the approximation which had to be made in order to obtain J_s^* in analytic form were relatively moderate, however, those required to derive the time-dependent rate are sufficiently severe so that achievement of an accurate, closed form solution is apparently not possible. A physically-based approach will have to be used to develop an accurate relationship for τ.

The differential equation describing time-dependent nucleation may be obtained by first performing the partial differentiation indicated in Eq. (11) and accepting the size dependence of J_n, β_n, C_n and C'_n:

$$J_n = -\beta \frac{\partial C'}{\partial n} + \frac{\beta C'}{C} \frac{\partial C}{\partial n} \tag{21}$$

Differentiating eqn. (5) with respect to n and substituting into Eq. (21) gives

$$J = -\beta \frac{\partial C'}{\partial n} - \frac{\beta C'}{kT} \frac{\partial (\Delta F^{\circ})}{\partial n} \tag{22}$$

The rate of change in the concentration of embryos of a given size is simply the difference between the fluxes in and out of that size:

$$\frac{\partial C'_n}{\partial t} = J_{n-1} - J_n \tag{23}$$

As was the case for concentrations, fluxes vary slowly with n and the difference may be replaced by a derivative:

$$\frac{\partial C'_n}{\partial t} = -\frac{\partial J_n}{\partial n} \tag{24}$$

Substituting Eq. (22) into Eq. (24) and ignoring the minor n-dependence of β:

$$\frac{\partial C'}{\partial t} = -\beta \frac{\partial^2 C'}{\partial n^2} - \frac{\beta}{kT} \frac{\partial \left[C' \cdot \partial (\Delta F^{0}) / \partial n \right]}{\partial n} \tag{25}$$

Eq. (25) is a form of the well known Master equation or Fokker-Planck equation for the behavior of a non-equilibrium system of particles. This equation thus describes time-dependent nucleation and could, in principle, be solved (subject to suitable initial and boundary conditions) for the total time-dependent nucleation behavior of a system. However, even with simple initial conditions (only monomers at $t = 0$) and boundary conditions (constant monomer concentration, and $C' = 0$ for $n \gg n^*$), Eq. (25) has not yet been solved.

There have been a number of attempts (10-16) at finding a solution, all of which involved severe approximations. The treatments agreed, however, in finding that the transient and steady state nucleation rates are related by:

$$J^*(t) = J_s^* \, e^{-\tau/t} \tag{26}$$

where with one exception the incubation time τ (also called a time lag, delay time, or induction period) was of the form:

$$\tau = \pm \frac{\kappa}{\dfrac{\beta^*}{kT} \left.\dfrac{\partial^2(\Delta F)}{\partial n^2}\right|_{n^*}} \tag{27}$$

where the dimensionless constant, κ, is found by various investigators to lie between 0.5 and 5.

To obtain a better relationship for τ, Feder et al. (17) resorted to a more physical approach. As shown in Fig. 1, they divided the evolution of an embryo from monomer to n^* into two regions: t', extending from monomer to $n^* - \delta/2$, and τ_δ, encompassing the region from $n^* - \delta/2$ to $n^* + \delta/2$. The kinetics of evolution through these two regions may be compared upon the basis of the principle of time reversal (17). It is thus legitimate to examine the kinetics of embryo evolution by considering the more readily visualized and microscopically spontaneous reverse process of dissolution. Passage in this manner through the τ_δ region occurs across rather small gradients in ΔF_n^o with respect to n. Thus, the drift term in Eq. (25) is unimportant and motion can be considered to take place by means of random walk in ΔF_n^o vs. n space. On the other hand, dissolution through the t' region takes place down a much steeper average gradient in ΔF_n^o and hence should occur more rapidly, i.e., the drift term in Eq. (25) is dominant. Feder at. al. demonstrated this point mathematically, and were thus led to approximate $\tau \approx \tau_\delta$. They calculated τ_δ by determining the value of δ at which $\Delta F_n^o = \Delta F^* - kT$ (cf. Fig. 1). Again employing Taylor's Series, Eq. (17), with $n^* = z_0$, $\delta/2 = z - z_0$ and $\Delta F_n^o = f(z_0)$ leads, upon rearrangement, to:

$$\delta = \left[-\frac{1}{8kT} \frac{\partial^2 \Delta F^0}{\partial n^2} \bigg|_{n^*} \right]^{-1/2} \qquad (28)$$

From the Einstein equation for Brownian motion kinetics, the time to random walk a root mean square distance δ is:

$$\tau_\delta = \frac{\delta^2}{2\beta^*} \qquad (29)$$

where δ is treated as the diffusion distance, β* as the diffusivity and τ_δ as the diffusion time. Substituting Eq. (28) into Eq. (29),

$$\tau \approx \tau_\delta = \frac{-4kT}{\beta^* \dfrac{\partial^2 (\Delta F^0)}{\partial n^2} \bigg|_{n^*}} \qquad (30)$$

Abraham (18) has evaluated Eq. (25) numerically. He found an exponential approach to steady state (Eq. (26)) and incubation times in good agreement with the time reversal value.

<u>β*, the Frequency Factor</u>

Because the critical nucleus is in equilibrium (albeit an unstable one) with the matrix, the composition of the matrix in contact with the nucleus is the average matrix composition (19). Even clusters moderately removed from critical size are also surrounded by approximately the average matrix composition (19). Evaluation of β* can be accomplished on this basis by deducing the rate at which solute atoms cross the matrix:nucleus interface (assuming, for convenience, that the precipitate is a solute-rich phase). This is simply the number of solute atoms within a single jump distance of an attachment site on the critical nucleus multiplied by the jump frequency toward the embryo. The number of appropriately situated solute atoms is approximately $S^* x_\beta / a^2$ where S* is the area of embryo which can accept atoms, a is the lattice constant, and x_β is the mole fraction of solute in the matrix. The atomic jump frequency, Γ, is related to the diffusivity, D, by (20):

$$\Gamma = \frac{6D}{\alpha^2} \tag{31}$$

where α = jump distance, approximately equal to the lattice constant. Multiplying these two factors:

$$\beta^* \approx \frac{DS^* x_\beta}{a^4} \tag{32}$$

where D = the diffusivity which controls mass transport to or from the precipitate (19). As such, different D's can obtain for nucleation taking place under various circumstances, e.g., whether nucleation takes place in an interstitial or a substitutional phase, whether nucleation occurs homogeneously (where the appropriate volume diffusivity should be used) or at dislocations or at grain boundaries, where the diffusivity along the imperfection catalyzing nucleation may predominate. We emphasize that in Eq. (32) the geometrical factors involved in S* are secondary compared to the importance of selecting the proper "D". One may thus reasonably well approximate S* by a^2 when making rough calculations.

D. the Diffusivity

Let us now consider in more detail the proper diffusivity for use in β^* during homogeneous nucleation in a substitutional solid solution. If the nucleus: matrix interface is fully coherent (but still mobile, a situation which should obtain only when the two phases have the same crystal structure and spatial orientation (21)), diffusion of one atom toward the nucleus implies movement of another atom, usually of another species, away from the nucleus. However, the almost inevitable difference in the diffusivities of the two species cannot be compensated by a flux of vacancies, as in bulk diffusion phenomena (22). Thus, in homogeneous nucleation a "constrained" or "coherent" diffusivity is appropriate (22):

$$D = \frac{D_A D_B}{D_A x_\beta + D_B(1 - x_\beta)} \tag{33}$$

where D_A and D_B are the intrinsic diffusivities (23) of solute and solvent, respectively.

Nucleation of an entirely incoherent precipitate at a disordered grain boundary is normally controlled by the boundary diffusivity. When a grain boundary-nucleated precipitate is coherent with one of the matrix grains, however, unless the nucleus can attain critical size without displacement of the coherent interface, solute must reach this interface by volume diffusion and hence the coherent diffusivity is again appropriate (22).

N, the Nucleation Site Density

When homogeneous nucleation occurs, all lattice sites are eligible to participate and thus N is simply:

$$N = N_o/V \tag{34}$$

where N_o = Avogadro's number and V = molar volume. When nucleation takes place at lattice imperfections, N is greatly reduced. Following Cahn (22), nucleation at grain faces reduces N by the factor $d^2\lambda/2$ divided by d^3, where d = length of a grain edge and λ = grain boundary thickness. Division by 2 is required to take account of the sharing of the grain face by the two grains forming it and $d^3 \approx$ grain volume. Similarly, for nucleation at grain edges, the reduction factor is $d\lambda^2/3$ divided by d^3 and for grain corners $\lambda^3/4$ divided by d^3. Hence:

$$N \approx \left(\frac{\lambda}{d}\right)^{3-m}\frac{N_o}{V} \tag{35}$$

where m = dimensionality of the nucleation site: 3 for homogeneous nucleation, 2 for grain faces, 1 for edges and 0 for corners. Since λ/d is usually within an order of magnitude of 10^{-6}, N/N_o decreases from 10^{-6} for m = 2 to 10^{-18} for m = 0. Cahn has shown, however, that the decrease in ΔF^*, as more grain boundary area is destroyed with decreasing dimensionality, can more than compensate for even these large reductions in N when the ratio of the grain boundary energy to the energy of nucleus:matrix boundaries is sufficiently high and the matrix grain size is small. For nucleation on or along dislocations, N equals length of dislocation per unit volume multiplied by the number of nucleation sites per unit length.

Influence of Critical Nucleus Shape upon Nucleation Kinetics

The Equilibrium Shape Problem

Critical nuclei are so very small--and transient--that the details of their shape are exceedingly difficult to discern experimentally. Nonetheless the Gibbs view that this shape must be the equilibrium one continues to be considered correct. In a process as thoroughly kinetic in character as nucleation, why should this condition obtain? The answer to this question lies in the circumstances that interfacial energy usually represents the principal barrier to nucleation and that the equilibrium shape is, in the absence of strain energy, the one having the smallest total interfacial energy for a given volume of precipitate phase. The problem of finding the equilibrium shape was solved by Wulff (25) a generation after Gibbs posed it. Herring (26) has reviewed and further contributed to efforts to prove Wulff's solution correct. The solution which Wulff provided to the equilibrium shape problem is based upon a polar plot of specific interfacial free energy, γ, as a function of boundary (or surface) orientation, constructed in three dimensions for all orientations. The envelope connecting these points is known as the polar γ-plot. A cross-section through a schematic plot of this type is given by the curved solid line in Figure 4 (26). Note that this particular plot contains a number of cusps, at which there is a discontinuous change in the derivative of γ with respect to interface orientation. In the case of an interphase boundary, cusps are to be expected at precisely those boundary orientations which yield particularly good matching of atom patterns and spacings across the boundary. Small deviations from exact cusp orientation can be accommodated by the introduction of ledges, and if necessary, of kinks in the edges of the ledges. Such defects, however, will increase the boundary energy. Hence the boundary orientation at which they are not required should lie at a sharp, or cusped minimum, with the cusp pointing toward the interior of the γ-plot. Herring (26) argues that cusps are blunted at temperatures near the melting temperature for crystal:vacuum interfaces. This seems physically less likely at interphase boundaries because the formation of thermally activated kinks, which help to remove crystal:vacuum interfaces from the cusp orientation, is energetically so much more difficult at a crystal:crystal interface.

The Wulff construction of the equilibrium shape is obtained by drawing a plane through each point on the γ-plot perpendicular to the vector connecting that point to the origin. (A few sample traces of such planes are shown as light dashed lines in Figure 4.) The body formed by all points reachable from the origin without crossing any of these planes is geometrically similar to the equilibrium shape. A two-dimensional section through the equilibrium shape so constructed

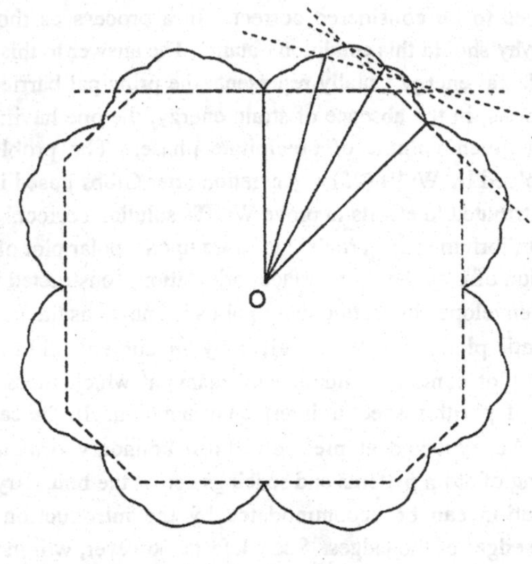

Figure 4: Schematic γ-plot and the Gibbs-Wulff construction based upon this plot. Herring (26).

is delineated in Figure 4 with heavier dashed lines forming an octagon in two dimensions.

Calculation of ΔF^*

Not only is ΔF^* the most widely variable and thus the most influential term in the general equation for J^* (Eq. (1)) but the mathematical apparatus used to calculate ΔF^* in various situations will also be seen to be directly applicable to the calculation of Z, β^* and τ. The approach to be used in calculating ΔF^* entails first the determination of the equilibrium shape. The volume strain energy, ΔF_ε, appropriate to that shape--or at least to the shape best approximating that of the critical nucleus--is then algebraically added to ΔF_v. Although simultaneous minimization of both interfacial energy and volume strain energy would obviously provide a more accurate evaluation of ΔF^*, only quite limited progress has as yet been made in developing a general solution to this problem. An effort made in this direction will be summarized in a later subsection.

For Homogeneous Nucleation in the Absence of Crystallography: Assume that the nucleus:matrix boundary is disordered (incoherent) and has the specific interfacial free energy $\gamma_{\alpha\beta}$ at all boundary orientations, where the matrix phase is designated β and the product phase α. The polar γ-plot is thus a sphere, and hence so also is the equilibrium shape. Following Gibbs (1), the standard free energy change, ΔF^o, associated with the formation of an embryo of this morphology is:

$$\Delta F^o = (\text{embryo volume})\cdot \Delta F_v + (\text{embryo interfacial area})\cdot \gamma_{\alpha\beta}$$
$$= \tfrac{4}{3}\pi r^3 \cdot \Delta F_v + 4\pi r^2 \gamma_{\alpha\beta} \qquad (36)$$

where r = radius of the spherical embryos. As implied in Figure 1, r* is found by differentiating ΔF^o respect to r and setting the resulting expression equal to zero. Solving for r, which becomes the critical radius:

$$r^* = -2\gamma_{\alpha\beta}/\Delta F_v \qquad (37)$$

Substituting this result into eqn. (36),

$$\Delta F^* = \frac{16\pi\gamma_{\alpha\beta}^3}{3\cdot \Delta F_v^2} \qquad (38)$$

The basis for a subsequently developed (27) alternative method for calculation of ΔF* will now be presented. This method is a little simpler than the foregoing one even in the case of a spherical nucleus and will be seen to become appreciably easier to use when more complex nucleus morphologies are considered.

This approach begins by re-writing Eq. (36) in generalized form:

$$\Delta F^o = K_v r^m \cdot \Delta F_v + K_a r^{m-1} \gamma_{\alpha\beta} \qquad (39)$$

where K_v = volume shape factor, K_a = interfacial energy factor and m = dimensionality of the nucleus = 3 for the cases discussed in this article. Proceeding as in the derivation of eqn. (37),

$$r^* = \frac{(1-m)\gamma_{\alpha\beta}K_a}{m \cdot \Delta F_v \cdot K_v} \tag{40}$$

Substituting into eqn. (39):

$$\Delta F^* = \frac{(1-m)^{m-1}K_a^m\gamma_{\alpha\beta}^m}{m^m \Delta F_v^{m-1} K_v^{m-1}} = \frac{(1-m)^{m-1}E^m}{m^m \cdot \Delta F_v^{m-1} \cdot v^{m-1}} \tag{41}$$

where E = total interfacial energy of the critical nucleus and v = volume of the critical nucleus. Herring (26) has used the Brunn-Minkowski inequality to derive the following relationship between E and the volume of the Wulff construction:

$$E = mv^{\left(\frac{m-1}{m}\right)} v_W^{\left(\frac{1}{m}\right)} \tag{42}$$

where v = volume of the equilibrium shape in physical space and v_W = volume of this shape in "Wulff space", wherein $\gamma_{\alpha\beta}$ replaces r as a coordinate. Substituting into eqn. (41) yields:

$$\Delta F^* = \frac{(1-m)^{m-1}}{\Delta F_v^{m-1}} \cdot v_W \tag{43}$$

For a sphere, $v_W = (4/3)\pi\gamma_{\alpha\beta}^3$. Substitution into Eq. (43) leads to immediate recovery of Eq. (38).

<u>For Homogeneous Nucleation in the Absence of Crystallography--Non-Classical Theory</u>: Nucleation theory so far described is known as "classical" theory and follows directly from that of Gibbs. The founding paper in this approach to nucleation theory, termed "non-classical"--

---

* Essentially all of the equation sequences presented for classical nucleation theory in this chapter should be readily followed with an algebra-calculus type of background. Following the derivations involved in Cahn-Hilliard non-classical theory, on the other hand, requires a more extensive mathematical base.

which can also be considered as implicit in the work of Gibbs--is that of Cahn and Hilliard (4). This work was based, in turn, upon their preceding paper on the free energy of an inhomogeneous binary solution (28). Both matrix and product phases are assumed to be incompressible fluids. Fluctuations within the matrix are taken to occur on a scale large with respect to the inter-atomic spacing. These assumptions are also reasonable for a solid→solid phase transformation in which the crystal structures of the matrix and precipitate are the same. e.g., fcc→fcc, the lattice parameters of the two phases are equal and independent of composition and their elastic constants are identical and isotropic, thus making $\Delta F_{\varepsilon} = 0$.

The Cahn-Hilliard (4,28) equation for the Helmholtz free energy of a binary solid solution of the type described is:

$$F = \int_V [f(x) + \kappa(\nabla x)^2]\, dV \tag{44}$$

where $f(x)$ = free energy/unit volume of homogeneous solution with composition (atom fraction) x, κ = gradient energy coefficient and $\kappa(\nabla x)^2$ = gradient energy. When a composition fluctuation develops in an initially homogenous solution, the barrier to nucleation created by the gradient energy (essentially equivalent to the interfacial energy term in classical theory) is eventually overcome by the integrated negative value of $f(x)$. Thus ΔF^* is the minimum height of the barrier to be overcome by statistical fluctuations. Applying the Euler equation to Eq. (44) yields the relationship which must be solved to describe this fluctuation; this relationship may be simplified in the present situation because under the assumptions used the critical nuclei are spherical :

$$2\kappa\frac{d^2x}{dr^2} + \frac{4\kappa}{r}\frac{\partial\kappa}{\partial x} + \frac{\partial\kappa}{\partial x}\left(\frac{dx}{dr}\right)^2 = \frac{\partial(\Delta f)}{\partial x} \tag{45}$$

where Δf is described in Fig. 5. The increase in free energy attending the formation of a spherical critical fluctuation then follows from Eq. (44) as:

$$\Delta F^* = 4\pi\int_0^{\infty} [\Delta f + \kappa(dx/dr)^2]\, r^2 dr \tag{46}$$

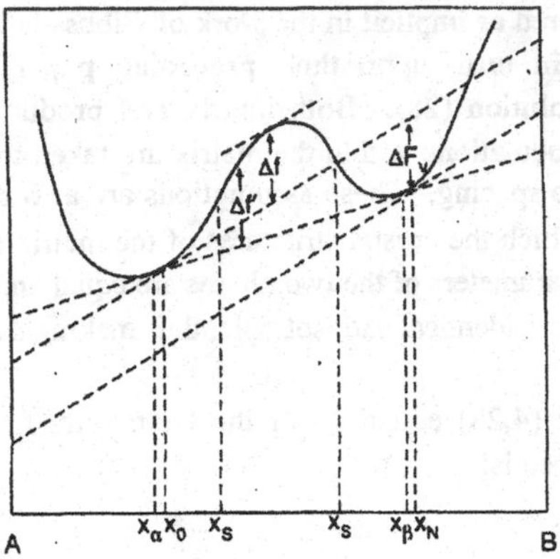

Figure 5: f-x diagram for the miscibility gap system A-B. x_o = bulk composition, x_α and x_β = miscibility gap compositions, x_s = spinodal compositions, x_N = composition of classical critical nucleus, Δf = free energy difference between the f-x curve tangent at x_o and the parallel tangent to the "hump" in the f-x curve, $\Delta f'$ = difference between the f-x curve and the tangent connecting the equilibrium compositions, and ΔF_v = difference between the tangent to the f-x curve at x_o and the curve itself. Cahn and Hilliard (4).

where r = radial distance from the center of the fluctuation. The composition-distance profile of critical fluctuations is obtained by solving Eq. (45) for x as a function of r and substituting the data obtained into Eq. (46). Typical results are shown in Fig. 6. Normalized concentrations are $X_o = (x_o - x_c)/2(x_\alpha - x_c)$, where x_c = composition of T_c for the miscibility gap, and normalized radial distance from the center of the nuclei, $t = r\{[2\xi(x_\alpha - x_c)^2]/\kappa_c\}^{1/2}$, where $\xi = (\partial^4 f/\partial x^4)_{crit}/4!$ and κ_c = value of κ at x_c. When the supersaturation is small, e.g., X_o = -0.45, the solute concentration well exceeds that of the bulk matrix in the region from the center of the nucleus to about one-half the reduced radius of the fluctuation. However, with increasing supersaturation, e.g., X_o = -0.30, even in the center of the critical nucleus the solute concentration is not much higher than that in the bulk matrix.

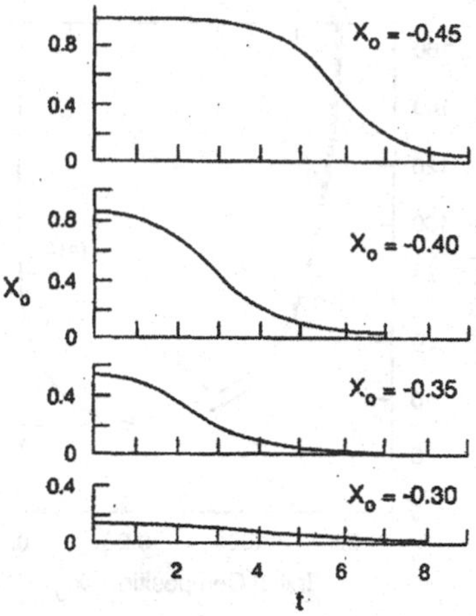

Figure 6: Normalized composition profiles through critical nuclei at constant temperature at various normalized compositions, X_o, and thus supersaturations. Cahn and Hilliard (4).

Gibbs (1) recognized that critical nuclei can be compositionally inhomogeneous but deliberately defined their properties so that they could be treated as having a constant composition up to a well defined nucleus:matrix interface (4). Defining with increasing arbitrariness the radius of non-classical critical nuclei as $r_{1/2}$, the radius at which the composition of the nucleus is reduced to one-half the difference between the composition of the nucleus center and c_o, Fig. 7 shows that $r_{1/2}$ passes toward infinity as the solvus temperature, x_α, and also as the spinodal composition, x_s, are approached. This figure also demonstrates that whereas the r^* of classical nucleation theory becomes coincident with $r_{1/2}$ as $x_o \rightarrow x_\alpha$, when $x_o \rightarrow x_s$ classical theory forecasts a continued slow diminution of r^*. Non-classical theory, on the other hand, appropriately predicts a smooth merger of nucleation (and growth) with spinodal decomposition under the latter circumstance. Similarly, Fig. 8 demonstrates that the behavior of ΔF^* is similar in classical and in non-classical theory as x_α is approached. However, the two diverge near x_s, with ΔF^* remaining finite on classical theory but (appropriately) diminishing to zero at x_s.

Cahn and Hilliard (4) defined the supersaturation region within which classical nucleation theory is applicable as $l \ll r_{1/2}$, where l is the thickness of the interface, defined as the radial distance over which the nucleus composition differs significantly from both the central region of the nucleus and the bulk composition. They estimated the maximum value of $\Delta F^*/kT$ at which nucleation may occur at detectable rates as ca. 60. On this basis, the applicability criterion for

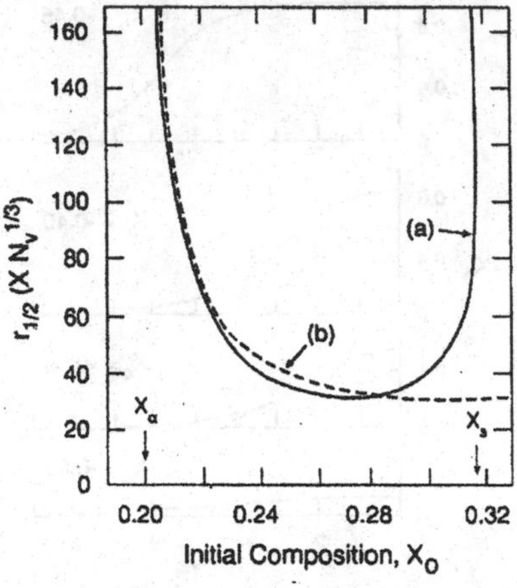

Figure 7: Variation of critical nucleus radius with x_o on (a) non-classical and (b) classical nucleation theory. Cahn and Hilliard (4).

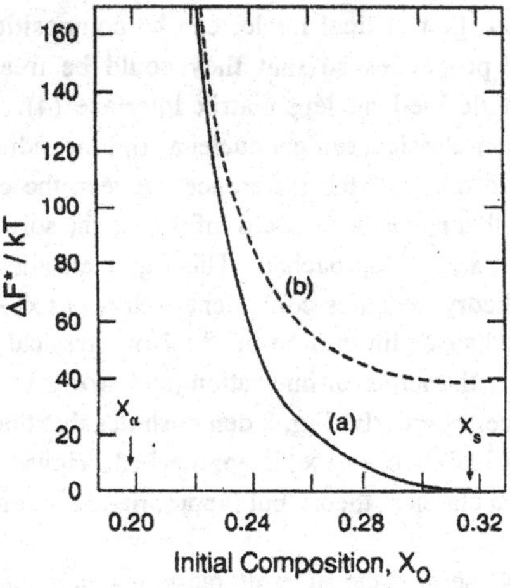

Figure 8: Variation of $\Delta F^*/kT$ with X_o on (a) non-classical and (b) classical nucleation theory. Cahn and Hilliard (4).

classical theory is:

$$\pi\gamma l^2/(45kT) << 1 \quad (47)$$

The terms other than ΔF* in Eq. (1) have yet to be derived for non-classical nucleation theory. The relationships from classical nucleation theory for these terms must therefore be used instead, though such usage is presumably decreasingly accurate as the spinodal region is approached.

Finally, it should be emphasized that in the continuum form provided by Cahn and Hilliard, non-classical theory must be modified in the presence of strain energy and considerably further altered when the crystal structures of the matrix and nucleus phases are different. Cahn (29-31) has incorporated strain energy into continuum nucleation (and spinodal decomposition) theory. In the nucleation context, however, continuum theory is rather difficult to use. Cook et al. (32) laid the basis for a more flexible approach to the problem by discretizing Cahn-Hilliard continuum theory. Cook and DeFontaine (33) then incorporated strain energy in this theory. LeGoues et al. (34) found that the combined model could be conveniently treated numerically, yielding an approach into which both crystalline and elastic anisotropy can be incorporated, there are no restrictions as to critical nucleus shape and both cubic and tetragonal distortion tensors can be accommodated. Fig. 9 shows the added complexity of shape displayed by the plate-like

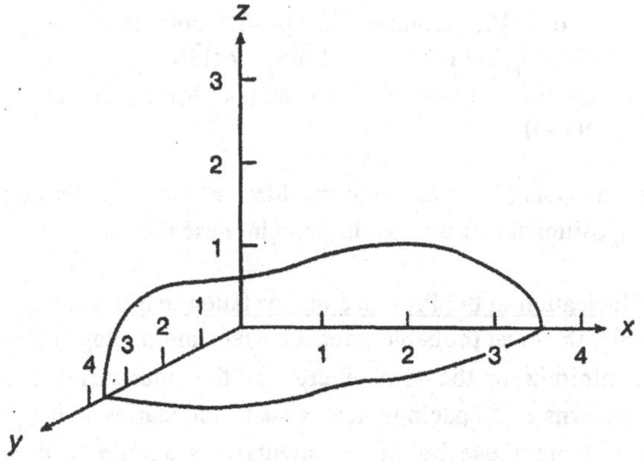

Figure 9: Approximate shape of one octant of the critical nucleus of a GP zone in Al-2 A/O Cu at 198 K. LeGoues et al (34).

critical nucleus of a (fcc) GP zone in an Al-Cu alloy, wherein a misfit of 12% is present, when the discretized approach is employed (34).

Exploration of the general case in which the crystal structures of the matrix and nucleus phases are sufficiently different that the stacking sequence changes across the interphase boundaries is presently in its earliest stages. The suggestion was made that in this situation a crystallographically sharp, i.e., classical interface, would be combined with a compositionally diffuse, i.e., non-classical interface (35). Fig. 10 (36) confirms the compositional portion of this prediction for the well defined interface $(0001)_{hcp}//\{111\}_{fcc}$, $<11\bar{2}0>_{hcp}//<1\bar{1}0>_{fcc}$ by means of a

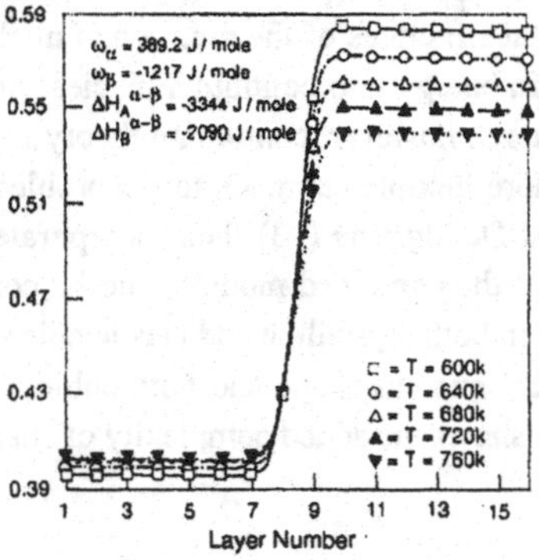

Figure 10: Composition profile normal to the $(0001)_{hcp}//\{111\}_{fcc}$, $<11\bar{2}0>_{hcp}//<1\bar{1}0>_{fcc}$ interface at various temperatures. Ramanujan et al (36).

discrete lattice plane analysis (36). Note that the different thermodynamic properties of the two phases make the compositional diffuseness different in these phases.

For Homogeneous Nucleation in the Presence of Crystallography: On Eq. (38), minimization of ΔF^* (in order to predict the most probable critical nucleus morphology) is promoted with special efficiency through minimizing the total energy of the boundaries enclosing the nucleus. Matching of atom patterns and spacings across such boundaries will therefore be especially good, and deviations from these boundary orientations should lead to rapid increases in interfacial energy. Hence these boundaries are likely to be of the energy cusp type.

Consider the particularly simple case in which an energy cusp is present at only one orientation of the $\alpha:\beta$ boundary (Fig. 11) (37). Heavy lines in this figure enclose the equilibrium shape. Light lines show the γ-plot where it differs from this shape. The dashed light lines show one of an infinite number of possible variations of the γ-plot in the vicinity of the cusp orientation which would yield the same equilibrium shape.

The Gibbs-type derivation of ΔF* for the faceted sphere is as follows. By construction, Figure 11 shows that the normal distance from the origin to a facet and to a curved surface is proportional to the energy of that interface. As such, in Fig. 11, $\cos\alpha = \gamma^c_{\alpha\beta}/\gamma_{\alpha\beta}$, where $\gamma^c_{\alpha\beta}$ = energy of a coherent (or partially coherent) α:β boundary. The volume of the truncated γ-sphere and the area of the various interfaces may be calculated from standard formulae (38) to obtain:

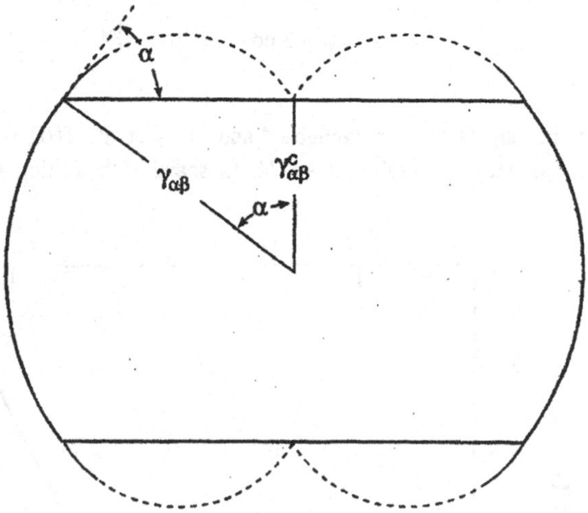

Figure 11: γ-plot and equilibrium shape of a critical nucleus faceted at one boundary orientation (37).

$$\Delta F^0 = \Delta F_v \cdot 2\pi r^3\left(\cos\alpha - \frac{\cos^3\alpha}{3}\right) + 2\gamma^c_{\alpha\beta}\pi r^2 \cdot \sin^2\alpha + 4\gamma_{\alpha\beta}\pi r^2\cos\alpha \tag{48}$$

where the first term in the r.h.s. is the product of the nucleus volume and ΔF_v, the second term gives the total interfacial energy needed by the two parallel facets and the third term takes account of the interfacial energy of the spherically curved portion of the interface, taken to have a disordered structure. It is not necessary to undertake the $\partial(\Delta F^o)/\partial r = 0$ operation to calculate r*. The facets must have the same chemical potential as the curved interface when the equilibrium shape is established. In turn, for a given value of $-\gamma_{\alpha\beta}/\Delta F_v$,, r* must be the same as for an unfaceted sphere. Substituting Eq. (37) into Eq. (48), with straightforward manipulation it is found that:

$$\Delta F^* = \frac{16\pi\gamma_{\alpha\beta}^3}{3 \cdot \Delta F_v^2} \cdot (1 - 2f\{\alpha\}) \equiv \Delta F_{hu}^* \cdot (1 - 2f\{\alpha\}) \tag{49}$$

where:

$$f\{\alpha\} = \left(2 - 3 \cos \alpha + \cos^3 \alpha\right)/4 \tag{50}$$

and the subscript "hu" signifies "homogeneous" and "unfaceted". f{α} is plotted as a function of α in Figure 12. Note that as α→90°, f{α}→1/2, the shape of the critical nucleus approaches

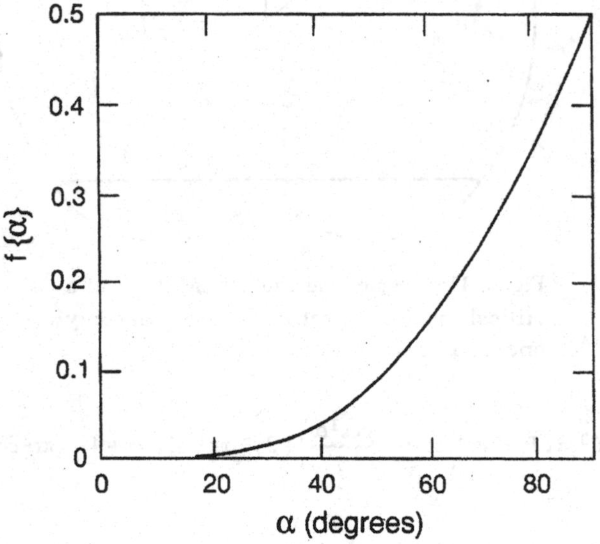

Figure 12: Dependence of f{α} upon α. This plot is also applicable to f{ψ} vs. ψ.

that of a thin disc and ΔF*→ 0. Homogeneous nucleation thus becomes a considerably more likely possibility when $\gamma_{\alpha\beta}^c/\gamma_{\alpha\beta}$ is sufficiently small.

Using the newer approach outlined in the previous sub-section, the volume of the equilibrium faceted sphere, given by the first term in the r.h.s. of Eq. (48), is converted to v_W simply by replacement of r and r cos α by $\gamma_{\alpha\beta}$ and $\gamma^c_{\alpha\beta}$, respectively. Substitution into Eq. (43) then yields Eq. (49) with considerably less effort.

Recalling from equation (43) that ΔF^* is proportional to v_W which is in turn proportional to the volume of the critical nucleus, the ratio of the ΔF^*'s of two nucleus morphologies for a given phase and specimen is equal to the ratio of their volumes as well as of their v_W's. Thus the influence of faceting upon ΔF^* may be assessed in terms of its effect upon either the physical or the Wulff volume of the critical nucleus.

Grain Boundary Nucleation in the Absence of Crystallography: In this and in the following sub-section, the grain boundary is assumed to be planar and to have a disordered structure. Derivation of the equilibrium shape at a grain boundary is considerably facilitated by a recently derived modification of the Wulff construction (39,40) which is valid even in some cases of faceting. (Winterbottom (41) earlier developed a similar construction for the equilibrium shape of a particle on an immobile vapor:solid boundary.) For simplicity, we first consider the case of isotropic and equal $\gamma_{\alpha\beta}$ values in both matrix grains. Spheres of radius $\gamma_{\alpha\beta}$ are constructed about points 0 and 0'. Figure 13 is a cross-section through diameters of these spheres. These points are separated by a distance proportional to $\gamma_{\beta\beta}$, the grain boundary energy. The shaded region in which the two spheres overlap is then geometrically similar to the equilibrium shape. The dashed line AA' represents the area of the grain boundary destroyed by the formation of this shape as the critical nucleus. Note that as $\gamma_{\beta\beta} \rightarrow 0$, the two spheres merge, and in the limit the single circle (sphere) appropriate to homogeneous nucleation is recovered. As $\gamma_{\beta\beta} \rightarrow 2\gamma_{\alpha\beta}$, on the other hand, the volume of the shaded region$\rightarrow$0 and complete "wetting", i.e., $\Delta F^* = 0$, is approached. From this modified Wulff plot, the governing dihedral angle of the nucleus morphology, ψ, is read directly as $\cos^{-1}(\gamma_{\beta\beta}/2\gamma_{\alpha\beta})$, without need to resort to force balancing at the root of the angle.

In writing ΔF^o for this nucleus, one new factor is introduced, i.e., the product of $\gamma_{\beta\beta}$ and the area of the grain boundary destroyed is subtracted from the interfacial energy required for the creation of α:β interfaces:

$$\Delta F^o = \Delta F_v \cdot \frac{2}{3}\pi r^3\left(2 - 3\cos\psi + \cos^3\psi\right) + \gamma_{\alpha\beta}4\pi r^2(1 - \cos\psi) - \gamma_{\beta\beta}\pi r^2 \sin^2\psi \tag{51}$$

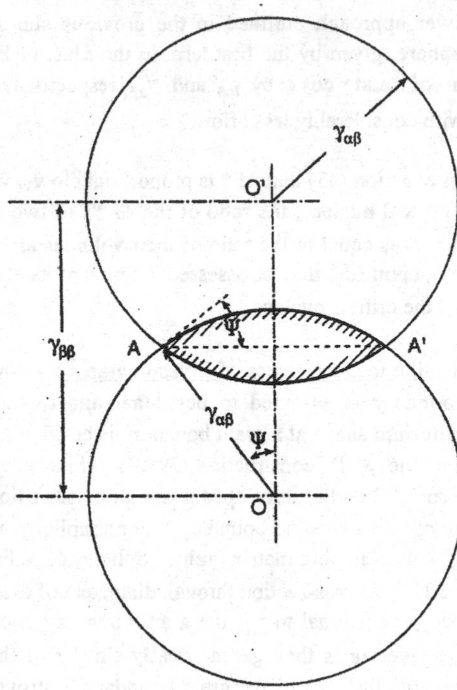

Figure 13: Modified γ-plot and Gibbs-Wulff construction for an unfaceted nucleus at a planar grain boundary (40).

Because this nucleus is made up from abutting caps of two spheres of equal radius, r* is again given by Eq. (37). Substituting this equation into Eq. (47) and rearranging yields:

$$\Delta F^* = \Delta F^*_{hu} \cdot 2 \cdot f\{\psi\} \qquad (52)$$

where f{ψ} is identical to f{α} when α's are replaced by ψ's. The newer approach, Eq. (43), yields the same result more readily.

<u>Grain Boundary Nucleation in the Presence of Crystallography</u>: Let us now consider the case when the α and β lattices match well enough and are spatially oriented with respect to each other

so that a facet appears at a singular orientation of the interface between the embryo and one of the matrix grains forming the grain boundary. The two matrix grains forming the grain boundary are taken to be irrationally oriented to the point where a facet will not be present at any orientation of the nucleus:matrix boundary with the other matrix grain. (The section on comparisons of nucleation theory with experiment at the end of this article casts considerable doubt upon the physical reality of this assumption, though it continues to be useful for the purpose of illustration.) The specific interfacial free energy is $\gamma_{\alpha\beta}^{c}$ at the facet orientation and $\gamma_{\alpha\beta}$ at all others. Initially, the further assumption will be made that the combination of the angle which the facet makes with respect to the grain boundary, ϕ, and the value of $\gamma_{\alpha\beta}^{c}/\gamma_{\alpha\beta}$ are such that the facet will not contact the grain boundary, as shown in the modified Wulff plot of Figure

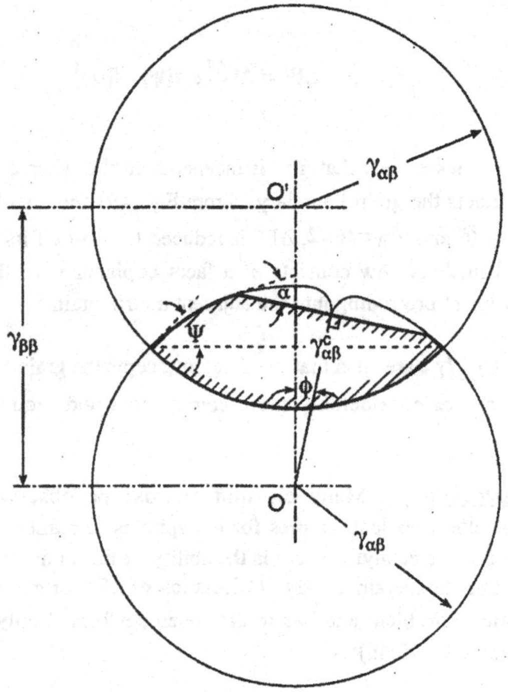

Figure 14: Modified γ-plot and Gibbs-Wulff construction for a nucleus formed at a grain boundary and faceted at a single boundary orientation in one matrix grain (40).

14 (30). Absence of contact between facet and grain boundary occurs over a finite range of ϕ

only when $\gamma_{\beta\beta} < 2\gamma^{c}_{\alpha\beta}$.

Once again, r* is given by Eq. (37). Now employing only the new method for obtaining ΔF*, and again equating r and $\gamma_{\alpha\beta}$:

$$V_W = \frac{2\pi\gamma^{3}_{\alpha\beta}}{3}\left(2 - 3\cos\psi + \cos^3\psi\right) - \frac{\pi\gamma^{3}_{\alpha\beta}}{3}\left(2 - 3\cos\alpha + \cos^3\alpha\right) \tag{49}$$

Hence from Eq. (43):

$$\Delta F^* = \Delta F^*_{hu}\left(2 \cdot f(\psi) - f(\alpha)\right) \tag{50}$$

Inspection of Fig. 14 makes clear that ΔF* is independent of ϕ when $\phi < \phi_{c1}$, the critical angle at which the facet contacts the grain boundary. From Eq. (49) and also from a result obtained by Gibbs (1), when $\phi = 0^\circ$ and $\gamma^{c}_{\alpha\beta} = \gamma_{\beta\beta}/2$, ΔF* is reduced to one-half its value in the absence of a facet. The critical nucleus now consists of a facet coplanar with the grain boundary and a spherical cap of radius r* protruding into the adjacent matrix grain.

When ϕ and $\gamma^{c}_{\alpha\beta}/\gamma_{\alpha\beta}$ are such that the facet intercepts the grain boundary, the shape of the critical nucleus becomes considerably more complicated and requires numerical evaluation (40,42).

<u>Nucleation at Dislocations</u>: Many electron microscope observations have shown that dislocations are excellent nucleation sites for precipitates in numerous alloy systems (43,44). The physical basis for this catalytic effect is the ability of the strain energy of the dislocation to diminish the transformation strain energy. Calculation of ΔF* for nucleation on dislocations is a very difficult elasticity problem and has to date been performed only approximately and then only for special situations (45-49).

Formation of an incoherent nucleus on a dislocation core relieves the strain energy (including the core energy) of the material replaced by the nucleus. The incoherent case has been analyzed (44,45) subject to somewhat different approximations. In both cases, however, ΔF* was much less for nucleation on a dislocation than for homogeneous incoherent nucleation in the matrix.

There is no net advantage to incorporating the dislocation core in a fully coherent nucleus because lattice continuity requires the dislocation line to thread the nucleus, thereby increasing its strain energy. Instead the coherent nucleus will lie away from, but very close to the core, so as to interact with the elastic stress field associated with the dislocation. The strain energy accompanying formation of an "oversized" nucleus (wherein the average volume per atom is less than that in the matrix phase) is reduced if nucleation occurs on the tension side of an edge dislocation. Similarly, an "undersized" nucleus favors the compression side of the dislocation. For a spherical nucleus with the same elastic constants as the matrix, Larche (49) has shown theoretically that ΔF* for nucleation at such dislocations may be written in a particularly convenient form through the following small formal modification of ΔF* for homogeneous nucleation in the absence of faceting (Eq. (38)):

$$\Delta F_d^* = \frac{16\pi(\gamma - \gamma_{eff})^3}{3(\Delta F_v + \Delta F_\varepsilon)^2} \qquad (55)$$

where:

$$\gamma_{eff} = \frac{\mu b(1+\nu)}{3\pi(1-\nu)}\varepsilon - \frac{b(1+2\nu)(1-2\nu)}{12\pi(1+\nu)(1-\nu)}\Delta F_v \qquad (56)$$

where μ = shear modulus, b = Burgers vector length, ν = Poisson's ratio and ε = stress-free transformation strain. The reduction in $\gamma_{\alpha\beta}$ for $\varepsilon = 0.02$ is of the order of 100 mJ/m^2 and thus represents a major decrease in ΔF* compared to that for coherent homogeneous nucleation, where in most cases $\gamma^c_{\alpha\beta}$ is of similar order (5).

A dislocation will also catalyze nucleation of a coherent precipitate without misfit strain ($\varepsilon = 0$) if the precipitate is softer than the matrix. The strain energy of the dislocation is proportional to the shear modulus, and as such is decreased by the presence of the soft particle (46).

<u>Calculation of the Geometrical Portion of ΔF*</u>: Adaptation of Eq. (32) to particular nucleus shapes now requires only the evaluation of the influence of the particular shapes upon S*. As discussed in the next section, solute atoms can join the nucleus only at disordered portions of the nucleus:matrix boundary when the crystal structures of the two phases are different. For the models under consideration, these boundaries are spherically curved. Consistently with the approach employed to express the various ΔF* relationships, S* is described in terms of the constant L (37):

$$S^* = LS_{hu}^* = 4\pi r^{*2}L \qquad (57)$$

S^* is simply the term which is multiplied by $\gamma_{\alpha\beta}$ in the various equations for ΔF^*. Table I (37) lists both L and K (the multiplying factor for ΔF^* equations) for each of the nucleus morphologies discussed.

Substituting Eq. (37) into this relationship and the result into Eq. (32):

$$\beta^* = \frac{16\pi\gamma_{\alpha\beta}^2 D x_\beta L}{a^4 \cdot \Delta F_v^2} \qquad (58)$$

<u>Calculation of Z and τ: Further Application of Time Reversal</u>

Application of both Eq. (20), for Z, and Eq. (30), for τ, to specific nucleus morphologies requires evaluation of $\left[\partial^2(\Delta F^o/\partial n^2)\right]_{n^*}$. This should be done along the morphological path which embryos travel in passing through critical nucleus size. When the nuclei are spherical, this calculation is straightforward. However, when a facet is present a complication develops (37). Facets formed by differently structured crystals cannot migrate normal to themselves because substitutional atoms would have to occupy, temporarily, interstitial sites, an energetically infeasible process, particularly in reasonably close packed crystal structures (50). During growth, ledges formed on facets provide the facets with some mobility (50,51). However, formation of ledges on the facets present during nucleation, whose lateral extent may be no more than a very few lattice parameters, is improbable. The manner in which a faceted nucleus evolves despite this restriction is most readily understood by employing the principle of time reversal, and examining the dissolution of a critical nucleus for the representative case of the faceted sphere shown in Fig. 15. The "point effect of diffusion" is seen to cause the facets to disappear early in the dissolution process. Now considering the "growth" process by looking at the sequence of morphology in reverse, the embryo is observed to "grow" as an unfaceted sphere until the precise positions for facet formation are reached. The facets then extend laterally until the equilibrium shape is achieved. Failure of facet formation to begin--purely through statistical fluctuations--in these positions will result in the generation of a higher ΔF^* nucleus morphology. Such a morphology is most unlikely to contribute to the overall kinetics of nucleation because of overwhelming competition from the minimum ΔF^*, i.e., equilibrium-shaped critical nuclei.

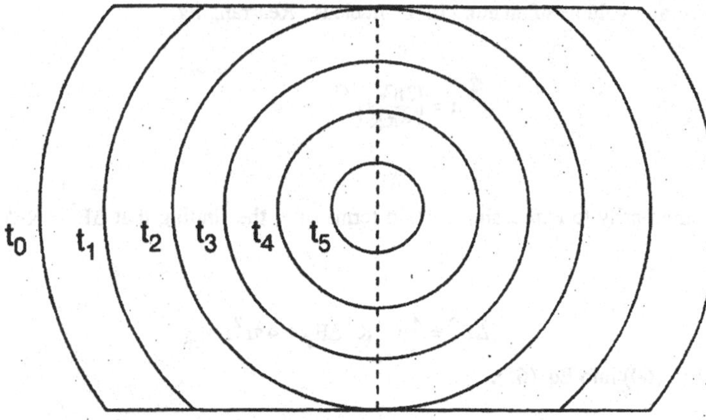

Figure 15: Successive stages in the dissolution ($t_o<t_1<t_2<t_3<t_4<t_5$) and formation ($t_o>t_1>t_2>t_3>t_4>t_5$) of the heavily outlined singly faceted nucleus. Johnson et al. (37)

$\left[\partial^2(\Delta F^o/\partial r^2)\right]_{n*}$ was calculated for both faceted spheres and faceted grain boundary allotriomorphs, using both the correct "growth" path and the mathematically more convenient one corresponding to the equilibrium shape at all stages of "growth", i.e., assuming the facets to be mobile (37). When evaluated for representative values of the various interfacial energies involved, the two results were found to be the same to within a factor of two. This represents but a small error in the context of Eqns. (1), (20) and (30). Since the correct solution is both more complicated and different for each nucleus morphology whereas the "shape preserving" solution is both simple and basically the same for all nucleus morphologies based upon spheres, only the latter is presented here.

As follows from Eq. (43) and as was shown by Gibbs (1), the constant K is the ratio of the volume of the critical nucleus to that of a sphere with the same radius. This result will now be used to calculate the second derivative of ΔF^* at n^* in terms of $\underline{n}$ instead of $\underline{r}$. The volume of a given sphere-based morphology is $(4/3)\pi r^3 K$. Hence:

$$(4/3)\pi r^3 K = nV_\alpha \tag{59}$$

where V_α = average volume of an atom in the nucleus. Rearranging,

$$r = \left(\frac{3nV_\alpha}{4\pi K}\right)^{1/3} \tag{60}$$

Reverting momentarily to expressing ΔF* in terms of r, the finding that $\Delta F^* = K \cdot \Delta F^*_{hu}$ shows that:

$$\Delta F^o = \frac{4}{3}\pi r^3 K \cdot \Delta F_v + 4\pi r^2 K \gamma_{\alpha\beta} \tag{61}$$

Substituting Eq. (60) into Eq. (61),

$$\Delta F^0 = nV_\alpha \cdot \Delta F_v + (4\pi K)^{1/3} \gamma_{\alpha\beta} (3nV_\alpha)^{2/3} \tag{62}$$

Differentiating with respect to n and setting the result equal to zero:

$$n^* = -\frac{32\pi K}{3V_\alpha}\left(\frac{\gamma_{\alpha\beta}}{\Delta F_v}\right)^3 \tag{63}$$

Twice differentiating Eq. (62) with respect to n:

$$\frac{\partial^2 (\Delta F^o)}{\partial n^2} = -\frac{2}{9}(4\pi K)^{1/3} \gamma_{\alpha\beta} (3V_\alpha)^{2/3} n^{-4/3} \tag{64}$$

Substituting Eq. (63) into Eq. (64) yields.

$$\left(\frac{\partial^2 (\Delta F^o)}{\partial n^2}\right)_{n^*} = -\frac{V_\alpha^2 \cdot \Delta F_v^4}{32\pi K \gamma_{\alpha\beta}^3} \tag{65}$$

Substituting this relationship for the second derivative into Eq. (20):

$$Z = \frac{V_\alpha \cdot \Delta F_v^2}{8\pi\left(kTK\gamma_{\alpha\beta}^3\right)^{1/2}} \tag{66}$$

Table I. Equations for K and L for Various Nucleus Shapes

Nucleus Shape	K	L
Sphere	1	1
Faceted Sphere	$1 - 2 \cdot f\{\alpha\}$	$\cos \alpha$
Unfaceted Grain Boundary Allotriomorph	$2 \cdot f\{\Psi\}$	$1 - \cos \psi$
Faceted Grain Boundary Allotriomorph	$2 \cdot f\{\Psi\} - f\{\alpha\}$	$1/2\,(1 + \cos \alpha) - \cos \psi$

Using Eqns. (58) and (64) in Eq. (30):

$$\tau = \frac{8kT\gamma_{\alpha\beta}a^4}{Dx_\beta V_\alpha^2 \cdot \Delta F_v^2} \cdot \frac{K}{L} \tag{67}$$

Eqns. (66) and (67) for Z and τ are applicable to all cases of homogeneous and grain boundary nucleation discussed herein.

The Ancillary Parameters

These parameters, which must be accurately evaluated in order to calculate J*, include D, ΔF_v, $\gamma_{\alpha\beta}$ and ΔF_ε.

ΔF_v: See the paper by Hillert in this book.

D: See the discussion in connection with Eq. (33).

$\gamma_{\alpha\beta}$: Calculation of $\gamma_{\alpha\beta}$ for disordered interphase boundaries has yet to be satisfactorily accomplished. For partially coherent interphase boundaries, it is a useful approximation (52) to add the chemical component of $\gamma_{\alpha\beta}$, which arises from changes in the chemical identity of some proportion of the nearer neighboring atoms at the interface, and the structural component of $\gamma_{\alpha\beta}$, which results from the presence of any structural disorder at the interface. As considered in the section on Quantitative Comparison of Homogeneous Nucleation Theory and Experiment, it now appears that the chemical component of $\gamma_{\alpha\beta}$ may well be the only one present at most (and perhaps even all) nucleus:matrix interfaces.

A discrete lattice plane (DLP) approach to calculation of the chemical interfacial energy for fully coherent interphase boundaries, $\gamma^{c}_{\alpha\beta}$, has been developed by Lee and Aaronson (53) on the basis of an earlier analysis reported by Wynblatt and Ku (54) for segregation to a free surface. The model used in both treatments is of the nearest neighbors, broken bond type. The similarly based earlier model of Becker (55) confined analysis to the pair of conjugate planes forming the coherent interface, using the assumption that the composition of these planes is identical to that of their bulk phases. The later models (53,54), on the other hand, allowed the chemical composition to vary gradually between those of the two bulk phases and to do so over as many atomic planes as necessary in order to minimize the total chemical interfacial energy of the system. DLP analyses thus yield the same type of result as the Cahn-Hilliard (CH) (28) continuum analysis of the chemical $\gamma^{c}_{\alpha\beta}$. The DLP analysis sketched here becomes mathematically equivalent to that of CH at sufficiently high temperatures when the regular solution model is employed (53). The DLP approach begins by assuming that each phase *initially* has the bulk composition right up to the interphase boundary. Segregation of atoms toward and away from the interface region is then allowed. This segregation produces both energy changes, ΔE, and configurational entropy changes, ΔS. These continue until the variation of the Helmholtz free energy change, $\Delta F = \Delta E - T \cdot \Delta S$, with any further change in composition, $\partial(\Delta F)/\partial x = 0$. Hence the overall problem is reduced to calculation of the energy and entropy changes occurring as the equilibrium distribution of solute atoms, B, in the region of the coherent boundary between the two phases is achieved.

The geometric basis for the ΔE calculation is shown in Fig. 16, a schematic illustration of the distribution of atoms about an atom P on the i'th plane within the region where the composition differs from that of the equilibrium compositions of the two participating phases, α and β. Z_l is the lateral coordination number within the i'th plane and Z_j is the coordination number with respect to the nearest neighboring atoms in the j'th plane. The total coordination number, Z, is accordingly:

$$Z = Z_l + 2 \cdot \sum_{j} Z_j \tag{68}$$

The bond energy required to break bonds between a B atom and its nearest neighbors in the bulk α phase (whose composition is taken to be that of the α/(α + β) phase boundary, $x_{\alpha}^{\alpha\beta}$) and to form bonds between this atom and its nearest neighbors in the i'th layer of the interface is thus:

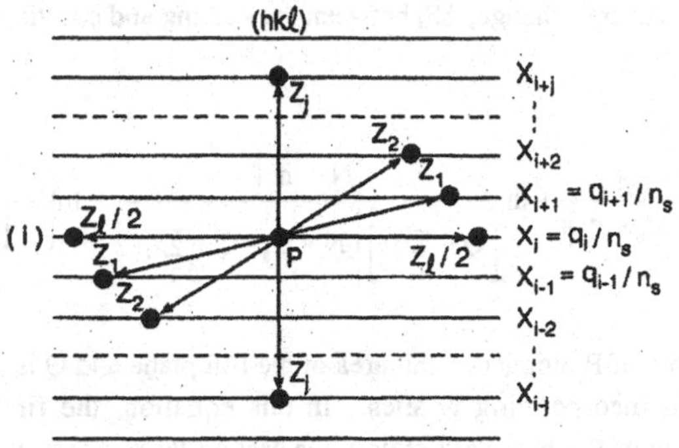

Figure 16: Geometry of the DLP Method (43).

$$\frac{1}{2}\left(\sum_j [x_{i+j} + x_{i-j}]Z_j + x_i Z_l - x_{\alpha}^{\alpha\beta} Z\right) (\varepsilon_{BB} - \varepsilon_{AB}) \tag{69}$$

where ε_{BB} and ε_{AB} are the energies of single B-B and A-B nearest neighbor bonds. A similar relationship can then be written for an A (solvent) atom transferred from the i'th plane to the bulk lattice of the α phase. The sum of these two relationships is thus the energy change associated with this change of positions, Δh_i.

$$\Delta h_i = \Delta\varepsilon\left[x_{\alpha}^{\alpha\beta} Z - x_i Z_l - \sum_j (x_{i+j} + x_{i-j}) Z_j\right] \tag{70}$$

where $\Delta\varepsilon = \varepsilon_{AB} - 1/2(\varepsilon_{AA} + \varepsilon_{BB})$ and ε_{AA} is the energy of single A-A nearest neighbor bonds. The total bond energy change required to reach the equilibrium state from the initial state is therefore:

$$\Delta H = n_s \sum_i \left(x_i - x_{\alpha}^{\alpha\beta}\right) \cdot \Delta h_i \tag{71}$$

where n_s is the number of atoms per unit area in the interface and the summation is over the entire region normal to the boundary in which the composition differs from that of either bulk phase.

The configurational entropy change, ΔS, between the starting and equilibrium states is:

$$\Delta S = k\left(\sum_i \ln\frac{n_s!}{q_i![n_s-q_i]!} + \ln\frac{[N-in_s]!}{\left[Q-\sum_i q_i\right]!\left[N-in_s-Q+\sum_i q_i\right]!} - \ln\left[\frac{N!}{Q![N-Q]!}\right]\right) \tag{72}$$

where q_i is the number of B atoms per unit area of the i'th plane and Q is the number of B atoms in the entire system incorporating N sites. In this equation, the first term in parentheses represents the entropy of the boundary region, the second term is that of the regions outside of the boundary region and the third term applies to the bulk phases or reference state. Approximating the logarithm of arguments consisting of N, Q or combinations thereof ± terms involving i, n_s and/or q_s as equal to N, Q, etc., this equation is reduced to:

$$\Delta S = -n_s k \sum_i \left(x_i \ln\left[\frac{x_i}{x_\alpha^{\alpha\beta}}\right] + (1-x_i)\ln\left[\frac{1-x_i}{1-x_\alpha^{\alpha\beta}}\right]\right) \tag{73}$$

The equilibrium solute distribution occurs when $\partial(\Delta F)/\partial x_i = 0$. Substituting Eqns. (71) and (73) into the relationship for ΔF yields a set of difference equations which must be solved simultaneously for x_i:

$$2\cdot\Delta\varepsilon\left\{x_\alpha^{\alpha\beta}Z - x_i Z_l - \sum\left[x_{i+j}+x_{i-j}\right]Z_j\right\} - kT\left\{\ln\left[\frac{1}{x_i}-1\right] - \ln\left[\frac{1}{x_\alpha^{\alpha\beta}}-1\right]\right\} = 0 \tag{74}$$

Cahn and Hilliard (28) define the interfacial energy as the difference between the free energies of an equilibrium mixture of α + β and of a homogeneous, continuous α or β phase at the composition of the alloy. Hence ΔF = ΔE - T·ΔS gives the interfacial energy of the system when the equilibrium concentration profile calculated from Eq. (74) is substituted into this relationship. After some manipulation to put the ΔF equation into usable form, the final result obtained is:

$$\gamma_c = n_s \sum_i \left\{ -\Delta\varepsilon\left(x_i - x_\alpha^{\alpha\beta}\right)^2 Z + \Delta\varepsilon \sum_j \left(x_i - x_{i+j}\right)^2 Z_j + kT\left[x_i \ln\frac{x_i}{x_\alpha^{\alpha\beta}} + (1 - x_i)\ln\frac{(1 - x_i)}{\left(1 - x_\alpha^{\alpha\beta}\right)} \right] \right\} \quad (75)$$

The quantities n_s, Z_l and Z_j in this equation are evaluated by a vectorial method described in detail in the original reference (53).

Fig. 17 shows the concentration profiles calculated at three different temperatures, expressed as fraction of T_c, the critical temperature of the symmetrical regular solution miscibility gap. In this figure, note that over the wide relative temperature range studied the concentration profiles are rather insensitive to the orientation of the interphase boundary. The increasing diffuseness of the interface with rising temperature is in good accord with the Cahn-Hilliard (28) treatment and with the expectation that the concentration profile will stretch out toward infinity as T_c is approached.

Fig. 18 compares γ_c values computed for an fcc lattice as a function of T/T_c from three different treatments. Curves (a), a (111) interface, and (b), a (100) interface, were obtained from Eq. (75), i.e., the DLP treatment. Curve (c) is the result of the Cahn-Hilliard (28) continuum analysis. The formal result for the latter analysis is:

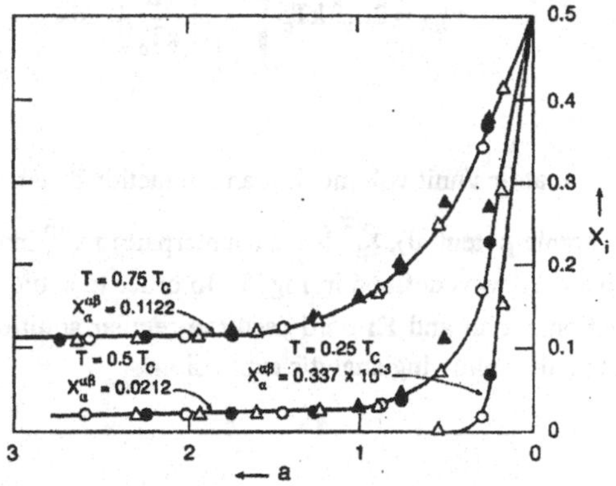

Figure 17: One-half the concentration-penetration curve for a coherent α:β boundary calculated from the DLP method at three reduced temperatures. • represents a (100) boundary, Δ the (111) boundary and ♦ the boundary resulting from the Cahn-Hilliard (28) continuum calculation.

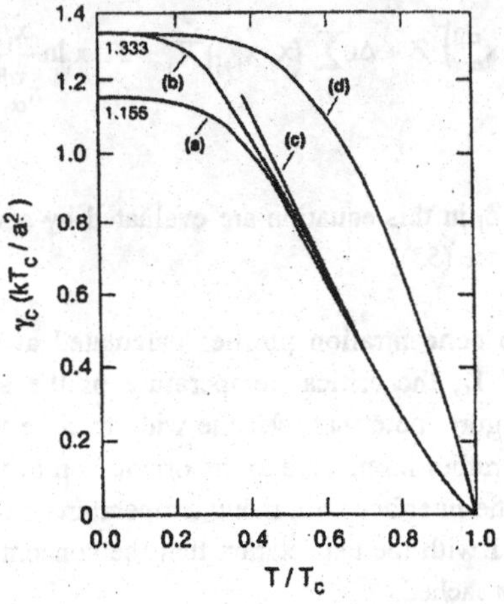

Figure 18: Comparison of various interfacial energy calculations (53).

$$\gamma^{c}_{\alpha\beta} = 2n_v\lambda kT_c \int_{x_\alpha^{\alpha\beta}}^{x_\beta^{\beta\alpha}} \left(\frac{\Delta f}{kT_c}\right)^{1/2} dx \tag{76}$$

where n_v is the number of atoms/unit volume, λ is an interaction distance (which is very sensitive to the choice of inter-atomic potential), $x_\beta^{\beta\alpha}$ is the counterpart to $x_\alpha^{\alpha\beta}$ in defining the compositions of the miscibility gap and Δf was defined in Fig. 5. In order to avoid the numerical integration required by this equation, Cahn and Hilliard used the regular solution model and some other approximations to obtain the following analytic expression:

$$\gamma^{e}_{\alpha\beta} = n_v\lambda[kT]^{1/2}\left[\pi\left(0.5 - x_\alpha^{\alpha\beta}\right)\left(\Delta f_{max}\right)^{1/2}\right]\left[1 - (\pi/2 - 4/3)(T/T_c)\right] \tag{77}$$

where:

$$\Delta f_{max} = -2kT_c\left(0.5 - x_\alpha^{\alpha\beta}\right)^2 + 0.5\,kT\left[\ln\left(0.5/x_\alpha^{\alpha\beta}\right) + \ln\left(0.5/\left(1-x_\alpha^{\alpha\beta}\right)\right)\right] \qquad (77A)$$

Curve (d) was obtained from the aforementioned Becker (55) equation:

$$\gamma_{\alpha\beta}^c = n_s Z_s \Delta\varepsilon\left(x_\alpha^{\alpha\beta} - x_\beta^{\beta\alpha}\right)^2 \qquad (78)$$

Fig. 18 shows that this equation markedly over-estimates $\gamma_{\alpha\beta}^c$ when $T > 0.2\ T_c$, doubtless as a result of its failure to take into account the compositional diffuseness of a coherent interface at finite temperatures. Hence, in the temperature range where this equation is accurate, diffusion may be too slow to support nucleation. This figure also shows that when $T > 0.7\ T_c$, the two DLP and the continuum curves converge; at lower temperatures, $\gamma_{(111)} < \gamma_{(100)}$ and both interfacial energies are less than their continuum counterpart, both as would be intuitively anticipated.

Dilatational ΔF_ε for Incoherent Ellipsoidal Precipitates with Isotropic Homogeneous Elasticity

The term "homogeneous" denotes in the present context that the matrix and precipitate have the same elastic constants. The ellipsoids referred to in this and in succeeding subsections are taken to be ellipsoids of revolution.

Nabarro (56) and later Kroner (57) derived a relationship for the volume strain energy, ΔF_ε, caused by uniform dilatational strains arising solely from a change in the average volume per atom attending a phase transformation in the situation where elasticity is isotropic and incompressible, and the elastic constants are the same in both phases, and showed that the strain energy per unit volume of the ellipsoid is:

$$\Delta F_\varepsilon = 6\mu\varepsilon^2\, f(\beta) \qquad (79)$$

where $f(\beta)$ is a function of the aspect ratio, β. (The use of β to denote the aspect ratio is now well established in the literature. Because this term is employed in a quite different context than β and β^*, the frequency factor for an embryo and the critical nucleus, respectively, one trusts that no confusion will result from this duplication.) Fig. 19 shows the results obtained. The strain energy is seen to be a maximum for a sphere, to pass to zero for an infinitesimally thin disc

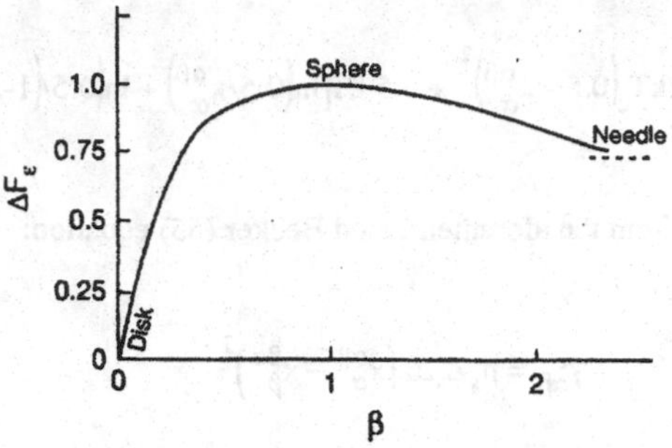

Figure 19: Volume strain energy vs. aspect ratio for an incoherent ellipsoid of revolution. Nabarro (56).

and to assume an intermediate value for a needle.

Although Eq. (79) has long been used for the volume strain energy associated with nucleation, the probability that incoherent nuclei can form will later be shown to be negligible. Further, Russell (58) has pointed out that the ambient supply of vacancies should migrate to or from such embryos until ΔF_ε is eliminated.

Dilatational ΔF_ε for Coherent Ellipsoidal Precipitates with Isotropic Homogeneous Elasticity

Eshelby (59,60) and later Lazlo (61) have shown that under these conditions ΔF_ε is independent of β, the aspect ratio of ellipsoids of revolution. The following is an elementary derivation of their relationship.

The area under a load (P) vs. elastic deformation (δ) curve is equal to the total strain energy (U) of a stressed specimen. Taking this relationship to be linear,

$$U = 1/2\, P\, \delta \tag{80}$$

Since the stress, $\sigma = P/A$, where A = the area over which the load is applied and the strain, $\varepsilon = \Delta L/L_o = \delta/L_o$, where L_o is the original length of a body and $\Delta L = \delta$ = the change in length resulting from the application of P,

$$U = 1/2(\sigma A)(\varepsilon L_o) = 1/2\, \sigma \varepsilon v_o \tag{81}$$

where $v_o = AL_o$, the original volume of the stressed body. Thus U/v_o is defined as the volume strain energy, $W \equiv \Delta F_\varepsilon = 1/2\ \sigma\varepsilon$. If a body is stressed only in the x-direction, $W_x = 1/2\ \sigma_x\varepsilon_x$; similarly, $W_y = 1/2\ \sigma_y\varepsilon_y$. When $\sigma_x = \sigma_y$,

$$\Delta F_\varepsilon = \sigma\varepsilon \tag{82}$$

Noting now that a strain in the x-direction requires a proportional strain in the y-direction, and vice versa, in order to maintain constancy of volume, and that the constant of proportionality is Poisson's Ratio ν, E is Young's modulus, $\varepsilon_x = 1/E(\sigma_x - \nu\sigma_y)$ and $\varepsilon_y = 1/E(\sigma_y - \nu\sigma_x)$. When $\varepsilon = \varepsilon_x = \varepsilon_y$,

$$\varepsilon = \frac{\sigma}{E}(1 - \nu) \tag{83}$$

Substituting into Eq. (82) for σ,

$$\Delta F_\varepsilon = \frac{E\varepsilon^2}{1 - \nu} \tag{84}$$

<u>Dilatational ΔF_ε for Coherent Ellipsoidal Precipitates with Isotropic Heterogeneous Elasticity</u>

The term "heterogeneous" is used here to indicate that the relevant elastic constants are different in the matrix and precipitate phases.

Laszlo (59) derived individual equations for fully coherent discs, spheres and needles. Using the "transformation strain" approach of Eshelby (60, 61), Barnett (62) verified these results. Subsequently, Barnett et al. (63) derived an expression for the dilatational strain energy for coherent ellipsoids of revolution as a function of their aspect ratio, thereby encompassing all of the special shapes studied by Laszlo (59). This expression is both lengthy and complex. However, the results, illustrated graphically for three different ratios of the shear modulus for the matrix (μ) relative to that of the precipitate (μ^*) in Fig. 20, are quite clear. This figure shows that when the shear moduli of the two phases are identical (Poisson's ratio is taken to be the same for all cases considered here), the strain energy is independent of β, in which case Eq. (84) applies without specification of the phases involved. On the other hand, the ratio of the strain energy for the case when $\mu^* \neq \mu$ to that for $\mu = \mu^*$ is seen in Fig. 20 to diminish with decreasing β when $\mu^* < \mu$ but to increase with decreasing β when $\mu^* > \mu$.

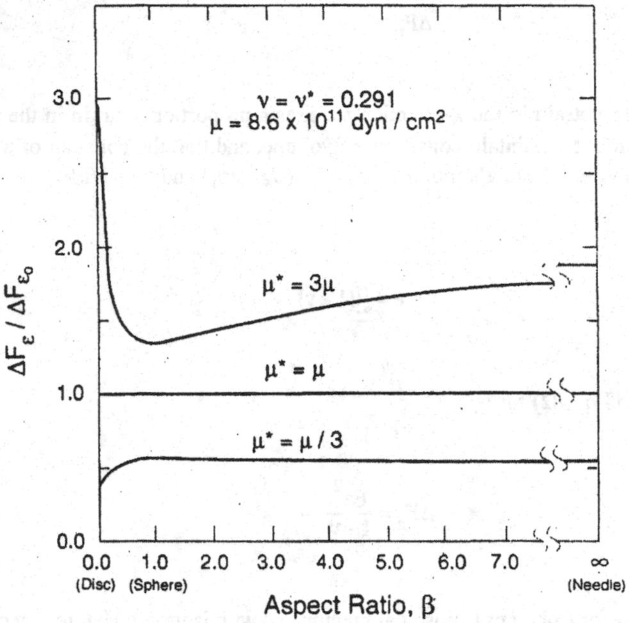

Figure 20: Variation of the volume strain energy, ΔF_ε, of an ellipsoid of revolution in which the elastic constants (μ = shear modulus, ν = Poisson's ratio, * denotes the nucleus phase) are unequal or equal relative to the volume strain energy of ellipsoid in which these constants are the same, $\Delta F_{\varepsilon 0}$, as a function of aspect ratio, β. Barnett et al. (46).

Shear ΔF_ε for Incoherent Ellipsoidal Precipitates with Isotropic Homogeneous Elasticity

It is unlikely that shear (i.e., martensite-like movement of interphase boundaries) plays a mechanistic role in any of the transformations to which the considerations of this paper apply. On the other hand, when a phase transformation is accompanied by a change in crystal structure and this change entails a change in the *shape* of the crystal lattice, then in addition to the volume-change or dilatational strain energy with which previous relationships have dealt there is also likely to be, in a geometric rather than a mechanistic sense, a shear strain energy involved. Even when the elastic constants are the same and are isotropic in both phases, Eshelby (60) has shown that the volume shear strain energy does depend upon the morphology of the product phase even when the elastic constants are homogeneous (unlike the dilatational strain energy: the $\mu = \mu^*$ plot in Fig. 20). For the case of an oblate ellipsoid of revolution (a reasonable approximation of a plate-shaped precipitate) (60):

$$\Delta F_{\varepsilon} = \frac{E\left(\varepsilon_{13}^{T}\right)^{2}}{1+\nu} \cdot \frac{\pi(2-\nu)c}{4(1-\nu)a} \tag{85}$$

where ε_{13}^{T} = stress-free tensor shear strain (equal to one-half the engineering shear). Thus making c/a sufficiently small will greatly reduce ΔF_{ε} due to shear strain. However, the dependence of ΔF^* upon $(\gamma_{\alpha\beta})^3$ but only on $(\Delta F_V + \Delta F_{\varepsilon})^2$ makes strain energy minimization in diffusional phase transformations decided less important than in martensitic transformation. In the latter type, strain energy minimization is doubtless the dominant factor (64).

Combined ΔF_{ε} and $\gamma_{\alpha\beta}$ Minimization

On the basis of the analysis by Barnett et al. (63) of the dilatational strain energy associated with coherent ellipsoids of revolution under the condition of isotropic heterogeneous elasticity, Lee et al. (65) incorporated strain energy as well as interfacial energy in a derivation of ΔF^*. They assumed that nucleation is homogeneous, $\gamma_{\alpha\beta}$ is independent of boundary orientation, isotropic elasticity applies but $\mu^* \neq \mu$ and the volume free energy change and the volume strain energy are additive. The standard free energy change for the formation of an ellipsoidal nucleus under these conditions is written:

$$\Delta F^{\circ} = \tfrac{4}{3}\pi a^{3}\beta\left[\Delta F_{V} + \Delta F_{\varepsilon}\right] + \pi a^{2}\gamma_{\alpha\beta}\left[2 + g(\beta)\right] \tag{86}$$

where:

$$g(\beta) = \frac{2\beta^2}{\sqrt{1-\beta^2}} \tanh^{-1}\sqrt{1-\beta^2} \text{ when } \beta < 1$$

$$g(\beta) = 2 \text{ when } \beta = 1$$

$$g(\beta) = \frac{2\beta}{\sqrt{1-\beta^{-2}}} \sin^{-1}\sqrt{1-\beta^{-2}} \text{ when } \beta > 1 \tag{87}$$

ΔF^* is thus a f(a^*, b^*), where a^* and b^* are semi-axes of the ellipsoid at the critical nucleus size (with $\beta = a^*/b^*$) satisfies simultaneously the conditions $[\partial(\Delta F_o)/\partial a]_\beta = 0$ and $[\partial(\Delta F_o)/\partial\beta]_a = 0$. The first of these differentiations yields:

$$a^* = \frac{\gamma_{\alpha\beta}[2 + g(\beta)]}{2\beta[\Delta F_v + \Delta F_\varepsilon]} \tag{88}$$

However, the differentiation required with respect to β cannot be performed analytically because of the very complex form of the relationship between ΔF_ε and β (63). Accordingly, β^* was numerically evaluated by substituting Eqs. (87) and (88) into Eq. (86) to obtain:

$$\Delta F^* = \frac{\pi\gamma_{\alpha\beta}^3[2 + g(\beta)]^3}{12\beta^2[\Delta F_v + \Delta F_\varepsilon]^2} \tag{89}$$

and this expression for ΔF^* was plotted against β. In order to avoid the need to use absolute values of ΔF_v and $\gamma_{\alpha\beta}$, the ΔF^* of Eq. (89) was divided by ΔF^*_{huo} (Eq. (38), in the absence of strain energy) to obtain:

$$\frac{\Delta F^*}{\Delta F^*_{huo}} = \frac{[2 + g(\beta)]^3}{[8\beta(1 + \{\Delta F_\varepsilon/\Delta F_v\})]^2} \tag{90}$$

Fig. 21 evaluates $\Delta F^*/\Delta F^*_{hu^o}$ as a function of β at five levels of $-\Delta F_\varepsilon/\Delta F_v$.

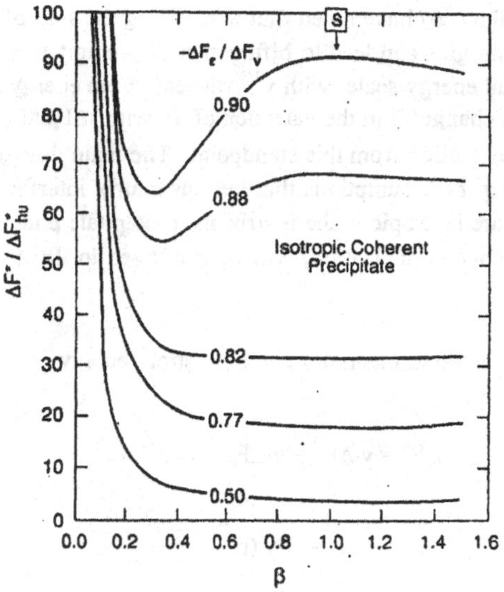

Figure 21: Variation of $\Delta F^*/\Delta F^*_{hu}$ with β at various values of $-\Delta F_\varepsilon/\Delta F_v$ (65).

Until $-\Delta F_\varepsilon/\Delta F_v$ reaches ~ 0.82, $\Delta F^*/\Delta F^*_{hu^o}$ is seen to pass through a minimum at $\beta = 1$, i.e., when the critical nucleus is a sphere. Hence only when $|\Delta F_v|$ is small will ΔF_ε control the critical nucleus shape and make an increasingly oblate ellipsoid the critical nucleus shape. However, when the volume strain energy consumes such a large portion of the volume free energy change, nucleation would probably be quite unlikely to occur under most circumstances.

In further studies, Lee et al. (65) found a similar result (the sphere ceased to be the minimum ΔF* shape when $-\Delta F_\varepsilon/\Delta F_v > 0.81$ for the particular case studied—a Ag precipitate in an Al matrix with a specified orientation relationship) when the elastic constants of the two phases were allowed to differ and anisotropic elasticity was introduced. Similarly when an incoherent nucleus was considered, returning to homogeneous isotropic elasticity, the critical nucleus was no longer spherical at a similar value of this ratio, i.e., $-\Delta F_\varepsilon/\Delta F_v > 0.75$.

Bifurcation of Critical Nucleus Shape Under the Combined Influence of ΔF_ε and $\gamma_{\alpha\beta}$ Minimization

Johnson and Cahn (66) have noted that minimizing the sum of the interfacial and elastic transformation strain energies can lead to bifurcation of precipitate shape during growth. This arises because interfacial energy scales with $v^{2/3}$ whereas strain energy varies as v, the precipitate volume. Hence abrupt change(s) in the variation of β^* with $-\Delta F_\varepsilon/\Delta F_v$ can occur. Lee and Yoo (67) have considered nucleation from this standpoint. The main thread of their treatment will be presented here, retaining the assumptions that nucleus:matrix interfacial energy is isotropic and that the elastic moduli are isotropic in the matrix and precipitate phases. One of their examples, for the case of homogeneous nucleation, will be employed to illustrate the application of this analysis.

Eq. (36) is first rewritten to incorporate volume strain energy:

$$\Delta F^o = v \cdot \Delta F_v + v \cdot \Delta F_\varepsilon + S\gamma_{\alpha\beta} \tag{91}$$

where:

$$\Delta F_\varepsilon = g \cdot w(\beta) \tag{92}$$

and:

$$S = v^{2/3} \cdot s(\beta) \tag{93}$$

in which g is defined as the strain energy "strength", of order $\mu\varepsilon^2$ where ε = misfit strain, S = interfacial area and $w(\beta)$ and $s(\beta)$ are respectively a strain energy function and an interfacial area function. Substituting Eqns. (92) and (93) into Eq. (91), and setting $(\partial(\Delta F)/\partial V)_\beta = 0$ yields:

$$v^{1/3} = \frac{-2\gamma_{\alpha\beta} \cdot s(\beta)}{3[\Delta F_v + g \cdot w(\beta)]} \tag{94}$$

Substituting Eq. (94) into Eq. (91):

$$\Delta F^o = \frac{4[\gamma_{\alpha\beta} \cdot s(\beta)]^3}{27[\Delta F_v + g \cdot w(\beta)]^2} \tag{95}$$

At a given value of ΔF_v, the critical value of the aspect ratio, β^*, is obtained from $\partial(\Delta F)/\partial\beta = 0$. Performing this operation on Eq. (95):

$$3\frac{d(s(\beta))}{d\beta}\left[\frac{\Delta F_v}{g}+w(\beta)\right]-2\cdot s(\beta)\frac{\partial(w(\beta))}{\partial\beta}=0 \tag{96}$$

Bifurcation arises because there are multiple solutions to Eq. (96). The critical aspect ratio, β^*, must satisfy the condition $\partial^2(\Delta F)/\partial\beta^2 > 0$ since the corresponding ΔF represents a minimum in the ΔF vs. β "surface". When β^* is utilized in Eqns. (94) and (95),

$$v^* = -\left(\frac{2\gamma_{\alpha\beta}\cdot s(\beta)}{3[\Delta F_v + g\cdot w(\beta)]}\right)^3 \tag{97}$$

$$\Delta F^* = \frac{4[\gamma_{\alpha\beta}\cdot s(\beta)]^3}{27[\Delta F_v + g\cdot w(\beta)]^2} \tag{98}$$

Turning attention now to Eq. (92), for both dilatational and shear strain energies associated with both coherent and incoherent critical nuclei, Lee and Yoo (52) concluded from consideration of the literature that when $0 \leq \beta \leq 1$, $n \approx 3$ in the relationship:

$$w(\beta) = 1 - (1-\beta)^n \tag{99}$$

The analysis thus developed will now be applied to the homogeneous nucleation of cuboidal prisms, also termed rectangular parallelepipeds. Despite the use of this shape, the assumption that γ is isotropic is retained. The dimensions of such a prism are 2a, 2a and $2a\beta$. Thus $v = 8\beta a^3$ and $S = 8a^2(1+2\beta)$. From Eq. (83),

$$s(\beta) = 2(1+2\beta)\beta^{-2/3} \tag{100}$$

Substituting Eqns. (89) and (90) into Eq. (91):

$$\Delta F = v\cdot\Delta F_v + gv[1-(1-\beta)^n] + 2\gamma_{\alpha\beta}(1+2\beta)(v/\beta)^{2/3} \tag{101}$$

Applying the foregoing relationships for cuboidal prisms to Eqns. (96) - (100),

$$(1 - \beta)[(\Delta F_v/g) + 1 - (1 - \beta)^n] + n\beta(1 + 2\beta)(1 - \beta)^{n-1} = 0 \quad (102)$$

$$v^* = -\left(\frac{2\gamma_{\alpha\beta}\cdot s(\beta^*)}{3[\Delta F_v + g\cdot w(\beta^*)]}\right)^3 \quad (103)$$

$$\Delta F^* = \frac{32[\gamma_{\alpha\beta}(1 + 2\beta^*)]^3}{27\{\beta^*\langle\Delta F_v + g[1 - (1 - \beta^*)^n]\rangle\}^2} \quad (104)$$

When n = 3, Eq. (102) becomes:

$$(1 - \beta)[(\Delta F_v/g) + 6\beta - 5\beta^3] = 0 \quad (105)$$

The three accessible solutions to Eq. (105) are:

$$\beta_1^* = 1 \quad (106)$$

$$\beta_2^* = (1.6)^{1/2} \cos\{(4\pi/3) + (1/3)\cos^{-1}[\Delta F_v/((6.4)^{1/2}g)]\} \quad (107)$$

$$\beta_3^* = (1.6)^{1/2} \cos^{-1} \{(1/3) \cos^{-1} [\Delta F_v/((6.4)^{1/2}g)]\} \quad (108)$$

These three solutions are plotted as a function of $-\Delta F_v/g$ in Fig. 22. Whereas β_1^* and β_3^* represent local minima, β_2^* describes a local maximum. β_3^* is delineated by the heavy horizontal

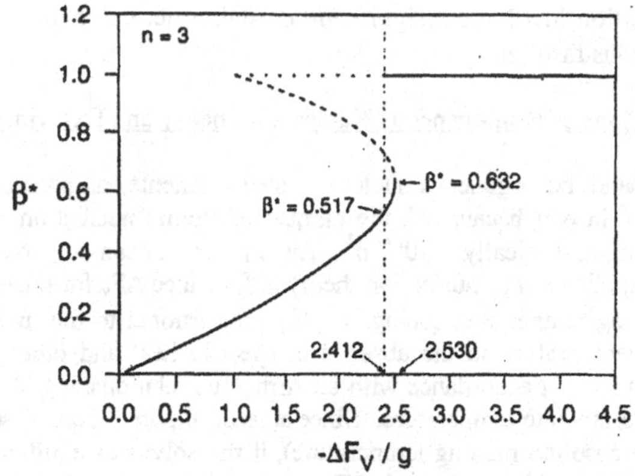

Figure 22: Homogeneous nucleation bifurcation diagram for cuboidal prism nucleus when n = 3. Lee and Yoo (67).

line, β_1^* is plotted as the heavy curve in this figure* and β_2^* is shown by the dashed curve. From the variation of ΔF* with $\Delta F_v/g$ for the three β*'s (which is not shown here), it is found that when $-\Delta F_v/g \le 2.412$, ΔF* is smaller when β* of the critical nuclei follows the solid β_1^* curve. When $-\Delta F_v/g \ge 2.412$, ΔF* is minimized when β* switches to the β_3^* horizontal line. Hence β* abruptly increases from 0.517 to 1 under this circumstance. Thus β* = 0.517 and $-\Delta F_v/g = 2.412$ define the bifurcation point in this system. As Johnson and Cahn (65) previously recognized, this discontinuity resembles a first order phase transformation.

Comparisons of Theory and Experiment

Introduction

Comparisons between the predictions of nucleation theory and experiment are best conducted when nucleation takes place homogeneously, since heterogeneous nucleation can involve complex interactions of the critical nuclei with the structure and chemistry of the nucleation site whose quantitative description is not readily accomplished. Additionally, determination of the critical nucleus shape has yet to be performed reliably for any diffusional transformation involving a change in crystal structure, even though this shape usually has by far the most powerful influence upon nucleation kinetics. Also, it has yet to be established that any

---

The thin solid curve, on which the maximum β = 0.632, is derived from β_3^*.

diffusional transformation involving a significant crystal structure change (e.g., fcc↔hcp) can nucleate in homogeneous fashion.

Quantitative Comparisons of Homogeneous Nucleation Theory and Experiment

On the other hand, homogeneous nucleation measurements can pose difficult experimental problems of their own, in part because N, the number of atomic nucleation sites per unit volume (Eqs. (1,34,35)) is so high, typically $\sim 10^{29}/m^3$. These problems can be solved, however, through the following simple application of nucleation theory (68). Since ΔF_v for nucleation becomes more negative with decreasing temperature (being roughly proportional to the undercooling below the relevant equilibrium temperature in the absence of specific heat and other anomalies) while D diminishes rapidly (usually in accordance with an Arrhenius relationship), J* must pass through a maximum at some intermediate temperature. Since at least in pure metals D scales well with T/T_m (69) (where T_m is the absolute melting temperature), if the solvus or equilibrium temperature for the transformation lies at a high value of T/T_m, the maximum J* is likely to be so high that transformation during cooling to and/or from the intended isothermal reaction temperature will be very difficult to prevent and transformation at this temperature will be too rapid to study in specimens quenched to room temperature. On the other hand, if the solvus temperature is so low that D is negligible even at moderate undercoolings, the kinetics of nucleation may be almost undetectably slow. Hence at some intermediate alloy composition the maximum in the nucleation rate will occur at a temperature where nucleation kinetics can be conveniently determined without requiring extraordinary precautions to avoid anisothermal transformation. As illustrated in Figure 23(a), if the slope of the solvus curve is relatively shallow, the "nucleation window" for making such measurements will encompass an appreciable range of composition. On the other hand, when the solvus curve has a relatively steep slope, e.g., Fig. 23(b), for practical purposes there may be only one alloy composition which furnishes a "nucleation window" within which kinetic measurements can be made. Nucleation kinetics in dilute Cu-Co alloys (70) provide a fairly good case of Fig. 23(a) behavior whereas those in an Al-0.11 A/O Sc alloy (71) represent a clear example of the situation described by Fig. 23(b).

Measurements of homogeneous nucleation kinetics and comparison of the results with nucleation theory have been made for vapor→liquid, liquid→liquid and solid→solid transformations. Excellent reviews of the first two types (72,73) and of the third type (5) have been published. Measurement of nucleation kinetics in the first two milieu has posed an exaggerated form of the type of difficulty just considered for the third case as a result of the orders of magnitude higher diffusivities in liquids and the still higher diffusivities in gases. Hence the view that many of the measurements made in the first two milieu contradicted nucleation theory was finally traced to the realization that the so-called "cloud point" (occurring at the supersaturation where the transparent or translucent matrix medium abruptly turned cloudy), by which the onset of rapid nucleation had been judged, is actually a composite of nucleation,

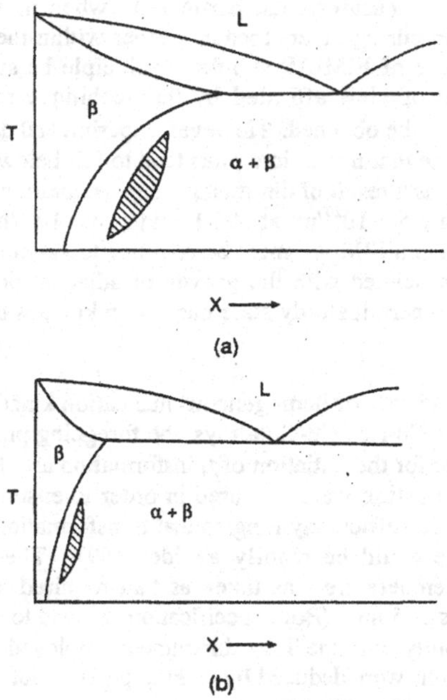

Figure 23: The "Nucleation Window" as a function of the slope of the solvus (68).

growth and coarsening (73,74). Such measurements were thus recognized as being a useful reflection of nucleation kinetics only at very small undercoolings (74). In a review of experimental studies on homogeneous nucleation within solids it was noted that most of these data had been reported where the coarsening regime had at least been entered, and in some cases had entirely encompassed the temperature-time range in which measurements were made (68).

Before considering a set of experimental data in which these problems were successfully avoided, it is appropriate to comment briefly upon experimental techniques used for nucleation kinetics measurements. A physical property measurement known to be sensitive to the amount of precipitate formed or to the concentration of solute remaining in solid solution in the matrix phase, e.g., electrical resistivity measured as a function of isothermal reaction or aging time, readily produces large amounts of data. However, a multi-layer analysis is required to convert data on resistivity to particle number density (n_v) vs. isothermal reaction time, e.g., (75), and at small particle sizes there is evidence that the connection between the solute concentration remaining in the matrix and the resistivity is non-linear (76). Small angle neutron scattering (SANS) has been used to advantage (77), but again requires a number of assumptions to derive data on n_v. TEM, both in conventional form and as HRTEM, is probably the most generally

useful technique for such measurements, particularly when n_v is small enough so that the probability of particles obscuring others located deeper within the foil is minimized. The still higher resolution technique of FIM-AP should in principle be still more useful. In practice, however, the small field of view afforded by this technique requires that $n_v \geq 10^{23}/m^3$ if statistically valid data are to be obtained. However, experimental data on nucleation kinetics in a Cu-1 A/O Co alloy indicate that n_v vs. time plots tend to fall below linearity (denoting departure from steady state kinetics as a result of diminishing supersaturation of the untransformed matrix) at number densities of only $5 \times 10^{20}/m^3$ at 893 K and at $8 \times 10^{21}/m^3$ at 873 K (Fig. 25). Hence the acquisition of J* data with FIM-AP must be confined to systems in which significant overlap of the diffusion fields associated with the growth of adjacent precipitates does not occur for sufficient time intervals to permit steady state nucleation kinetics to be established at n_v's of the order of $10^{23}/m^3$.

In an experimental study of homogeneous nucleation kinetics of fcc Co-rich precipitates in the fcc matrix of dilute Cu-rich Cu-Co alloys, the foregoing problems were circumvented by calculating the TTT-curve for the initiation of transformation as a function of A/O Co before the alloys used for the investigation were procured in order to ensure that the "nose" of the TTT-curve fell at a reaction time sufficiently long so that transformation during quenching to or from the reaction temperature could be readily avoided (69). The time for the "initiation" of transformation at each temperature was taken as that required to reach $n_v = 10^{17}/m^3$ and an average precipitate radius of 5 nm. (Both specifications proved to be pessimistic estimates of the conditions for easy visibility with the TEM instrument employed.) The critical nucleus shapes central to the J* calculations were deduced by making polar γ-plots of γ_c calculated from Eq. (75) as a function of T/T_c. The Wulff construction was then performed on these plots. Representative results are shown in Fig. 24. At 0°K (Fig. 24 (a)), the particles were completely faceted, with {111} facets predominating and smaller {100} facets enclosing the remainder of the nuclei. With increasing temperature, both facets shrank, the {100} facets disappearing first and the {111} facets vanishing not far above $T/T_c = 0.50$.

The predictions of transformation kinetics thus made proved to be a suitable guide for the experimental measurements of n_v. As illustrated in Fig. 25 for a Cu-1 A/O Co alloy, within the usual scatter accompanying such measurements this quantity increased linearly with time over a long enough interval of isothermal reaction time at each of the reaction temperatures studied to permit the steady state nucleation rate (simply the slope of the linear regions) to be determined.

Fig. 26 compares the measured nucleation rates in Cu-1% Co with those calculated from classical theory and the DLP model, from the Cahn-Hilliard (CH) (4) continuum non-classical model and from the Cook-DeFontaine (33) discrete lattice point form of the continuum non-

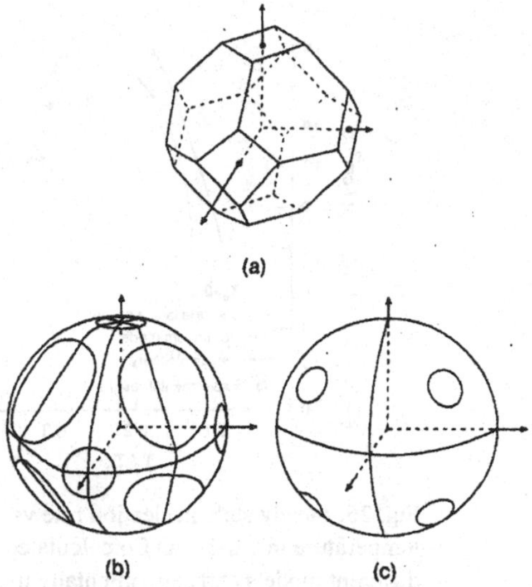

Figure 24: Critical nucleus shape in Cu-Co at: (a) $T/T_c = 0$, (b) $T/T_c = 0.25$, (c) $T/T_c = 0.50$ (78).

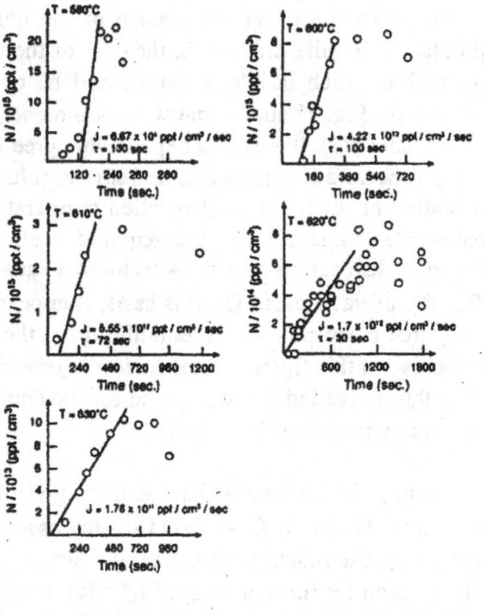

Figure 25: Particle number density, n_v, vs. reaction time at 5 temperatures in Cu-1 A/O Co (70).

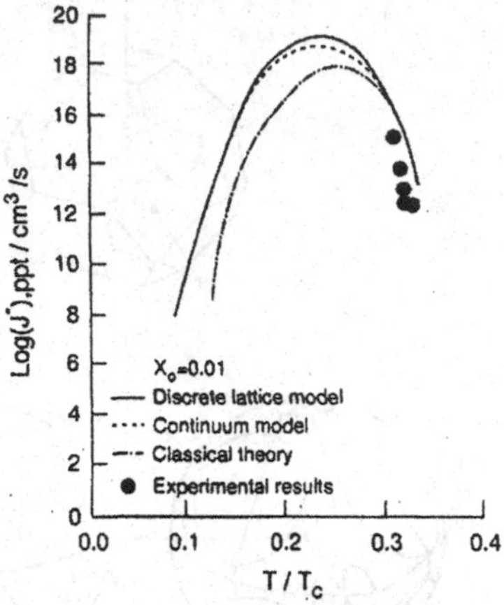

Fig. 26: Steady state nucleation rate vs. reduced temperature in Cu-1 A/O Co calculated from three different models and experimentally measured (70).

classical model. These comparisons were made in the absence of strain energy because incorporation of this factor is quite difficult in the case of the CH theory. Note that within the temperature region within which usable n_v data could be obtained, the three theories yield essentially identical results. Fig. 27 shows that when strain energy is taken into account through the Cook-DeFontaine model (33), theory and experiment agree very well when the assumption is made that the vacancy concentration quenched in from the solution annealing temperature is still present during nucleation at the isothermal reaction temperature (70). Russell (79) suggests, however, that a reasonable estimate of the time required to eliminate excess vacancies is simply l^2/D_v, where l is the inter-dislocation spacing ≈ reciprocal square root of the dislocation density and D_v is the diffusivity of vacancies. On this basis, it appears that the vacancy concentration present was probably that corresponding to equilibrium at the isothermal reaction temperature (68), i.e., the dashed curve in this figure. In the three alloys studied, 0.5, 0.8 and 1.0 A/O Co, the experimental J_s^* data thus exceeded that calculated for the equilibrium vacancy concentration at the reaction temperature by factors of 10^2 - 10^6.

On the other hand, Fig. 28 shows that the measured incubation time, τ, is in accurate agreement with that calculated in the Cu-1 A/O Co alloy assuming that the equilibrium vacancy concentration is present at the reaction temperature. Similar results were obtained in the two lower Co alloys. The reason for the matching of J_s^* with the quenched-in vacancy concentration and of τ with the equilibrium vacancy concentration at the reaction temperature is probably

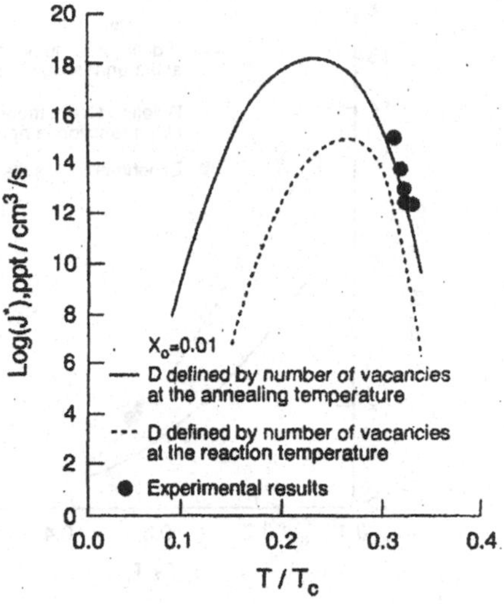

Fig. 27: Comparison of experimental data on J_s* in Cu-1 A/O Co with calculated values in the presence (—) and absence (---) of quenched-in vacancies (70).

associated with the circumstance that $\Delta F^* \propto \gamma_{\alpha\beta}^3$ (Eq. (38)), whereas $\tau \propto \gamma_{\alpha\beta}$ (Eq. (67)). Hence a small error in the calculation of $\gamma_{\alpha\beta}$ exerts far more influence upon J_s* than upon τ. Calculation of $\gamma_{\alpha\beta}$ was performed by the nearest neighbor DLP method (Eqn. (75)). However, the observation by FIM-AP (79) and SANS (e.g., refs, 81, 82) that there are marked fluctuations in composition prior to nucleation cannot be explained by such short-range interactions (83-85). Hence a much larger value of the Cahn-Hilliard interaction parameter λ (Eq. (76)) must be operative, requiring a different formulation of the relationship for $\gamma_{\alpha\beta}$. These considerations also indicate that comparison between theory and experiment for homogeneous nucleation theory is more reliably based upon τ than upon J_s*.

Quantitative Comparison of the Theory of Nucleation at Dislocations with Experiment

The Larche (49) theory of coherent nucleation at edge dislocations (Eqns. (55) and (56)) has recently been quantitatively tested by Wang and Shiflet (86) in an Al-8.28 A/O Li alloy, taking advantage of the circumstance that the $L1_0$ transition phase, δ', has been shown by Cassada et al. (87), and later more rigorously confirmed by Wang and Shiflet (86), to nucleate

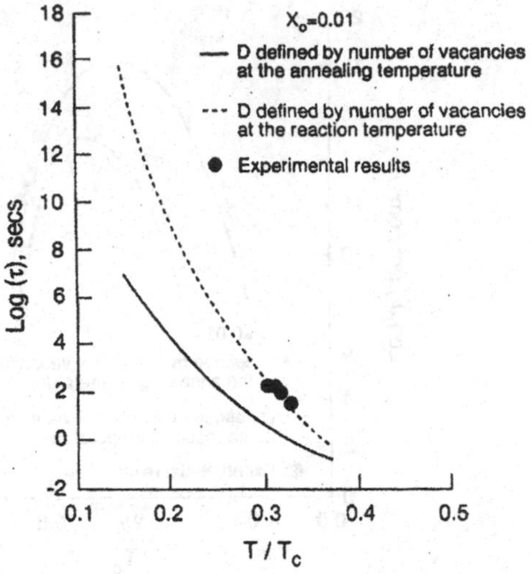

Figure 28: Incubation time, τ, as a function of reduced temperature in Cu-1 A/O Co, calculated and experimentally measured (70).

with pronounced preference at edge dislocations. In agreement with theory, δ' was also shown by these workers to nucleate on the compressive or half-plane side of edge dislocations, consistently with the negative misfit strain associated with this precipitate.

Eqs. (55) and (56) were used to compute ΔF^*. To adapt the other components of the homogeneous nucleation rate equation to the Larche theory, Wang and Shiflet (86) also modified them only in respect of replacing γ with $\gamma - \gamma_{eff}$. They measured the particle number density per unit length of edge dislocation line. A typical data plot is shown in Fig. 29. As in the homogeneous nucleation case (Fig. 24), steady state nucleation kinetics appeared to develop immediately. The fall-off in number density at later reaction times is again caused by coarsening. J_s^* was measured at five temperatures from 483° to 523°K. From the Larche modification of nucleation theory, the nucleus:matrix interfacial energy was back-calculated from the measured nucleation kinetics and found to decrease from 16 mJ/m^2 at 483 K to 8 mJ/m^2 at 523 K. The temperature dependence of γ thus determined was in good agreement with that calculated from Cahn-Hilliard (28) continuum theory. Also, these results agree well with the value of 14 mJ/m^2 obtained at 473 K by Baumann and Williams (88) from the Gibbs-Thompson equation and TEM measurements of the smallest stable particle size at this temperature. These results thus provide

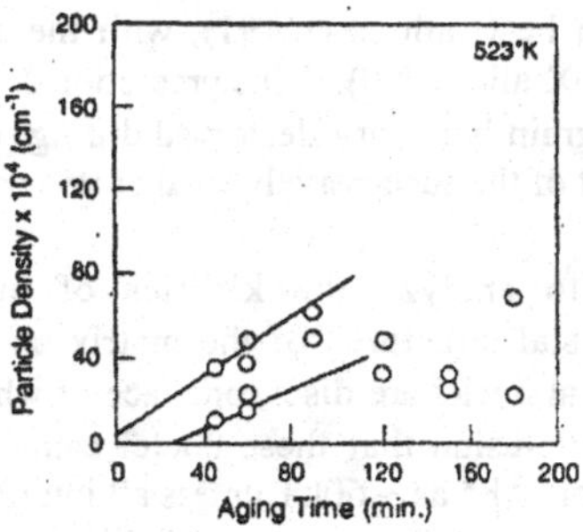

Figure 29: Number density of δ′ precipitates vs. aging time in Al-A/O 8.28 Li. Wang and Shiflet (86).

further support for the essential correctness of the theory of solid→solid nucleation.

<u>Qualitative Comparisons of Nucleation Theory with Experiment</u>

In addition to quantitative support for nucleation theory, there are also a number of observations which provide important qualitative affirmation for aspects of the theory which cannot yet be treated analytically with sufficient accuracy. Most of the problems arising in quantitatively connecting theory and experiment are associated with the shape of critical nuclei and the energy of the various interfaces comprising them in situations where the crystal structures of the matrix and precipitate phases are distinctly different.

(i) Homogeneous nucleation kinetics have so far been confirmed only when the crystal structure of the matrix and product phase are either identical or the product is a long-range ordered version of the matrix (68). In these situations it should be feasible to have full coherency at all boundary orientations of the critical nuclei, thus limiting the energy of nucleus:matrix boundaries to the chemical component, i.e., to differences in the chemical identity of atoms at equivalent sites on opposite sides of the interphase boundary. Hence these interfaces should have much lower energies than their counterparts in transformations where the change in crystal structure is sufficient to produce changes in stacking sequence across the nucleus:matrix boundaries. The result is in qualitative accord with $\Delta F^* \propto \gamma_{\alpha\beta}^3$.

(ii) The prediction that nucleation at grain boundaries, particularly at relatively small undercoolings below the solvus temperature, is most rapid at grain corners and then successively less so at grain edges and grain faces (24) has been qualitatively confirmed for the proeutectoid ferrite reaction in Fe-C alloys (89-91), with the edge vs. face differences also having been confirmed in Fe-C-X alloys (90). This prediction follows from the more important influence of the larger area of grain boundary destroyed during progression along the sequence faces→edges→corners than that of the successively smaller values of N applicable (24).

(iii) Attempts to analyze the kinetics of nucleation at grain faces in transformations where the crystal structures of the matrix and precipitate are significantly different by assuming that critical nuclei are discs, one face of which is coplanar with grain face boundaries, have led to the conclusion that these nuclei cannot fulfill the Cahn-Hilliard (4) evaluation of the maximum viable ΔF^* as ~ 60kT unless all interfaces on them are of low energy type, i.e., are fully or partially coherent (92-96). This deduction has been supported experimentally by the TEM finding that the broad face of grain boundary allotriomorphs nucleated at grain faces which are irrationally oriented with respect to the matrix grain they abut are nonetheless partially coherent (97,98).

(iv) Although ΔF_v for the nucleation of transition phases is normally less negative than that for the equilibrium phase in a given alloy, the formation of transition phases in many alloy systems is consistent with the general experimental observation that the matching of the transition phase lattice to that of the matrix phase tends strongly to be much better than that of the equilibrium phase (100). Thus the interfacial energies associated with transition phase nuclei should be lower that those of equilibrium phase nuclei. This observation supports again the greater importance of γ relative to ΔF_v in the context of ΔF^* as described, for example, by Eq. (38).

(v) Reproducible lattice orientation relationships between precipitate and matrix phases nucleated intragranularly, which usually correspond to well matched pairs of conjugate planes and directions in the two lattices, are consistent with the reduced ΔF^* for the lower $\gamma_{\alpha\beta}$ values associated with faceted critical nucleus shapes, e.g., Eq. (49). Deviations from such relationships when nucleation takes place at grain boundaries (97) are presumed to be associated with the need to take up compromise orientation relationships in order to minimize the total interfacial energy of critical nuclei under these circumstances.

(vi) Even though the Cahn (45) theory for nucleation at dislocations was formulated (for the sake of generality) for incoherent nuclei, this theory nonetheless predicts a number of important experimental observations concerning such nucleation.* The central result of this theory may be written for edge dislocations as:

---

*The Larche (49) theory for coherent nucleation at dislocations is not used here because it does not include an analysis for nucleation at screw dislocations.

$$\frac{\Delta F_d^*}{\Delta F_{hu}^*} \approx 1 + \frac{\mu b^2 \cdot \Delta F_v}{2\pi^2 \gamma_{\alpha\beta}^2 (1 - \nu)} \tag{109}$$

and for screw dislocations as:

$$\frac{\Delta F_d^*}{\Delta F_{hu}^*} \approx 1 + \frac{\mu b^2 \cdot \Delta F_v}{2\pi^2 \gamma_{\alpha\beta}^2} \tag{110}$$

where ΔF_d^* is the work of critical nucleus formation at a dislocation, ΔF_{hu}^* = work for homogeneous nucleation (here Eq. (38)) and **b** = Burgers vector of the dislocation at which nucleation occurs; ν and μ are taken to be the same in both phases. In agreement with experiment, these equations predict that at a given ΔF_v nucleation occurs more readily at edge than at screw dislocations (86,87,100,101), and also at dislocations with the longest Burgers vector (101).

The preference for only one of several crystallographically equivalent habit planes exhibited by precipitate plates nucleated at dislocations (102,103) can also be understood on this general basis. This preference is for the particular habit plane whose "misfit vector"--taken as orthogonal to the habit plane--makes the smallest angle with respect to the Burgers vector of the dislocation at which nucleation occurs (103,104). This geometry makes the maximum use of the dislocation **b** by permitting the largest reduction in dilatational transformation strain energy. The absence of nucleation at dislocations at small undercoolings when misfit between the precipitate and the matrix phase is small (105) is consistent with the hindrance offered by the extra free volume associated (particularly with edge) dislocations under this circumstance.

Nucleation occurring more readily at isolated dislocations than at those grouped in sub-boundaries (106,107) evidently results from the partial mutual neutralization of strain fields of dislocations comprising well formed sub-boundaries (108). The readier occurrence of precipitate nucleation at jogs in dislocations (109) and at dislocation nodes (101,107,110) than at straight dislocations may be due in good part simply to the larger length of dislocation likely to be in contact with critical nuclei under these circumstances.

Concluding Remarks

Despite doubts often expressed, as reviewed in ref. (35), the theory of diffusional solid→solid nucleation has good predictive capability. However, the theory is presently useful in a quantitative sense only for phase transformations in which the crystal structure of the precipitate is the same as that of the matrix. The extension of the calculation of a polar γ-plot of coherent interphase boundary energy and thus of critical nucleus shape to the case where one phase is a long-range ordered version of the other (but their spatial orientations are the same) is in principle readily accomplished. However, calculating this energy as a function of boundary orientation has yet to be accomplished when the crystal structures of the matrix and product phases differ sufficiently so that there is a change in stacking sequence across interphase boundaries; this accomplishment is now badly needed. With the advent of the position-sensitive atom probe (111), however, it may now be possible to determine critical nucleus shapes experimentally for this situation. Unfortunately, the inability of such transformations to take place homogeneously--at least under conditions so far investigated--requires that these shapes be determined for particular nucleation sites. The work of Wang and Shiflet (86) on nucleation at edge dislocations suggests, though, the possibility that these might be particularly simple sites at which to begin conducting such experiments, particularly in the sense that the shapes of critical nuclei may not be significantly changed by the dislocation strain field (though the growth shape is drastically altered thereby (86)).

Acknowledgments

Appreciation is expressed to the National Science Foundation for supporting preparation of the original version of this paper and to Prof. K. C. Russell (M.I.T.) for valuable discussions of that manuscript as well as numerous individual topics in nucleation theory over a period of many years.

References

1. J. W. Gibbs. On the Equilibrium of Heterogeneous Substances, Collected Works, Vol. I, Longmans Green and Co., New York, 1928.
2. A. C. Zettlemoyer (ed.): Nucleation, Marcel Dekker, New York, 1969.
3. J. W. Cahn: Trans. Met. Soc. AIME, 1968, vol. 242, p. 166.
4. J. W. Cahn and J. E. Hilliard: Jnl. Chem. Phys., 1959, vol. 31, p. 539.
5. K. C. Russell: Advances in Colloid and Interface Science, 1980, vol. 132, p. 205; Phase Transformations, p. 219, ASM, Materials Park, OH, 1970.
6. L. Farkas: Zeit. Physik. Chem., 1927, vol. 125, p. 239.
7. R. Becker and W. Doering: Ann. Physik, 1935, vol. 24, p. 719.
8. J. B. Zeldovich: Acta Physicochim, URSS, 1943, vol. 18, p. 1.
9. L. Tisza and M. I. Manning: Phys. Rev., 1957, vol. 105, p. 1695
10. H. Wakeshima: J. Chem. Phys., 1954, vol. 22, p. 1614;
J. Phys. Soc. Japan, 1954, vol. 9, p. 400; 1955, vol. 10, p. 374.
11. H. Wakeshima and T. Takata: Jap. J. App. Phys., 1963, vol. 2, p. 792.
12. A. Kantrowitz: J. Chem. Phys., 1951, vol. 19, p. 1097.

13. F. C. Collins: Zeit. Elektrochem, 1955, vol. 59, p. 404.
14. J. F. Frisch J. Chem. Phys., 1957, vol. 27, p. 90.
15. R. F. Probstein: J. Chem. Phys., 1951, vol.19, p. 619.
16. B. K. Chakraverty: Absorption at Croissance Cristalline, p. 375. Colloques Internationaux du C.N.R.S., Paris, 1965.
17. J. Feder, K. C. Russell, J. Lothe and G. M. Pound, Advan. Phys., 1966, vol. 15, p. 111.
18. F. F. Abraham: J. Chem. Phys., 1969, vol. 51, p. 1632.
19. K. C. Russell: Acta Metall., 1968, vol. 16, p. 761.
20. P. G. Shewmon: Diffusion in Solids, p. 42, McCraw-Hill, New York, 1963.
21. H. I. Aaronson: Phase Transformations, p. II-1, Chameleon Press, London, 1979.
22. K. C. Russell: Acta Metall., 1969, vol. 17, p. 1123.
23. L. S. Darken: Trans. AIME, 1948, vol. 175, p. 184.
24. J. W. Cahn: Acta Metall., 1955, vol. 4, p. 449.
25. G. Wulff: Zeit. Kristallog., 1901, vol. 34, p. 449.
26. C. Herring: Structure and Properties of Solid Surfaces, p. 5, Univ. of Chicago Press, 1953.
27. J. K. Lee and H. I. Aaronson: Scripta Met., 1974, vol. 8, p. 1451.
28. J. W. Cahn and J. E. Hilliard: Jnl. Chem. Phys., 1958, vol. 28, p. 258.
29. J. W. Cahn: Acta Metall., 1962, vol. 10, p. 907.
30. J. W. Cahn: Acta Metall., 1961, vol. 9, 795.
31. J. W. Cahn: The Mechanism of Phase Transformations in Crystalline Solids, p. 1, Inst. of Metals, London, 1969.
32. H. E. Cook, D. deFontaine and J. E. Hilliard: Acta Metall., 1969, vol. 17, p. 765.
33. H. E. Cook and D. deFontaine: Acta Metall., 1969, vol. 17, p. 915; 1971, vol. 19, p. 607.
34. F. K. LeGoues, H. I. Aaronson and Y. W. Lee: Acta Metall., 1984, vol. 32, p. 1845.
35. H. I. Aaronson and K. C. Russell: Proc. of an Int. Conf. on Solid→Solid Phase Transformations, p. 371, TMS, Warrendale, PA, 1982.
36. Raju V. Ramanujan, J. K. Lee, F. K. LeGoues and H. I. Aaronson: Acta Metall., 1989, vol. 37, p. 3051.
37. W. C. Johnson, C.L. White, P. E. Marth, P. K. Ruf, S. H. Tuorninen, K. D. Wade, K. C. Russell and H. I. Aaronson: Met. Trans., 1975, vol. 6A, p. 911.
38. S. M. Selby: Standard Mathematical Tables, pp. 12, 17, The Chemical Rubber Co., Cleveland, OH, 1973.
39. J. W. Cahn and D. W. Hoffman: Acta Metall., 1974, vol. 22, p. 1205.
40. J. K. Lee and H. I. Aaronson: Acta Metall., 1975, vol. 23, p. 799.
41. W. L. Winterbottom: Acta Metall., 1967, vol. 15, p. 303.
42. J. K. Lee and H. I. Aaronson: Acta Metall., 1975, vol.23, p. 809.
43. R. B. Nicholson: Phase Transformations, p 269, ASM, Metals Park, OH, 1970.
44. H. B. Aaron and H. I. Aaronson: Met. Trans., 1971, vol. 2, p. 23.
45. J. W. Cahn: Acta Metall., 1957, vol. 5, p. 169.
46. R. Gomez-Ramirez and G. M. Pound: Met. Trans., 1973, vol. 4, p. 1563.
47. C. C. Dollins: Acta Metall., 1970, vol. 18, p. 1209.
48. B. Ya. Lyubov and V. A. Solovyev: Phys. Met. Metallog., 1965, vol. 19, no. 3, p. 13.
49. F. C. Larche: Dislocations in Solids, ed. by F. R. N. Nabarro, North-Holland Publishing Co., Amsterdam, vol. 4, p. 135.
50. H. I. Aaronson, C. Laird and K. P. Kinsman: Phase Transformations, p. 313, ASM, Materials Park, OH, 1970.

51. H. I. Aaronson: Decomposition of Austenite by Diffusional Processes, p. 387, Interscience, New York, 1962.
52. D. Tumbull: Impurities and Imperfections, p. 121, ASM, Materials Park, OH, 1955.
53. Y. W. Lee and H. I. Aaronson: Acta Metall., 1980, vol. 28, p. 539.
54. P. P. Wynblatt and R. Ku: Surface Science, 1977, vol. 65, p. 5511.
55. R. Becker: Ann. Phys.1938, vol. 32, p. 128.
56. F. R. N. Nabarro: Proc. Roy. Soc. A, 1940, vol. 175, p. 519.
57. E. Kroner: Acta Metall., 1954, vol. 2, p. 302.
58. K. C. Russell: Scripta Metall., 1969, vol. 3, p. 313.
59. F. Laszlo: Jnl. Iron Steel Inst., 1950, vol. 5, p. 164.
60. J. D. Eshelby: Proc. Roy. Soc A., 1957, vol. 241, p. 376
61. J. D. Eshelby: Prog. in Solid Mechanics, 1961, vol. 2, p. 289.
62. D. M. Barnett: Scripta Metall., 1971, vol. 5, p. 261.
63. D. M. Barnett, J. K. Lee, H. I. Aaronson and K. C. Russell: Scripta Metall., 1974, vol. 8, p.1447.
64. J. W. Christian: The Theory of Phase Transformations in Metals and Alloys, p. 802, Pegamon Press, Oxford, 1965.
65. J. K. Lee, D. M. Barnett and H. I. Aaronson: Metall. Trans., 1977, vol. 8A, p. 963.
66. W. C. Johnson and J. W. Cahn: Acta Metall., 1984, vol. 32, p. 1925.
67. J. K. Lee and M. H. Yoo: Metall. Trans., 1992, vol. 23A, p. 1891.
68. H. I. Aaronson and F. K. LeGoues, Metall. Trans., 1992, vol. 23A, p. 1915.
69. P. G. Shewmon: Diffusion in Solids, 2nd Ed., p. 86, TMS, Warrendale, PA, 1989.
70. F. K. LeGoues and H. I. Aaronson: Acta Metall., 1984, vol. 32, p. 1855.
71. R. W. Hyland, Jr., Metall. Trans., 1991, vol. 23A, p. 1947.
72. C. M. Knobler: Decomposition of Alloys: The Early Stages, p. 55, P. Haasen. V. Gerold. R. Wagner and M. F. Ashby, ed., Pergamon Press, N.Y., 1984.
73. K. Kinder and D. Stauffer: Adv. in Physics, 1976, vol. 25, p. 343.
74. J. S. Langer and A. J. Schwartz: Phys. Rev. Lett., 1980, vol. 21, p. 948.
75. I. Servi and D. Turnbull: Acta Metall., 1966, vol. 14, p. 161.
76. A. J. Hillel, J. T. Edwards and P. Wilkes: Phil. Mag., 1980, vol. 21, p. 948.
77. R. Kampmann and R. Wagner: Atomic Transport and Defects in Metals by Neutron Scattering, p. 73, Springer-Verlag, Berlin, 1986.
78. F. K. LeGoues, H. I. Aaronson, Y. W. Lee and G. J. Fix: Proc. of an Int. Conf. on Solid→Solid Phase Transformations, p. 427, TMS, Warrendale, PA, 1982.
79. K. C. Russell: priv. comm., M.I.T., 1992.
80. X. Xiang, W. Wagner and H. Wollenberger: Z. Metallkunde, 1991. vol. 92, p. 192.
81. M. F. Chisholm and D. E. Laughlin: Phase Transformations '87, G. W. Lorimer, ed., Inst. of Metals, London, 1988, p. 1.
82. W. Wagner and W. Petry: Phys. B, 1989, vols. 156-157, p. 65.
83. P. C. Clapp and S. C. Moss: Phys. Rev., 1966, vol. 142, p. 418.
84. M. Kalos, J. L. Leibowitz, O. Penrose and A. Sur: J. Stat. Phys., 1978, vol. 18, p. 39.
85. K. Binder and M. H. Kalso: J. Stat. Phys., 1980, vol. 22, p. 363.
86. Z. M. Wang and G. J. Shiflet: Metall. Trans., 1996, vol. 27A, p. 1599.
87. W. A. Cassada, G. J., Shiflet and W. A. Jesser: Acta Mater. Metall., 1992, vol. 40, p. 2101.
88. S. F. Baumann and D. B. Williams: Scripta Metall., 1984, vol. 18,p. 611.
89. W. F. Lange III and H. I. Aaronson: Metall. Trans., 1979, vol. 10A, p. 1951.

90. M. Enomoto, W. F. Lange III and H. I. Aaronson: Metall. Trans., 1986, vol. 17A, p. 1399.
91. Weiming Huang and M. Hillert: Metall. Trans., 1996, vol. 27A, p. 480.
92. W. F. Lange III, M. Enomoto and H. I. Aaronson: Metall. Trans., 1988, vol. 19A, p. 427.
93. M. Enomoto and H. I. Aaronson: Metal. Trans., 1986, vol. 17A, 1385.
94. T. Tanaka, H. I. Aaronson and M. Enomoto: Metall. Trans., 1995, vol. 26A, 547.
95. M. R. Plichta, J. H. Perepezko, H. I. Aaronson and W. F. Lange III: Acta Metall., 1980, vol. 28, 1031.
96. E. S. K. Menon and H. I. Aaronson: Metall. Trans., 1986, vol. 17A, 1703.
97. T. Furuhara and H. I. Aaronson: Acta Metall., 1991, vol. 39, 2887.
98. T. Furuhara and T. Maki: Mater. Trans. J.I.M., 1992, vol. 33, p. 734.
99. K. C. Russell and H. I. Aaronson: Jnl. Mater. Science, 1975, vol. 10, 1991.
100. W. C. Dash: Dislocations and Mechanical Properties of Crystals, John Wiley & Sons, New York, 1957, p. 57
101. S. Amelinckx: Acta Metall., 1958, vol. 6, 34.
102. H. Wilsdorf and D. Kuhlman-Wilsdorf: Defects in Crystalline Solids, The Physical Society, London, 1955, p. 175.
103. G. Thomas and J. Nutting: The Mechanism of Phase Transformations in Metals, Inst. of Metals, London, 1956, p. 57.
104. R. B. Nicholson: Proc. European Conf. on Electron Microscopy, Delft, 1961, p.375.
105. E. Hornbogen and M. Roth: Z. Metallk., 1967, vol. 58, p. 842.
106. A. S. Keh and H. A. Wreidt: Trans. TMS-AIME, vol. 224, p. 561.
107. E. Hornbogen: Trans. ASM, 1962, vol. 55, p. 719.
108. W. T. Read, Jr. and W. Shockley: Phys. Rev., 1950, vol. 78, p. 275.
109. J. M. Hedges and J. W. Mitchell: Phil. Mag., 1953, vol. 44, p. 223.
110. G. Thomas: Phil. Mag., 1959, vol. 4, p. 606.
111. A. Cerezo, D. Gibuoin, S. Kim, S. J. Sijbrandij, F. M. Venker, P. J. Warren, J. Wilde and G. D. W. Smith: J. de Phys.: IV, 1996, vol. 6-C5, p. 205.

MOVING PHASE BOUNDARY PROBLEMS

Robert F. Sekerka and Shun-Lien Wang*
Carnegie Mellon University
Pittsburgh, Pennsylvania 15213-3890

Introduction

This article is an update and augmentation of an article originally published in the first edition of this volume [1] with title THE MOVING BOUNDARY PROBLEM and authors R. F. Sekerka, C. L. Jeanfils and R. W. Heckel. With permission of the former coauthors Jeanfils and Heckel, this revised version is now being published by Sekerka and a new coauthor, Shun-Lien Wang. Section **Formalism**, which deals with the formalism of diffusion in multiphase media, has been completely revised, with considerable simplification of the cumbersome notation of the previous version. The specific examples in section **Analytical Solutions to Planar Problems** are not changed much from those of the original article, except for corrections of misprints, clarifications, and improvement of notation. Section **Numerical Methods for Solution of Planar Problems** has been revised to include a more detailed discussion of expanding and contracting coordinate systems. Section **Analytical Solutions to Non-Planar Problems** has been expanded considerably, based in part on the doctoral thesis of Wang [2]. Examples of non-planar problems are presented in the context of solidification of a pure material by diffusive transport of heat, and shown to be isomorphous to the growth of stoichiometric precipitates by solute diffusion. New references have also been added.

The movement of the boundary between two phases of different composition and possibly different structure under the influence of diffusional fluxes is of importance in a wide range of processes. Moving boundaries of this type are of fundamental scientific importance when considering the growth or dissolution kinetics of a second phase or the interdiffusion processes that occur in a multiphase diffusion couple [3–16]. In addition, moving boundaries are also of great importance in establishing the kinetics of engineering processes such as the formation of alloy layers during galvanizing, the homogenization of compacted blends of powders, the development of oxidation resistant coatings via diffusion, and the decarburization of steel [17–20]. Furthermore, moving boundaries play a significant role in thermal degradation processes such as the formation of brittle intermediate phases at the interface between dissimilar metals that have been bonded together [21].

Moving boundaries associated with phase growth or dissolution are thus a significant feature of many scientific and engineering problems in the materials field, and have therefore attracted a great deal of attention from research workers. We anticipate that interest in this topic will continue to grow in the future, especially since our understanding of this topic

*Current address: Avant! Corporation, 46871 Bayside Parkway, Fremont, California 94538

continues to develop and the availability of parameters needed to solve moving boundary problems (diffusion coefficients and phase solubilities) continues to expand [22–28].

Our intention in this paper is to present moving boundary problems in a tutorial manner suitable for presentation as part of a graduate course on phase transition kinetics. The presentation is made for boundaries that have a specific geometry (e.g, planar, spherical, etc.) and the approach taken is the same as one would use for multiphase diffusion couples. Moving boundary problems that occur during phase transformations, such as growth during precipitation or dissolution during solution treatment, can obviously involve complexities in addition to those considered in the present treatment [4, 5, 29, 30]. Such complexities might include capillarity effects on compositions existing at the boundaries, non-equilibrium phase transformation, boundary movement controlled by interfacial structure characteristics, irregular boundary geometries, and ternary and higher order diffusion effects [31]. However, the current presentation permits the diffusional aspects of boundary migration to be considered in detail and also allows for the explicit statement or removal of assumptions which have commonly been made in the literature and textbooks. Furthermore, the authors look upon phase transformations as covering a broad range of topics including diffusional coating reactions and diffusional effects that occur during solid state joining of dissimilar materials. The formalism to be presented is thus applicable to a spectrum of diffusional effects in binary alloys.

Another situation in which the movement of the boundary between two phases is important is solidification, in which one phase is a solid (crystal) and the other is a liquid (melt). In this situation, it is possible for convection to occur in the melt, in which case transport in the melt takes place by both convection and diffusion (of heat and/or solute). For the special case in which convection is negligible, however, the problem of solidification of a pure material is very similar to the problem of phase boundary motion in a binary alloy solid. We will take advantage of this similarity to present examples in the context of solidification in the section **Analytical Solutions to Non-Planar Problems.**

The presentation made in this article treats a) the fundamentals of the formalism for multiphase diffusion, including boundary conditions and the flux balance equation that is basic to the moving boundary problem and b) the application of the flux balance equation to various geometries and boundary conditions. Analytical solutions to planar and non-planar moving boundary problems will be developed, the latter in the context of the solidification problem. Other methods of solution will be mentioned but not treated comprehensively.

Formalism

The motion of a phase boundary in a solid binary alloy is caused by a net flux of chemical species across the boundary (interface) that separates the phases, with due consideration of the differences in partial molar volumes between adjacent phases. This section of the paper presents the fundamental formalism that underlies these considerations, and indicates when, for practical reasons, the differences in partial molar volumes may or may not be neglected.

General description of diffusion

Observed fluxes We first define observed fluxes of the species (atoms or molecules) of a multicomponent system. We consider a system with κ components which we number

$(1, 2, \ldots, \kappa)$; later we shall specialize to a binary system and refer to the components as A and B.

The *observed flux* of a species, i, measured in units of moles $\mathrm{m}^{-2}\ \mathrm{s}^{-1}$, is given by[1]

$$\mathbf{J}_i := c_i \mathbf{v}_i \tag{1}$$

where c_i is the concentration of species i, expressed in moles/m^3, and $\mathbf{v}_i$ is the average velocity of species i, measured in m/sec with respect to an arbitrary observer. Such a flux depends on the state of motion of the observer, or for an observer fixed in the laboratory, the relative motion of the sample, via motion of the whole sample and/or local flow within the sample. We emphasize that c_i is *concentration* ("stuff" per unit volume), not *composition*, which would be measured in terms of mole fractions, $X_i = c_i/c_{\mathrm{tot}}$, where the total concentration (reciprocal of the molar volume) is $c_{\mathrm{tot}} = \sum_{i=1}^{\kappa} c_i$.

Alternatively, we could define observed mass-based fluxes

$$\tilde{\mathbf{J}}_i := \rho_i \mathbf{v}_i, \tag{2}$$

measured in kg $\mathrm{m}^{-2}\ \mathrm{s}^{-1}$, where ρ_i is the partial density of species i, expressed in kg/m^3. Here again, we emphasize that ρ_i is a type of *concentration* in the sense of "stuff" per unit volume; the corresponding *composition* would be expressed in terms of mass fractions, $\omega_i = \rho_i/\rho$, where the total density $\rho = \sum_{i=1}^{\kappa} \rho_i$.

Provided that each species has a non-vanishing molecular mass (molecular "weight") m_i, one has simply $\rho_i = m_i c_i$ and $\tilde{\mathbf{J}}_i = m_i \mathbf{J}_i$. If one treats a lattice model in which vacancies are considered to be a hypothetical species with zero molecular weight, the connection between a mole-based description and a mass-based description is more complicated. In the following, we limit ourselves to descriptions for which $m_i \neq 0$, in which case the only difference in these observed fluxes is the units in which they are measured. In the examples to be described in the section **Analytical Solutions to Planar Problems**, we shall use mole-based concentrations and fluxes, but for the moment we retain both descriptions in order to emphasize some fundamental points.

<u>**Conservation laws**</u> Provided that there are no chemical reactions (which we assume throughout this article), each species is conserved, as expressed by equations of the form

$$\frac{\partial c_i}{\partial t} + \nabla \cdot \mathbf{J}_i = 0 \tag{3}$$

for the mole-based description, or

$$\frac{\partial \rho_i}{\partial t} + \nabla \cdot \tilde{\mathbf{J}}_i = 0 \tag{4}$$

for the mass-based description, where t is time and ∇ is the gradient operator in an arbitrary coordinate system. These equations resemble what is referred to as Fick's second law, but we emphasize that they are in terms of observed fluxes, not the diffusive fluxes that we shall define subsequently.

[1]The notation := means "is defined to be equal to." We use it to distinguish definitions from results.

Diffisive fluxes We now define diffusive fluxes for each phase of a multicomponent system [3,6,7,32–37]. Such fluxes must be consistent with the fact that diffusion is independent of the observer. This requirement is met if the diffusive fluxes are expressed in terms of *differences* of velocity fields [38]

$$\begin{aligned} \mathbf{j}_i^F &:= c_i(\mathbf{v}_i - \mathbf{v}^F) \\ &= \mathbf{J}_i - c_i\mathbf{v}^F \end{aligned} \tag{5}$$

where $\mathbf{v}^F$ is a velocity field, a vector quantity *common to all species* that can vary with position and time. The quantity $\mathbf{j}_i^F$ is the diffusive flux, measured in moles m^{-2} s^{-1}, of species i with respect to the field $\mathbf{v}^F$.

We can also define a mass-based diffusive flux,

$$\begin{aligned} \tilde{\mathbf{j}}_i^F &:= \rho_i(\mathbf{v}_i - \mathbf{v}^F) \\ &= \tilde{\mathbf{J}}_i - \rho_i\mathbf{v}^F. \end{aligned} \tag{6}$$

Here, $\tilde{\mathbf{j}}_i^F$ is the diffusive flux, measured in kg m^{-2} s^{-1}, of species i with respect to the field $\mathbf{v}^F$.

Diffusion equations To get diffusion equations, we must introduce diffusive fluxes by means of Eq(5) or Eq(6), which requires specification of the field $\mathbf{v}^F$, and then substitute into Eq(3) or Eq(4). This gives:

$$\frac{\partial c_i}{\partial t} + \nabla\cdot(c_i\mathbf{v}^F) + \nabla\cdot\mathbf{j}_i^F = 0 \tag{7}$$

$$\frac{\partial \rho_i}{\partial t} + \nabla\cdot(\rho_i\mathbf{v}^F) + \nabla\cdot\tilde{\mathbf{j}}_i^F = 0 \tag{8}$$

There are many ways to choose $\mathbf{v}^F$, but the general strategy is to make a choice that will simplify the problem. Usually, $\mathbf{v}^F$ is chosen to be some weighted average of the velocities of the individual species. For example, if we were dealing with a fluid that is undergoing convection, the usual choice would be to work with Eq(8) and to choose $\mathbf{v}^F$ to be the barycentric velocity [36]

$$\mathbf{v} := \sum_{i=1}^{\kappa} \omega_i\mathbf{v}_i, \tag{9}$$

where the weighting factors are the mass fractions, $\omega_i = \rho_i/\rho$. This barycentric velocity (also known as the velocity of the local center of mass) is governed by the conservation of momentum (e.g., as expressed by the Navier-Stokes equation [36]) subject to the continuity equation. In fact, the continuity equation just expresses the conservation of total mass, so it can be obtained by summing Eq(4) over all species to obtain

$$\frac{\partial \rho}{\partial t} + \nabla\cdot(\rho\mathbf{v}) = 0. \tag{10}$$

We denote the corresponding diffusive fluxes by $\tilde{\mathbf{j}}_i = \rho_i(\mathbf{v}_i - \mathbf{v})$. Eq(8) can then be simplified by means of Eq(10) to the form

$$\rho\left[\frac{\partial \omega_i}{\partial t} + \mathbf{v}\cdot\nabla\omega_i\right] + \nabla\cdot\tilde{\mathbf{j}}_i = 0. \tag{11}$$

Eq(11) applies to the $\kappa - 1$ independent mass fractions and contains $\kappa - 1$ independent diffusive fluxes (since by summing Eq(6) when $\mathbf{v}^F$ is chosen to be the barycentric velocity, one has $\sum_{i=1}^{\kappa} \tilde{\mathbf{j}}_i = 0$). In this manner, one deals with the diffusion problem for $\kappa - 1$ species, but at the expense of having to determine the fluid velocity $\mathbf{v}$ self-consistently.

A similar treatment of Eq(7) can be made by choosing $\mathbf{v}^F$ to be the velocity of the local center of moles,

$$\mathbf{v}^* := \sum_{i=1}^{\kappa} X_i \mathbf{v}_i, \tag{12}$$

where the weighting factors are the mole fractions, $X_i = c_i/c_{tot}$. This choice is particularly useful for isothermal ideal gases that attain mechanical equilibrium (constant pressure, and therefore constant molar concentration) on a time scale short compared to that of diffusive processes. Under these conditions, c_{tot} is uniform throughout the system, resulting in $\nabla \cdot \mathbf{v}^* = 0$, so $\mathbf{v}^*$ can be uniform throughout the system. The molar diffusive fluxes relative to the local center of moles, namely

$$\mathbf{j}_i^* := c_i(\mathbf{v}_i - \mathbf{v}_i^*), \tag{13}$$

obey the relationship $\sum_{i=1}^{\kappa} \mathbf{j}_i^* = 0$, so again only $\kappa - 1$ of them are independent.

Center of volume For nearly incompressible solids and liquids, under certain circumstances to be described subsequently, there is a very useful choice of $\mathbf{v}^F$ that can greatly simplify the problem, namely the center-of-volume velocity

$$\mathbf{v}^{\square} := \sum_{i=1}^{\kappa} c_i \bar{V}_i \mathbf{v}_i = \sum_{i=1}^{\kappa} \rho_i \tilde{V}_i \mathbf{v}_i \tag{14}$$

where $\bar{V}_i$ is the partial molar volume of species i and $\tilde{V}_i$ is the partial specific volume of species i; these are related by $\bar{V}_i = m_i \tilde{V}_i$. Specifically, $\bar{V}_i = \partial V/\partial N_i$, where V is the volume and N_i is the number of moles of species i, with temperature T, pressure P, and the other mole numbers held constant in the differentiation. Since V is a homogeneous function of degree one in the N_i, the Euler theorem of homogeneous functions[2] gives $\sum_{i=1}^{\kappa} N_i \bar{V}_i = V$. Division by V then gives

$$\sum_{i=1}^{\kappa} c_i \bar{V}_i = 1, \tag{15}$$

so the weighting factors $c_i \bar{V}_i$ that appear in Eq(14) are normalized *volume fractions*. For the mass-based description, the corresponding equation is $\sum_{i=1}^{\kappa} \rho_i \tilde{V}_i = 1$. We assume that these equations hold locally, even for small departures from equilibrium [40, 41].

In general, determination of $\mathbf{v}^{\square}$ is not a simple matter, but for the special but still fairly general [42, 43] case that the $\bar{V}_i$ are *constants* (possibly different from one another), a great simplification is possible. Multiplication of Eq(3) by $\bar{V}_i$ and summing over i gives

$$\nabla \cdot \mathbf{v}^{\square} = 0, \tag{16}$$

so the velocity field $\mathbf{v}^{\square}$ is solenoidal[3]. This means that for constant $\bar{V}_i$, there are no sources or sinks of volume. Thus, volume is neither created nor destroyed *locally* as multicomponent

[2] If $f(\lambda x_1, \lambda x_2, \ldots, \lambda x_n) = \lambda^n f(x_1, x_2, \ldots, x_n)$, the function f is said to be homogeneous of degree n in the variables $x_1, x_2, \ldots, x_n$. Then Euler's theorem [39] states that $nf = \sum_{i=1}^{n} x_i(\partial f/\partial x_i)$.

[3] A "solenoidal" vector field forms closed loops, as do the lines of the magnetic field $\mathbf{B}$ for a solenoid.

diffusion progresses. Denoting the mole flux with respect to the center of volume by

$$\mathbf{j}_i^{\square} := c_i(\mathbf{v}_i - \mathbf{v}_i^{\square}), \tag{17}$$

Eq(7) takes the form

$$\frac{\partial c_i}{\partial t} + \mathbf{v}^{\square} \cdot \nabla c_i + \nabla \cdot \mathbf{j}_i^{\square} = 0. \tag{18}$$

Eq(18) still contains $\mathbf{v}^{\square}$ but it applies to the diffusion problem for $\kappa - 1$ species, since the diffusive fluxes, with respect to the local center of volume, obey the relation

$$\sum_{i=1}^{\kappa} \bar{V}_i \mathbf{j}_i^{\square} = 0. \tag{19}$$

Thus for a binary system, for example, the fluxes of A and B are not equal in magnitude and oppositely directed, unless $\bar{V}_A = \bar{V}_B$, which corresponds to the simple case for which the molar volume is independent of composition.

The general solution of Eq(16), familiar from electromagnetism, is that $\mathbf{v}^{\square}$ is the curl of some vector field, say $\mathbf{A}$:

$$\mathbf{v}^{\square} = \nabla \times \mathbf{A} \tag{20}$$

Under certain circumstances, however, $\mathbf{v}^{\square}$ is simply uniform throughout a phase. Such is the case for problems in one spatial dimension, say x, in which case Eq(16) takes the form

$$\frac{\partial v_x^{\square}}{\partial x} = 0, \tag{21}$$

which readily integrates to give

$$v_x^{\square} = g(t), \tag{22}$$

where $g(t)$ is some function of time. Under these circumstances, $\mathbf{v}^{\square}$ can used to define a new reference *frame*. Indeed, if we measure diffusive fluxes in a coordinate system that is moving such that $g(t)$ in Eq(22) is equal to zero, Eq(18) takes the simple form

$$\frac{\partial c_i}{\partial t} + \frac{\partial j_{ix}^{\square}}{\partial x} = 0. \tag{23}$$

Eq(23) is the common form of Fick's second law [44], and holds for diffusive fluxes measured in a unique (except for choice of origin) center-of-volume reference frame. This is the same reference frame as is used for the Boltzmann-Matano analysis [45,46], as will become clear when we take up constitutive laws (see the discussion of Eq(38)). Physically, this is the frame in which the "volume of the phase" is at rest. Note, however, that if more than one phase is being considered, there can be relative motion of the "volume" of one phase as compared to the "volume" of another. In the examples to be described in detail later, we will assume that Eq(23) holds in each phase and take the possible relative motion of the "volumes" of these phases into account explicitly. For cases in which the $\bar{V}_i$ are not constants, $\mathbf{v}^{\square}$ cannot be used to define a reference frame, and one must resort to special methods such as convected coordinates [7] or the method of Prager [47].

For problems in more than one spatial dimension, the establishment of a reference frame that leads to a simplification of Eq(18) is nontrivial, even for constant partial molar volumes, because Eq(16) admits solutions of the form of Eq(20), which allow $\mathbf{v}^{\square}$ to be nonuniform throughout a phase. Such solutions correspond to rigid rotations or material deformations that are pure shear. To insure that $\mathbf{v}^\square$ depends only on time, all of its spatial derivatives must vanish, which implies that the tensor

$$\dot{e}_{\mu\nu} := \frac{\partial v^\square_\mu}{\partial x_\nu} = 0, \quad (\mu, \nu = x, y, z). \tag{24}$$

Eq(24) is a much stronger condition than Eq(16), which only requires the sum of the diagonal components of $\dot{e}_{\mu\nu}$ to vanish. Under some conditions, Eq(24) might be derivable from fundamental considerations such as symmetry; otherwise it must be taken as a working "ansatz " or an approximation. If Eq(24) holds, then $\mathbf{v}^\square = \mathbf{g}(t)$ and one can choose a coordinate system which moves such that $\mathbf{g}(t) = 0$. For such a reference frame, which is actually stationary with respect to the rigid volume of the phase under consideration, Eq(18) becomes

$$\frac{\partial c_i}{\partial t} + \nabla \cdot \mathbf{j}_i^\square = 0. \tag{25}$$

__Constitutive laws__ In order to complete a description of diffusion, it is necessary to introduce constitutive laws for the fluxes. It is usually assumed that the diffusive fluxes are related linearly to gradients of suitable potentials. Irreversible thermodynamics suggests that such potentials are the reciprocal of the absolute temperature and the chemical potentials divided by the absolute temperature [37, 48]. Such potentials depend on temperature, pressure, and composition, so diffusive fluxes would be expected to depend on gradients of all of these quantities. For isothermal diffusion, which is the case we treat in later sections, gradients of pressure and composition still remain. In samples of laboratory size, except for unusual circumstances, it turns out that pressure gradients make a negligible contribution to diffusive fluxes [49]. Therefore, it is consistent with irreversible thermodynamics to adopt constitutive relations in which the diffusive fluxes are linearly related to composition gradients. For a binary system, one would therefore write constitutive equations of the form [36]

$$\mathbf{j}_A^\star = -(c_A + c_B)D^\star \nabla X_A, \quad \mathbf{j}_B^\star = -(c_A + c_B)D^\star \nabla X_B \tag{26}$$

where the factor of $(c_A + c_B)$ is consistent with the correct units of flux and $\mathbf{j}_A^\star + \mathbf{j}_B^\star = 0$ is satisfied automatically because $X_A + X_B = 1$. The quantity $D^\star$ is a chemical diffusivity, measured in units of m^2/s. For binary systems, Eq(15) becomes

$$c_A \bar{V}_A + c_B \bar{V}_B = 1 \tag{27}$$

and for constant partial molar volumes, differentiation gives

$$\bar{V}_A \nabla c_A = -\bar{V}_B \nabla c_B. \tag{28}$$

Therefore, the gradients of the concentrations are linearly related to one another. Thus the gradient of composition is

$$\nabla X_A = \frac{1}{\bar{V}_B (c_A + c_B)^2} \nabla c_A. \tag{29}$$

By multiplying $\mathbf{j}_A^\star$ by $\bar{V}_A$ and $\mathbf{j}_B^\star$ by $\bar{V}_B$, adding the results, and using Eq(14) and Eq(26), we find

$$\mathbf{v}^\star - \mathbf{v}^\square = (c_A + c_B)D^\star(\bar{V}_A - \bar{V}_B)\nabla X_A = (c_A + c_B)D^\star(\bar{V}_B - \bar{V}_A)\nabla X_B. \tag{30}$$

Thus,

$$\begin{aligned} \mathbf{j}_A^{\square} &= c_A(\mathbf{v}_A - \mathbf{v}^*) = c_A(\mathbf{v}_A - \mathbf{v}^*) + c_A(\mathbf{v}^* - \mathbf{v}^{\square}) \\ &= -(c_A + c_B)^2 D^* \bar{V}_B \nabla X_A = -D^* \nabla c_A. \end{aligned} \tag{31}$$

Similarly, $\mathbf{j}_B^{\square} = -D^*\nabla c_B$. Thus, the flux of one component of a *binary* solution can be taken to be proportional to its own concentration gradient, a form known as Fick's first law [44].

Another approach to constitutive laws is purely empirical, and leads to the same conclusion as above. We follow Darken [40, 50] and select the velocity field $\mathbf{v}^F$ to be that of inert markers. The latter, if they exhibit relative motion, are believed to do so in a manner independent of the marker material (operational definition of "inert") but closely related to the atomistics of the mixing process. For a binary solution, to which we specialize at this stage, the basic form of Fick's first law is taken to be

$$\mathbf{j}_i^M := c_i(\mathbf{v}_i - \mathbf{v}^M) = -D_i \nabla c_i \quad (i = A, B) \tag{32}$$

where $\mathbf{v}^M$ is the marker velocity field, $\mathbf{j}_i^M$ is the molar diffusive flux relative to the markers, and D_i is the *intrinsic* diffusion coefficient of species i, measured in m^2/s.

For a binary solution, Eq(32) is actually two equations, one for each species, because in general, $D_A \neq D_B$. Since we usually have little or no information about the marker velocity field, $\mathbf{v}^M$, we proceed to relate it to the center-of-volume velocity, $\mathbf{v}^{\square}$, previously defined by Eq(14). By multiplying Eq(32) for $i = A$ by $\bar{V}_A$, Eq(32) for $i = B$ by $\bar{V}_B$, adding the results, and using Eq(14), we find

$$\mathbf{v}^M - \mathbf{v}^{\square} = D_A \bar{V}_A \nabla c_A + D_B \bar{V}_B \nabla c_B. \tag{33}$$

Substituting into Eq(33) and using Eq(28) yields

$$\mathbf{v}^M - \mathbf{v}^{\square} = (D_A - D_B)\bar{V}_A \nabla c_A = (D_B - D_A)\bar{V}_B \nabla c_B. \tag{34}$$

Hence

$$\begin{aligned} \mathbf{j}_A^{\square} &= c_A(\mathbf{v}_A - \mathbf{v}^{\square}) = c_A(\mathbf{v}_A - \mathbf{v}^M) + c_A(\mathbf{v}^M - \mathbf{v}^{\square}) \\ &= -(c_A \bar{V}_A D_B + c_B \bar{V}_B D_A)\nabla c_A. \end{aligned} \tag{35}$$

Similarly

$$\mathbf{j}_B^{\square} = -(c_A \bar{V}_A D_B + c_B \bar{V}_B D_A)\nabla c_B. \tag{36}$$

Comparison of Eq(35) and Eq(36) shows that they contain a common diffusivity

$$D := c_A \bar{V}_A D_B + c_B \bar{V}_B D_A, \tag{37}$$

in terms of which they become simply

$$\mathbf{j}_A^{\square} = -D\nabla c_A, \quad \mathbf{j}_B^{\square} = -D\nabla c_B. \tag{38}$$

Then by using Eq(28) we see explicitly that

$$\mathbf{j}_A^{\square} = -(\bar{V}_B/\bar{V}_A)\mathbf{j}_B^{\square}, \tag{39}$$

in agreement with the general Eq(19). Eqs(38) are the common forms of Fick's first law for a binary system [44]. They contain a single interdiffusion coefficient D, sometimes referred to as the chemical diffusion coefficient. Note also that $D = D^*$, consistent with the form of Eq(31).

For the very special but often assumed (for the sake of tractability) case for which $\bar{V}_A = \bar{V}_B$, Eq(37) simplifies to $D = X_A D_B + X_B D_A$ and the fluxes are equal and opposite, i.e., $\mathbf{j}_A^{\square} = -\mathbf{j}_B^{\square}$. This was actually the case considered by Darken [40, 50].

Regardless of their origin, we shall regard Eqs(38) to hold. Moreover, in the examples to follow, we shall assume that D is a constant, irrespective of the details of Eq(37). This simplification allows us to exhibit some simple analytical solutions to diffusion problems. In the section **Numerical Methods for Solution of Planar Problems**, we examine briefly the case of interdiffusion coefficients that depend on concentration.

Multiphase diffusion

The formulae of the preceding section can be used to describe diffusion within a single phase. We now seek to relate the diffusional fluxes in several phases, which we denote by superscripts $\alpha, \beta, \gamma \ldots$. We confine our treatment to binary (A, B) systems, although the principles that we employ can be extended to higher order systems. In each phase, we assume the partial molar volumes to be constants, but not necessarily equal to each other in the same phase and possibly different for different phases.

For example, for two phases, α and β, there can be four distinct partial molar volumes, $\bar{V}_A^\alpha$, $\bar{V}_B^\alpha$, $\bar{V}_A^\beta$, and $\bar{V}_B^\beta$. If $\bar{V}_A^\alpha = \bar{V}_A^\beta$ and $\bar{V}_B^\alpha = \bar{V}_B^\beta$, the α and β phases have the same dependence of molar volume on composition, which leads to considerable simplification. If $\bar{V}_A^\alpha = \bar{V}_B^\alpha$, then the molar volume of the α phase is independent of composition, and similarly for β. Such special cases are often assumed in the literature, but we shall develop the general case, which can be simplified later if desired.

In the previous section, we saw that the assumption of constant (but possibly different) partial molar volumes leads to Eq(16) which guarantees that there are no sources or sinks of volume *within* that phase. But exchange of atoms between two phases can lead to sources or sinks of volume. These may be visualized by considering the transfer of δN_A moles of A and δN_B moles of B from the α phase to the β phase, across the $\alpha - \beta$ phase boundary; the change in volume will be $\delta V = (\bar{V}_A^\beta - \bar{V}_A^\alpha)\delta N_A + (\bar{V}_B^\beta - \bar{V}_B^\alpha)\delta N_B$. This volume change will not ordinarily vanish, except for the special case mentioned above, for which the α and β phases have the same dependence of molar volume on composition. In the general case, there will be relative motion of the α and β phases, as well as motion of the $\alpha\beta$ phase boundary relative to either one of them.

If there are no external fluxes of species A or B, it is possible to show that the *total* volume of a two-phase system remains constant, even as the $\alpha\beta$ boundary moves, provided that the constants $\bar{V}_A^\alpha = \bar{V}_A^\beta$ and $\bar{V}_B^\alpha = \bar{V}_B^\beta$. In this case, the mole numbers N_A of A and N_B of B are constants, given by the following expressions:

$$N_A = \int_{V^\alpha} c_A^\alpha \, d^3x + \int_{V^\beta} c_A^\beta \, d^3x \tag{40}$$

$$N_B = \int_{V^\alpha} c_B^\alpha \, d^3x + \int_{V^\beta} c_B^\beta \, d^3x \tag{41}$$

Multiplication of Eq(40) by $\bar{V}_A$ and Eq(41) by $\bar{V}_B$ and adding the results gives

$$\bar{V}_A N_A + \bar{V}_B N_B = \int_{V^\alpha} (\bar{V}_A c_A^\alpha + \bar{V}_B c_B^\alpha)\, d^3x + \int_{V^\beta} (\bar{V}_A c_A^\beta + \bar{V}_B c_B^\alpha)\, d^3x, \tag{42}$$

where we have dropped the superscripts on the partial molar volumes because they are the same for α and β. Thus by using Eq(27) for each phase, the integrands in Eq(42) are equal to one, so it becomes

$$\bar{V}_A N_A + \bar{V}_B N_B = V^\alpha + V^\beta. \tag{43}$$

Since the left hand side is a constant, the total volume on the right hand side is also a constant. If $\bar{V}_A^\alpha \neq \bar{V}_A^\beta$ or $\bar{V}_B^\alpha \neq \bar{V}_B^\beta$, the preceding demonstration that the total volume is constant would not be possible. The resulting change in total volume will then necessitate motion of one phase relative to the other, as discussed in the next paragraph for a planar system.

Planar $\alpha\beta$ system We treat a planar system consisting of two phases, α and β, for which x is the only relevant spatial variable. When two phases are involved, an equation of the form of Eq(22) applies for each phase

$$v^{\square\alpha} = g^\alpha(t), \quad v^{\square\beta} = g^\beta(t) \tag{44}$$

where we have dropped the subscript x for simplicity. We therefore choose a *master reference frame* by taking $g^\beta = 0$. In general, however, we cannot also take $g^\alpha = 0$ or we will not be able to allow for relative motion of α with respect to β. We denote the x component of the velocity of the $\alpha\beta$ boundary with respect to the β phase by $v^{\alpha\beta}$, considered to be positive when moving toward α from β.

We proceed to develop a pair of simultaneous equations to determine the two unknown velocities, $v^{\square\alpha}$ and $v^{\alpha\beta}$. To do this, we employ the conditions that the flux of each species i, measured with respect to the moving boundary, is continuous. This is based on the principle that the boundary itself is neither a source nor a sink for either A or B, but moves in such a way as to insure that this requirement is met. Thus (dropping the x subscripts on the fluxes)

$$j_i^{\square\beta} - c_i^{\beta\alpha} v^{\alpha\beta} = j_i^{\square\alpha} - c_i^{\alpha\beta}(v^{\alpha\beta} - v^{\square\alpha}) \quad (i = A, B) \tag{45}$$

where $c_i^{\beta\alpha}$ is the concentration in the β phase adjacent to the α phase, and $c_i^{\alpha\beta}$ is the concentration in the α phase adjacent to the β phase.[4] Note in Eq(45) that only the flux $j_i^{\square\alpha}$ gets corrected by the additional term $c_i^{\alpha\beta} v^{\square\alpha}$ because the master reference frame is that for which $v^{\square\beta} = 0$.

We solve the simultaneous equations Eq(45) for $v^{\square\alpha}$ and $v^{\alpha\beta}$. This procedure involves some algebraic manipulation and the use of Eq (15) and Eq (19) for each phase. The results are:

$$v^{\alpha\beta} = -Q^\beta \frac{D^\beta}{(c_B^{\beta\alpha} - c_B^{\alpha\beta})} \frac{\partial c_B^\beta}{\partial x} + Q^\alpha \frac{D^\alpha}{(c_B^{\beta\alpha} - c_B^{\alpha\beta})} \frac{\partial c_B^\alpha}{\partial x} \tag{46}$$

and

$$v^{\square\alpha} = \epsilon^\beta \frac{D^\beta}{(c_B^{\beta\alpha} - c_B^{\alpha\beta})} \frac{\partial c_B^\beta}{\partial x} + \epsilon^\alpha \frac{D^\alpha}{(c_B^{\beta\alpha} - c_B^{\alpha\beta})} \frac{\partial c_B^\alpha}{\partial x} \tag{47}$$

[4]This nomenclature will be used throughout the remainder of the paper.

The Q and ϵ coefficients, which are dimensionless, are given by:

$$Q^{\beta} := (\bar{V}_A^{\beta} c_A^{\alpha\beta} + \bar{V}_B^{\beta} c_B^{\alpha\beta})/(\bar{V}_B^{\beta}\Delta) \tag{48}$$

$$Q^{\alpha} := 1/(\bar{V}_A^{\alpha}\Delta) \tag{49}$$

$$\epsilon^{\beta} := (1 - \bar{V}_A^{\beta} c_A^{\alpha\beta} - \bar{V}_B^{\beta} c_B^{\alpha\beta})/(\bar{V}_B^{\beta}\Delta) \tag{50}$$

$$\epsilon^{\alpha} := (1 - \bar{V}_A^{\alpha} c_A^{\beta\alpha} - \bar{V}_B^{\alpha} c_B^{\beta\alpha})/(\bar{V}_B^{\alpha}\Delta) \tag{51}$$

$$\Delta := \frac{c_B^{\beta\alpha} c_A^{\alpha\beta} - c_B^{\alpha\beta} c_A^{\beta\alpha}}{c_B^{\beta\alpha} - c_B^{\alpha\beta}} = \frac{(c_A^{\beta\alpha} + c_B^{\beta\alpha})(c_A^{\alpha\beta} + c_B^{\alpha\beta})(X_B^{\beta\alpha} - X_B^{\alpha\beta})}{c_B^{\beta\alpha} - c_B^{\alpha\beta}} \tag{52}$$

Here, the X_B's are mole fractions, e.g., $X_B^{\beta\alpha} = c_B^{\beta\alpha}/(c_A^{\beta\alpha} + c_B^{\beta\alpha})$. Note that Eq(27) does not apply to quantities such as $\bar{V}_A^{\beta} c_A^{\alpha\beta} + \bar{V}_B^{\beta} c_B^{\alpha\beta}$ and $\bar{V}_A^{\alpha} c_A^{\beta\alpha} + \bar{V}_B^{\alpha} c_B^{\beta\alpha}$ because they involve the partial molar volumes of one phase and the concentrations of the other.

For the special case in which the molar volumes of α and β have the same dependence on composition, i.e., $\bar{V}_A^{\alpha} = \bar{V}_A^{\beta}$ and $\bar{V}_B^{\alpha} = \bar{V}_B^{\beta}$, we have $\epsilon^{\alpha} = \epsilon^{\beta} = 0$ and $Q^{\alpha} = Q^{\beta} = 1$. Under these conditions, $v^{\square\alpha} = 0$ (no relative motion of the phases) and Eq(46) reduces to

$$v^{\alpha\beta} = -\frac{D^{\beta}}{(c_B^{\beta\alpha} - c_B^{\alpha\beta})}\frac{\partial c_B^{\beta}}{\partial x} + \frac{D^{\alpha}}{(c_B^{\beta\alpha} - c_B^{\alpha\beta})}\frac{\partial c_B^{\alpha}}{\partial x}, \tag{53}$$

which is the simplified form of the flux balance equation that is commonly used in describing the movement of heterophase boundaries. In some cases, the values of the Q's will be close to unity and those of the $|\epsilon|$'s will be much less than unity. Since, however, Eq(46) contains the products $Q^{\alpha}D^{\alpha}$ and $Q^{\beta}D^{\beta}$, the use of appropriate values of Q^{α} and Q^{β} is necessary to relate accurately interface motion to measured diffusivities.

<u>**Planar $\alpha + \beta, \beta$ system**</u> In certain applications (see Figures 2, 3 and 8) it is desirable to treat the motion of a boundary that separates a single phase region, say β, from a two-phase region, say $\alpha+\beta$. Strictly speaking, such a system is not a planar system because the morphology of α and β phases in the two-phase region can be quite complicated. However, from the point of view of the driving force for diffusion, the $\alpha + \beta$ region is a region of uniform[5] chemical potential of each atomic species. Hence, there will be no diffusion in this region. Furthermore, we shall presume that such a two-phase region has a uniform average concentration given by $c_i^{\alpha+\beta} := f^{\alpha} c_i^{\alpha\beta} + f^{\beta} c_i^{\beta\alpha}$ where f^{α} and f^{β} are the volume fractions of the α and β phases in the two-phase region. Then the joining conditions at the moving $\beta - (\alpha + \beta)$ boundary that are analogous to Eqs. (46) and (47) can be obtained by simply modifying Eq(45) by omitting the terms $j_i^{\square\alpha}$ and introducing $\alpha + \beta$ subscripts whenever the various quantities pertain to the $\alpha + \beta$ region. Specifically,

$$j_i^{\square\beta} - c_i^{\beta\alpha} v^{\alpha+\beta,\beta} = -c_i^{\alpha+\beta}(v^{\alpha+\beta,\beta} - v^{\square,\alpha+\beta}) \ \ (i = A, B). \tag{54}$$

Simultaneous solutions to Eq(54) then yield

$$v^{\alpha+\beta,\beta} = -Q^{\alpha+\beta}\frac{D^{\beta}}{(c_B^{\beta\alpha} - c_B^{\alpha+\beta})}\frac{\partial c_B^{\beta}}{\partial x} \tag{55}$$

[5]If the α and β phases are very finely dispersed, considerations of interfacial energy will lead to a non-uniform chemical potential and concomitant "ripening" of the microstructure; such a problem would be intrinsically non-planar.

and

$$v^{\square,\alpha+\beta} = \epsilon^{\alpha+\beta} \frac{D^\beta}{(c_B^{\beta\alpha} - c_B^{\alpha+\beta})} \frac{\partial c_B^\beta}{\partial x} \tag{56}$$

where

$$Q^{\alpha+\beta} := (\bar{V}_A^\beta c_A^{\alpha+\beta} + \bar{V}_B^\beta c_B^{\alpha+\beta})/(\bar{V}_B^\beta \Delta') \tag{57}$$

$$\epsilon^{\alpha+\beta} := (1 - \bar{V}_A^\beta c_A^{\alpha+\beta} - \bar{V}_B^\beta c_B^{\alpha+\beta})/(\bar{V}_A^\beta \Delta') \tag{58}$$

$$\Delta' := \frac{c_B^{\beta\alpha} c_A^{\alpha+\beta} - c_B^{\alpha+\beta} c_A^{\beta\alpha}}{c_B^{\beta\alpha} - c_B^{\alpha+\beta}} = \frac{(c_A^{\beta\alpha} + c_B^{\beta\alpha})(c_A^{\alpha+\beta} + c_B^{\alpha+\beta})(X_B^{\beta\alpha} - X_B^{\alpha+\beta})}{c_B^{\beta\alpha} - c_B^{\alpha+\beta}} \tag{59}$$

The special case for which $\bar{V}_A^\beta = \bar{V}_A^\alpha$ and $\bar{V}_B^\beta = \bar{V}_B^\alpha$ leads to $Q^{\alpha+\beta} = 1$ and $\epsilon^{\alpha+\beta} = 0$.

Planar β system with free surface The final condition to be developed for a planar system pertains to a boundary that is actually a free surface of a phase, such as the β phase (see Figures 4-6). Let this boundary be located in the plane $x = \xi$. Then, by the same principles that led to Eq(45), the fluxes of A and B atoms, measured in a reference frame moving with the free surface $x = \xi$, must be continuous at $x = \xi$:

$$j_i^{\square\beta} - c_i^{\beta\xi} v^\xi = J_i^{\text{external}} \quad (i = A, B) \tag{60}$$

Here, $c_i^{\beta\xi}$ is the concentration of species i in the β phase at the free surface, J_i^{external} is the flux of species i in the external medium at the free surface of the β phase, and $v^\xi = d\xi/dt$ is the x component of the velocity of the free surface, measured with respect to the center-of-volume frame of the β phase.

A case of possible importance arises when only one species is supplied by, or lost to, the external medium. If we suppose that this species is B, then:

$$\begin{aligned} j_A^{\square\beta} - c_A^{\beta\xi} v^\xi &= 0 \\ j_B^{\square\beta} - c_B^{\beta\xi} v^\xi &= J_B^{\text{external}} \end{aligned} \tag{61}$$

Then by eliminating c_A^β and its derivative by using $c_A^\beta = (1 - \bar{V}_B^\beta c_B^\beta)/\bar{V}_A^\beta$, we obtain

$$v^\xi = \frac{\bar{V}_B^\beta}{(1 - \bar{V}_B^\beta c_B^{\beta\xi})} D^\beta \frac{\partial c_B^\beta}{\partial x} \tag{62}$$

and

$$J_B^{\text{external}} = -\frac{1}{(1 - \bar{V}_B^\beta c_B^{\beta\xi})} D^\beta \frac{\partial c_B^\beta}{\partial x}. \tag{63}$$

If it is presumed that $c_B^{\beta\xi}$ is specified (perhaps imposed by some local equilibrium with the external medium), Eq(62) is a convenient boundary condition at $x = \xi$ and Eq(63) simply gives the corresponding value of the external flux. If B is an interstitial solute (e.g., C in Fe), $\bar{V}_B^\beta$ will be small and the motion of the free surface, according to Eq(62), will be negligible. But if $\bar{V}_B^\beta$ is appreciable, as it would be expected to be if B were a substitutional element in the β phase, then the motion according to Eq(62) is substantial and the external flux J_B^{external} given by Eq(63) is increased in magnitude by a factor $(1 - \bar{V}_B^\beta c_B^{\beta\xi})^{-1}$ relative to the magnitude $|D^\beta(\partial c_B^\beta/\partial x)|$ of the diffusive flux in the center-of-volume frame of the β phase. It is precisely this difference of fluxes that leads to motion of the free boundary.

Other boundary conditions at a free surface are also possible, the correct choice in any physical problem being dependent upon the details of the kinetic processes at that surface. For example, if J_A^{external} is still presumed to be zero but J_B^{external} is known, for some reason, to be constant, then Eqs. (62) and (63) still hold, but $v^\xi = -\bar{V}_B^\beta J_B^{\text{external}}$ is now constant. Furthermore, $c_B^{\beta\xi}$ will now vary according to Eq(63) as the quantity $D^\beta(\partial c_B^\beta/\partial x)$ varies; i.e., the concentration at the free surface will not be constant.

Alternatively, it is possible that the external fluxes will be in the same ratio as the diffusive fluxes in the β phase, namely

$$J_A^{\text{external}} = -(\bar{V}_B^\beta/\bar{V}_A^\beta)\, J_B^{\text{external}} \tag{64}$$

In such a case, Eqs(60) are satisfied by $v^\xi = 0$ and the free surface does not move. This lack of motion is a common assumption in the literature, but is true only if the external fluxes are controlled very carefully! One obvious condition for which Eq(64) will be satisfied, however, is when both J_A^{external} and J_B^{external} are zero; this will certainly be true if the "free" surface is actually a plane of symmetry.

<u>**Non-planar systems**</u> In the treatment of multiphase nonplanar problems, we will always assume that $\bar{V}_i^\alpha = \bar{V}_i^\beta = \bar{V}_i^\gamma \ldots (i = A, B)$. These assumptions were not essential for planar problems, but are essential for non-planar situations in order to avoid intractable problems. Otherwise, volume changes, such as $\delta V = (\bar{V}_A^\beta - \bar{V}_A^\alpha)\delta N_A + (\bar{V}_B^\beta - \bar{V}_B^\alpha)\delta N_B$ which we mentioned previously, would have to be accommodated by either elastic or plastic deformation in solids or flow in fluids. In solids, the accompanying stresses could also affect the diffusion fluxes. The analysis of such problems would require, for each phase, a self-consistent solution for diffusion accompanied by deformation or flow.

Analytical Solutions to Planar Problems

A number of moving boundary problems can be solved by analytical methods provided that certain simplifying assumptions are made [6, 51, 52]. The main assumption is constancy of the interdiffusion coefficient within each phase, which renders the diffusion equation linear with constant coefficients. This assumption is not especially realistic but is very useful because such a large variety of solutions can be obtained with ease and the general dependence of these solutions on the parameters of the problem is readily apparent. Another simplifying assumption is constancy of concentrations at phase boundaries or free surfaces. Constancy of concentrations at phase boundaries would be a consequence of the assumption of local equilibrium, but local equilibrium is not a necessary assumption provided that the phase boundary concentrations are maintained constant for some other reason. We shall, however, draw all figures as if the assumption of local equilibrium were true. Another simplification is infinite or semi-infinite spatial domains. This is an idealization which is valid so long as the actual length of a phase is large compared to its so-called diffusion distance, $\sim \sqrt{Dt}$. Satisfying a boundary condition at infinity rather than at a definite large distance results in considerable simplification of the results, with negligible error. Finally, we only treat a restricted class of initial conditions that is compatible with the form of the solution and the boundary conditions.

All of these assumptions can be removed in principle, but at the expense of complications and cumbersome results. Solutions for finite samples and more general initial conditions can be treated by numerical techniques, as discussed briefly in the section **Numerical Methods for Solution of Planar Problems.**

Infinite extent planar problems

In this section, we consider problems in which two or more phases occupy the region $-\infty < x < \infty$. In each single phase region, the concentration[6] c will satisfy the equation

$$\frac{\partial c}{\partial t} = D\frac{\partial^2 c}{\partial x^2} \tag{65}$$

which is a combination of Eqs. (23) and (38). A well-known solution of Eq(65) is

$$c = \mathcal{A} + \mathcal{B}\,\mathrm{erf}\left(\frac{x}{\sqrt{4Dt}}\right) \tag{66}$$

where $\mathcal{A}$ and $\mathcal{B}$ are constants, D is the constant interdiffusion coefficient appropriate to the phase under consideration, and

$$\mathrm{erf}(u) := \frac{2}{\sqrt{\pi}}\int_0^u \exp(-\lambda^2)\,d\lambda \tag{67}$$

defines the error function [3, 6, 7] of argument u. We note that $\mathrm{erf}(u)$ is an odd function of u which increases monotonically with u from $\mathrm{erf}(0) = 0$ to $\mathrm{erf}(\infty) = 1$; intermediate values and other useful properties may be found in standard references [7, 52, 53]. Since

$$\frac{d}{dx}\mathrm{erf}(u) = \frac{2}{\sqrt{\pi}}\exp(-u^2)\frac{du}{dx} \tag{68}$$

with a similar expression for the time derivative, one can easily verify that Eq(66) is a solution by direct substitution into Eq(65).

The fact that the error function solution, Eq(66), is the key to the solution of planar problems involving moving boundaries can be appreciated by noting that Eq(66) becomes independent of t if x is evaluated at a position ξ that is proportional to $\sqrt{t}$. The constant of proportionality can be written in any convenient manner and is eventually determined by a complete solution to the problem. Since it is often desirable to deal with dimensionless quantities, one frequently writes

$$\xi = K\sqrt{4Dt} \tag{69}$$

where K is a dimensionless quantity, sometimes called the parabolic rate constant. The factor of four in Eq(69) is introduced so that when ξ is substituted for x in Eq(66), one gets simply $c = \mathcal{A} + \mathcal{B}\mathrm{erf}(K)$. Special cases of Eq(69) are $K = 0$, which corresponds to a fixed boundary at $x = 0$, or $K = \pm\infty$ which corresponds to boundaries located so far away from $x = 0$ that their precise distance does not matter. If Eq(66) is presumed to hold between any pair of boundaries of the type specified by Eq(69), c can always be made to take on constant specified values at these two boundaries by choice of the two disposable constants, $\mathcal{A}$ and $\mathcal{B}$.

---

[6] To simplify the notation, we shall use the symbol c without a subscript to denote the concentration of B atoms. The concentration of A atoms can be obtained from the formula $c_A = (1 - c\bar{V}_B)/\bar{V}_A$. Superscripts such as α, β, and γ will continue to refer to phases.

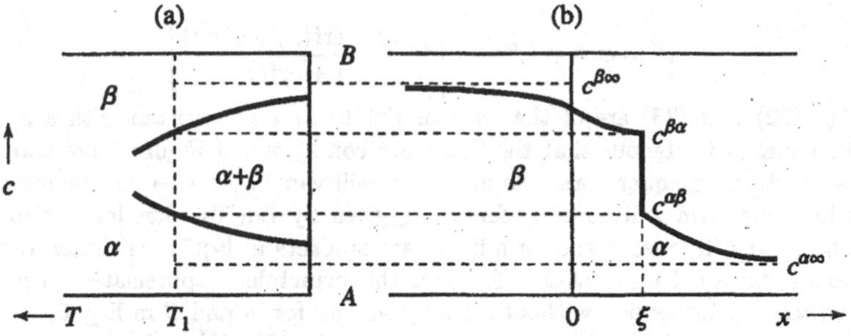

Figure 1: Concentration profile for a system of infinite extent with initial concentrations $c^{\alpha\infty}$ and $c^{\beta\infty}$ in the α and β fields of the phase diagram.

It is also easy to see that Eq(66) will lead to satisfaction of equations having the forms of Eq(46) and Eq(47), which involve the fluxes. For by Eq(68), the gradients of c, when evaluated at ξ given by Eq(69), will be proportional to $1/\sqrt{t}$, but $d\xi/dt$ will also be proportional to $1/\sqrt{t}$. Therefore, the dependence on time of both sides of Eq(46) or Eq(47) will cancel, provided that $v^{\alpha\beta}$ and $v^{\square\alpha}$ correspond to boundaries that move according to a law of the form of Eq(69). The resulting equation (or equations) will determine the appropriate constants, such as K.

At a boundary that separates two phases, say α and β, it might not be particularly clear which interdiffusion coefficient, D^α or D^β, should be used in Eq(69). In the subsequent examples, we have used

$$\xi = K\sqrt{4D^\beta t}, \tag{70}$$

whereas choices such as $\xi = K'\sqrt{4D^\alpha t}$ or $\xi = K''\sqrt{2(D^\alpha + D^\beta)t}$ will do just as well. Of course the actual value of the parabolic rate constant will depend on the choice, but these values will be related by

$$K = K'\sqrt{D^\alpha/D^\beta} = K''\sqrt{(D^\alpha + D^\beta)/(2D^\beta)}. \tag{71}$$

In any practical problem, it will be expedient to use the interdiffusion coefficient for the phase in which diffusion controls the rate of phase boundary motion. However, the third form involving K'' is always noncommital.

Finally, there are initial conditions implied by Eqs. (66) and (69). From Eq(69), it is apparent that all boundaries must be located initially at $x = 0$ (or $x \pm \infty$, which are idealized locations of far boundaries). For times $t > 0$, a number of phases can emanate and grow from $x = 0$. Moreover, the limit of Eq(66) for $t = 0$ is $\mathcal{A} + \mathcal{B}$ if $x > 0$ and $\mathcal{A} - \mathcal{B}$ if $x < 0$. Therefore, the semi-infinite regions $-\infty < x < 0$ and $0 > x > \infty$ must each have uniform concentrations initially, and these concentrations must be the same as the respective far-field concentrations $c^{\beta\infty}$ and $c^{\alpha\infty}$.

Infinite $\beta\alpha$ problems For a phase diagram and corresponding concentration-distance profile such as depicted in Figure 1, the concentrations in the α and β phases are:

$$c^\alpha = c^{\alpha\infty} + (c^{\alpha\beta} - c^{\alpha\infty})\left[\frac{1 - \mathrm{erf}(x/\sqrt{4D^\alpha t})}{1 - \mathrm{erf}(K\sqrt{D^\beta/D^\alpha})}\right] \tag{72}$$

and

$$c^\beta = c^{\beta\infty} + (c^{\beta\alpha} - c^{\beta\infty})\left[\frac{1+\mathrm{erf}(x/\sqrt{4D^\beta t})}{1+\mathrm{erf}(K)}\right] \tag{73}$$

Both Eqs. (72) and (73) are of the form of Eq(66) and are written with a grouping of terms that makes it obvious that the boundary conditions of Figure 1 are satisfied. For instance, the term in square brackets in Eq(72) will vanish for $x \to \infty$, leaving $c^\alpha = c^{\alpha\infty}$. In addition, this term will be unity for $x = \xi$ given by Eq(70), thus leading to $c^\alpha = c^{\alpha\beta}$ at that boundary. Similarly, the term in square brackets in Eq(73) is chosen to vanish at $x = -\infty$ and to equal unity at $x = \xi$. Once this principle is appreciated, such solutions can be written by inspection without actually solving for $\mathcal{A}$ and $\mathcal{B}$ in Eq(66).

Differentiation of Eqs. (72) and (73) with the aid of Eq(68) and substitution into the boundary condition, Eq(46), with $v^{\alpha\beta} = d\xi/dt$ yields

$$K = -\frac{1}{\sqrt{\pi}}\sqrt{D^\alpha/D^\beta}Q^\alpha S^\alpha \frac{\exp(-K^2 D^\beta/D^\alpha)}{1-\mathrm{erf}(K\sqrt{D^\beta/D^\alpha})} + \frac{1}{\sqrt{\pi}}Q^\beta S^\beta \frac{\exp(-K^2)}{1+\mathrm{erf}(K)} \tag{74}$$

where[7] $S^\alpha := (c^{\alpha\beta} - c^{\alpha\infty})/(c^{\beta\alpha} - c^{\alpha\beta})$ and $S^\beta := (c^{\beta\infty} - c^{\beta\alpha})/(c^{\beta\alpha} - c^{\alpha\beta})$. Values of K can be determined by successive approximations and computer techniques. In cases for which S^α and S^β are small, K will also be small[8] and an approximate solution can be obtained by expansion in powers of K. To first order in K, both exponentials in Eq(74) can be set equal to unity, the error functions can be expanded according to $\mathrm{erf}(K) = (2/\sqrt{\pi})[K + \ldots]$, and factors such as $1/[1 + (2/\sqrt{\pi})K + \ldots]$ can be rewritten as $[1 - (2/\sqrt{\pi})K + \ldots]$. The resulting approximation to Eq(74) can then be solved to yield

$$K \approx \frac{(-1/\sqrt{\pi})(\sqrt{D^\alpha/D^\beta})Q^\alpha S^\alpha + (1/\sqrt{\pi})Q^\beta S^\beta}{1 + (2/\pi)Q^\alpha S^\alpha + (2/\pi)Q^\beta S^\beta} \tag{75}$$

where, as a further approximation, the entire denominator can be taken equal to unity to first order in the quantities S^α and S^β.

Once K is determined, it can be substituted into Eq(72) and Eq(73) to obtain concentration profiles. Thus, the planar moving boundary problem has been solved for the situation of two phases, each of semi-infinite extent, with the assumptions of constant boundary concentrations and constant interdiffusion coefficients. We could also use Eq(47) to determine the relative motion of the phases, but we omit details.

Infinite $\beta, \alpha + \beta$ problems We now treat the case corresponding to Figure 2 in which the boundary moves from β toward a two-phase region $\alpha+\beta$ of average concentration $c^{\alpha+\beta}$. Since chemical potentials are uniform throughout such a two-phase region, there will be no diffusion to the right of $x = \xi$. The appropriate boundary condition at the $\beta - (\alpha + \beta)$ boundary is Eq(55). Moreover, Eq(73) is still appropriate for the β phase. Substitution of Eq(73) into Eq(55) yields

$$K = \frac{1}{\sqrt{\pi}}Q^{\alpha+\beta}\left(\frac{c^{\beta\infty} - c^{\beta\alpha}}{c^{\beta\alpha} - c^{\alpha+\beta}}\right)\frac{\exp(-K^2)}{1+\mathrm{erf}(K)}. \tag{76}$$

[7]We have corrected sign errors and missing factors of π in Eq(54) and Eq(55) of the original article.

[8]If $\sqrt{D^\alpha/D^\beta}$ is large, K will not necessarily be small for small S^α and S^β. In such a case, the problem could be set up in terms of K' or K'', which would then be small.

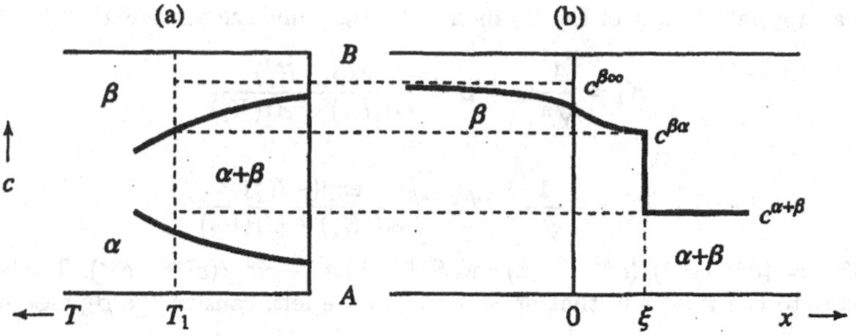

Figure 2: Concentration profile for a system of infinite extent with initial concentrations $c^{\alpha+\beta}$ and $c^{\beta\infty}$ in the $\alpha+\beta$ and β fields of the phase diagram.

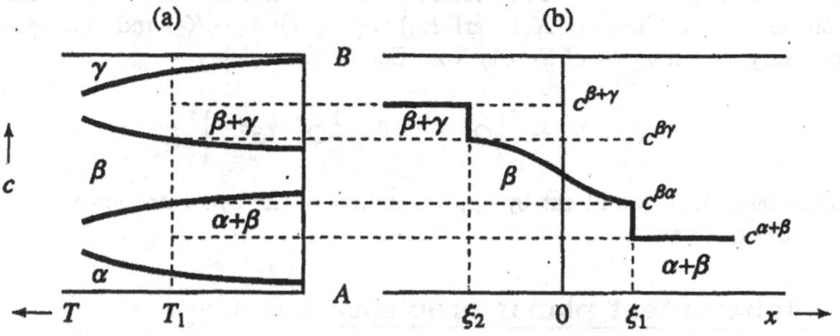

Figure 3: Concentration profile for a system of infinite extent with initial concentrations $c^{\alpha+\beta}$ and $c^{\beta+\gamma}$ in the $\alpha+\beta$ and $\beta+\gamma$ fields of the phase diagram.

For the concentration profile depicted in Figure 2, the boundary moves toward the $\alpha+\beta$ region (i.e., $K > 0$). If K is small, one can use the same reasoning that led to Eq(75) to obtain an approximate solution to Eq(76).

Infinite $\beta+\gamma$, β, $\alpha+\beta$ problems For two two-phase regions separated by a single-phase β region, as depicted in Figure 3, Eq(55) must be applied at each boundary. We take:

$$\xi_1 = K_1\sqrt{4D^\beta t} \tag{77}$$

$$\xi_2 = K_2\sqrt{4D^\beta t} \tag{78}$$

The concentration in the β phase,

$$c^\beta = c^{\beta\gamma}\left[\frac{\operatorname{erf}(K_1) - \operatorname{erf}(x/\sqrt{4D^\beta t})}{\operatorname{erf}(K_1) - \operatorname{erf}(K_2)}\right] + c^{\beta\alpha}\left[\frac{\operatorname{erf}(x/\sqrt{4D^\beta t}) - \operatorname{erf}(K_2)}{\operatorname{erf}(K_1) - \operatorname{erf}(K_2)}\right] \tag{79}$$

where the bracketed terms in Eq(79) are chosen alternately to vanish or equal unity at ξ_1 and ξ_2, thus making it obvious that the boundary conditions in β are satisfied. Substitution into the appropriate forms[9] of Eq(55) then yields the simultaneous equations

$$K_1 = \frac{1}{\sqrt{\pi}} Q^{\alpha+\beta} S^{\alpha+\beta} \frac{\exp(-K_1^2)}{\mathrm{erf}(K_1) - \mathrm{erf}(K_2)} \tag{80}$$

and

$$K_2 = -\frac{1}{\sqrt{\pi}} Q^{\gamma+\beta} S^{\gamma+\beta} \frac{\exp(-K_2^2)}{\mathrm{erf}(K_1) - \mathrm{erf}(K_2)} \tag{81}$$

where $S^{\alpha+\beta} := (c^{\beta\gamma} - c^{\beta\alpha})/(c^{\beta\alpha} - c^{\alpha+\beta})$ and $S^{\gamma+\beta} := (c^{\beta\gamma} - c^{\beta\alpha})/(c^{\gamma+\beta} - c^{\beta\gamma})$. The boundary at ξ_1 moves to the right and that at ξ_2 moves to the left, causing the β-phase region to thicken. Its width is just

$$\Delta\xi := \xi_1 - \xi_2 = (K_1 - K_2)\sqrt{4D^\beta t}. \tag{82}$$

An approximation for $K_1 - K_2$ can be found when $S^{\alpha+\beta}$ and $S^{\beta+\gamma}$ are small such that K_1 and K_2 are also small. Then $\mathrm{erf}(K_1) - \mathrm{erf}(K_2) \approx (2/\sqrt{\pi})(K_1 - K_2)$ and both exponentials are nearly unity. Subtraction of Eq(81) from Eq(80) then yields

$$K_1 - K_2 \approx \left[\frac{1}{2} Q^{\alpha+\beta} S^{\alpha+\beta} + \frac{1}{2} Q^{\beta+\gamma} S^{\beta+\gamma}\right]^{\frac{1}{2}}, \tag{83}$$

which is the result one would get by approximation of the concentration profile in the β phase by a straight line.

Semi-infinite extent planar problems

We next consider problems in which one or more phases occupy the infinite space between a free surface located at $x = \xi$, and $|x| \to \infty$. In general, ξ will be a function of time so the free surface, as well as other phase boundaries, will move relative to the center-of-volume frame of the β phase, for which $v^{\square\beta} = 0$.

Semi-infinite β problems As an introduction to the general considerations that underlie the motion of a free surface, we consider the case of a semi-infinite single phase located between $x = \xi$ and $x \to \infty$ where ξ can depend on t. We shall further assume that the concentration at $x = \xi$ has the constant value $c^{\beta\xi}$ while that at $x \to \infty$ has the constant value $c^{\beta\infty}$. Then the concentration in the β phase is

$$c^\beta = c^{\beta\infty} + (c^{\beta\xi} - c^{\beta\infty}) \left[\frac{1 - \mathrm{erf}\left(x/\sqrt{4D^\beta t}\right)}{1 - \mathrm{erf}(K)}\right] \tag{84}$$

where ξ is given by Eq(70). If the conditions at $x = \xi$ are such that there is no flux of A atoms (i.e., only B atoms) through this moving surface, then Eq(62) is applicable; substitution of Eq(84) into that boundary condition yields

$$K = -\frac{1}{\sqrt{\pi}} \frac{\bar{V}_B^\beta \left(c^{\beta\xi} - c^{\beta\infty}\right)}{1 - \bar{V}_B^\beta c^{\beta\xi}} \frac{\exp(-K^2)}{1 - \mathrm{erf}(K)}. \tag{85}$$

If $c^{\beta\xi} > c^{\beta\infty}$, as shown in Figure 4, B atoms are added to the solid and $K < 0$, indicating

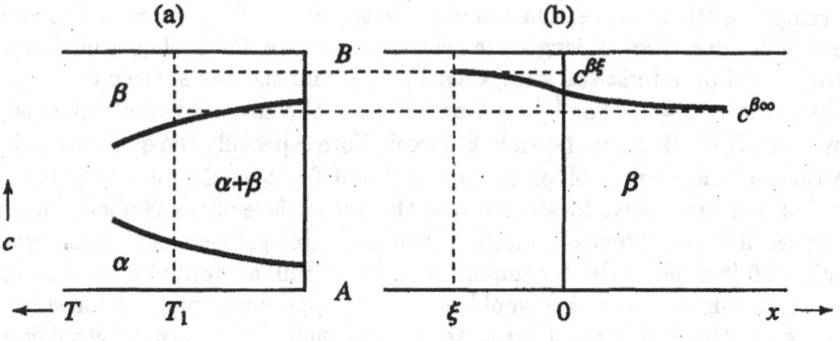

Figure 4: Concentration profile for a system of semi-infinite extent entirely in the β field of the phase diagram. The surface concentration $c^{\beta\xi}$ is greater than the initial concentration $c^{\beta\infty}$, resulting in an influx of species B and lengthening of the system.

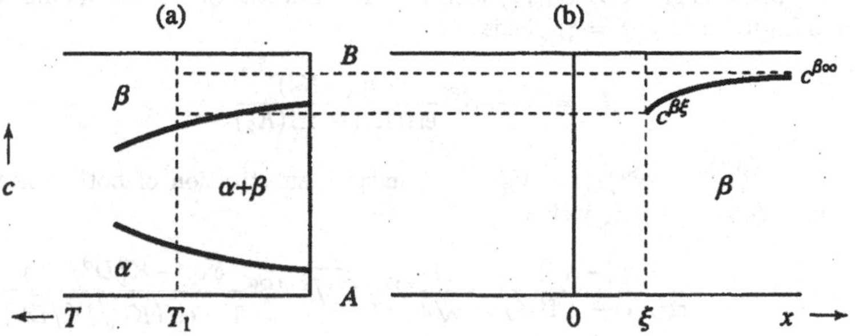

Figure 5: Concentration profile for a system of semi-infinite extent entirely in the β field of the phase diagram. The surface concentration $c^{\beta\xi}$ is less than the initial concentration $c^{\beta\infty}$, resulting in an outflux of species B and shortening of the system.

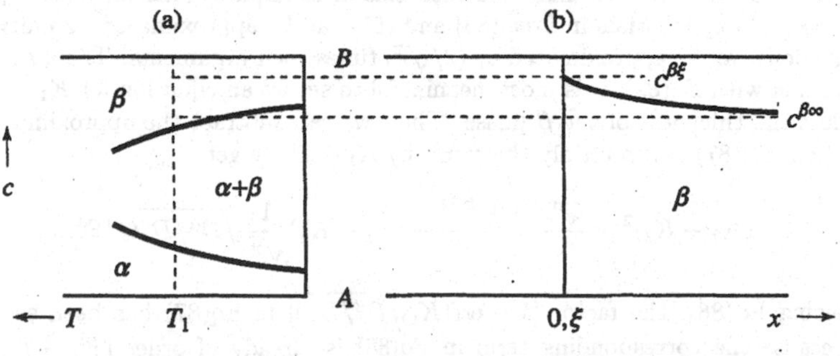

Figure 6: Concentration profile for a system of semi-infinite extent entirely in the β field of the phase diagram. B atoms in this instance are assumed to be interstitial with $\bar{V}_B^\beta \to 0$, resulting in a negligible change in the size of the system.

a lengthening[10] of the β phase. On the other hand, $c^{\beta\xi} < c^{\beta\infty}$ results in $K > 0$ in which case the β phase decreases in length; the corresponding profile is shown in Figure 5. If B atoms are located on interstitial sites, then $\bar{V}_B^\beta \to 0$ and the free surface of the β phase is practically static at $x = 0$. For $\bar{V}_B^\beta \to 0$ and $K \to 0$, the concentration profile is given by Eq(84) with $\text{erf}(K) = 0$ and is shown in Figure 6. Note especially the qualitative differences between the concentration profiles in Figures 4 and 6. It would be a serious mistake to presume that one can always locate $x = 0$ at the free surface of the β phase, presume that Eq(65) applies in that reference frame, and then invariably get a concentration profile of the form of Figure 6 [54,55]. If the B component is interstitial, as would be the case for carbon in iron, a static surface at $x = 0$ would be a good approximation, but for substitutional diffusion, e.g., in the dezincing of brass, the dimensional change would be substantial.

Semi-infinite $\beta\alpha$ problems This case is depicted in Figure 7. The free surface of the β phase is located at $\xi_2 = K_2\sqrt{4D^\beta t}$ and the $\alpha\beta$ boundary is located at $\xi_1 = K_1\sqrt{4D^\beta t}$. The concentration profile in the β phase is given by Eq(79) with $c^{\beta\gamma}$ replaced by $c^{\beta\xi}$, while that in the α phase is given by Eq(72) with $K = K_1$. Substitution of the modified form of Eq(79) into Eq(62) at $x = \xi = \xi_2$ yields

$$K_2 = -\frac{1}{\sqrt{\pi}} S^{\beta 2} \frac{\exp(-K_2^2)}{\text{erf}(K_1) - \text{erf}(K_2)} \tag{86}$$

where $S^{\beta 2} := \bar{V}_B^\beta(c^{\beta\xi} - c^{\beta\alpha})/(1 - \bar{V}_B^\beta c^{\beta\xi})$. Similarly, substitution of both concentration profiles into Eq(46) at $x = \xi_1$ yields

$$K_1 = \frac{1}{\sqrt{\pi}} Q^\beta S^{\beta 1} \frac{\exp(-K_1^2)}{\text{erf}(K_1) - \text{erf}(K_2)} - \frac{1}{\sqrt{\pi}} Q^\alpha \sqrt{D^\alpha/D^\beta} S^\alpha \frac{\exp(-K_1^2 D^\beta/D^\alpha)}{1 - \text{erf}(K_1\sqrt{D^\beta/D^\alpha})} \tag{87}$$

where $S^{\beta 1} := (c^{\beta\xi} - c^{\beta\alpha})/(c^{\beta\alpha} - c^{\alpha\beta})$ and $S^\alpha := (c^{\alpha\beta} - c^{\alpha\infty})/(c^{\beta\alpha} - c^{\alpha\beta})$ as before. Note that as $K_1 \to \infty$, Eq(86) is in agreement with Eq(85) with $K = K_2$.

In the general case, Eqs. (86) and (87) must be solved simultaneously to determine the motion of the free surface of β and the $\alpha - \beta$ phase boundary. Considerable simplification occurs if $S^{\beta 2}$, $S^{\beta 1}$ and S^α are small and their smallness implies small values of K_1 and K_2. In this case, all exponentials in Eqs. (86) and (87) can be approximated by unity and the error functions can be approximated by $(2/\sqrt{\pi})$ times their arguments. Then, as discussed in connection with Eq(82), it is most meaningful to set up an equation for $K_1 - K_2$ which determines the thickness of the β phase. Thus, we can subtract the approximate form of Eq(86) from Eq(87) and multiply the result by $K_1 - K_2$ to get

$$(K_1 - K_2)^2 \approx \frac{Q^\beta S^{\beta 1} + S^{\beta 2}}{2} - (K_1 - K_2)\frac{1}{\sqrt{\pi}}\sqrt{D^\alpha/D^\beta} Q^\alpha S^\alpha. \tag{88}$$

In obtaining Eq(88), the factor $[1 - \text{erf}(K_1\sqrt{D^\beta/D^\alpha})]$ in Eq(87) has been set equal to unity because the corresponding term in Eq(88) is already of order $(K_1 - K_2)S^\alpha$. An

[9] The form of Eq(55) at the $\gamma + \beta, \beta$ boundary is found by letting $\alpha \to \gamma$ and $v^{\alpha+\beta,\beta} \to d\xi_2/dt$.

[10] Since the β phase is infinite, this "lengthening" must be measured with respect to a fiducial mark that is stationary in the center-of-volume frame of the β phase.

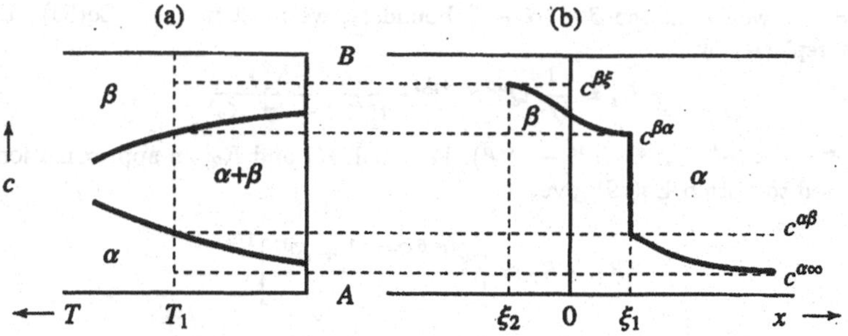

Figure 7: Concentration profile for a system of semi-infinite extent with surface concentration $c^{\beta\xi}$ in the β field and initial concentration $c^{\alpha\infty}$ in the α field of the phase diagram.

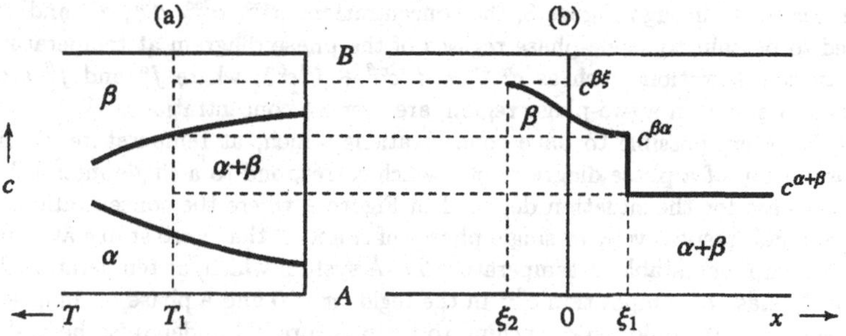

Figure 8: Concentration profile for a system of semi-infinite extent with surface concentration $c^{\beta\xi}$ in the β field and initial concentration $c^{\alpha+\beta}$ in the $\alpha+\beta$ field.

approximate[11] solution of Eq(88) is

$$K_1 - K_2 \approx \left[\frac{Q^\beta S^{\beta 1} + S^{\beta 2}}{2}\right]^{1/2} - \frac{1}{2}\frac{1}{\sqrt{\pi}}\sqrt{D^\alpha/D^\beta}Q^\alpha S^\alpha. \tag{89}$$

The first term arises from diffusion entirely in the β phase and dominates because of the small extent of that phase; the second term is a higher order correction for diffusion in α. Care should be taken in the use of Eq(89), however, since it has been assumed implicitly that $S^{\beta 1}$, $S^{\beta 2}$ and S^α are comparable in magnitude and that D^α is comparable to D^β. If the second term dominates in Eq(88), then $K_1 - K_2 \approx -(1/\sqrt{\pi})\sqrt{D^\alpha/D^\beta}Q^\alpha S^\alpha$ which is not a special case of Eq(89).

Semi-infinite $\beta, \alpha+\beta$ problems This case is depicted in Figure 8. Insofar as the concentration profile in the β phase and the motion of the free surface at ξ_2 are concerned, the situation is the same as for the case of semi-infinite $\beta\alpha$ problems; in particular, Eq(86)

[11]Eq(88) is of the form $(K_1 - K_2)^2 = a - b(K_1 - K_2)$ with roots $K_1 - K_2 = -(b/2) \pm \sqrt{(b/2)^2 + a}$. If b is smaller or even comparable to a, we take the + sign and Eq(89) results; but, if $a = 0$, we must take the − sign to get the non-zero root $K_1 - K_2 = -b$.

still holds. However, at the $\beta-(\alpha+\beta)$ boundary, we must now use Eq(55). Eq(87) is therefore replaced by

$$K_1 = \frac{1}{\sqrt{\pi}} Q^{\alpha\beta} S^{\alpha+\beta,1} \frac{\exp(-K_1^2)}{\mathrm{erf}(K_1) - \mathrm{erf}(K_2)} \tag{90}$$

where $S^{\alpha+\beta,1} := (c^{\beta\xi} - c^{\beta\alpha})/(c^{\beta\alpha} - c^{\alpha+\beta})$. For small K_1 and K_2, an approximation similar to that used to obtain Eq(89) gives

$$K_1 - K_2 \approx \left[\frac{Q^{\alpha+\beta} S^{\alpha+\beta,1} + S^{\beta 2}}{2}\right]^{1/2}. \tag{91}$$

Planar boundary motion following quenching

In the last two sections, the planar systems for both infinite and semi-infinite cases have been discussed relative to systems that have been prepared as multiphase diffusion couples. Thus, in Figure 1 through Figure 8, the concentrations $c^{\alpha\infty}$, $c^{\beta\infty}$, $c^{\alpha\beta}$, $c^{\beta\alpha}$ and $c^{\beta\xi}$ are all presumed to pertain to single-phase regions of the phase diagram at temperature T_1. In particular, concentrations such as $c^{\alpha+\beta} = f^{\alpha}c^{\alpha\beta} + f^{\beta}c^{\beta\alpha}$, where f^{α} and f^{β} are volume fractions of α and β in a two-phase region, are average concentrations.

It is, however, possible to have concentrations which, at temperature T_1, lie in the two-phase region of a phase diagram, but which correspond to a single metastable phase. Such is the case for the situation depicted in Figure 9 where the concentrations $c^{\alpha\infty}$ and $c^{\beta\infty}$ correspond, respectively, to single phases of α and β that were stable at temperature T_2, but that are metastable at temperature T_1. A system which, at temperature T_2, would consist of β phase of composition $c^{\beta\infty}$ in the region $x < 0$ and a phase of composition $c^{\alpha\infty}$ in the region $x > 0$ would, on quenching to temperature T_1, undergo isothermal diffusion with a concentration profile as shown in Figure 9. Insofar as the formal mathematics of the diffusion process and boundary motion are concerned, this problem is the same as that treated previously under infinite $\beta\alpha$ problems. Indeed, Eq(72) through Eq(75) are still applicable, but the quantities $(c^{\alpha\beta} - c^{\alpha\infty})$ and $(c^{\beta\infty} - c^{\beta\alpha})$ are now both negative, resulting in the differences in appearance between Figure 1 and Figure 9. In fact, the problem described by Figure 1 can be regarded as resulting from an upquench of a system that is prepared at a temperature much less than T_1. A second example of boundary motion following quenching is depicted in Figure 10 where the concentration $c^{\alpha\infty}$ corresponds to the α phase prepared at temperature T_2, but quenched to temperature T_1 where it is metastable. Insofar as isothermal diffusion at temperature T_1 is concerned, the mathematics is similar to that under semi-infinite $\beta\alpha$ problems to which Figure 7 applies. The only modification is that $c^{\beta\xi}$ must be taken equal to $c^{\beta\alpha}$ if $\xi_2 = x = 0$ is to be a plane of symmetry. This problem corresponds physically to the nucleation, at $x = 0$ and $t = 0$, and subsequent growth of a precipitate of β phase of composition $c^{\beta\alpha}$. The concentration profile in the α phase (for $x > 0$) is still given by Eq(72) with, of course, $K = K_1$. Similarly, Eq(87) applies with $S^{\beta 1} = 0$. These equations formally contain D^{β}; however, since there is no diffusion in the β phase, the appearance of D^{β} is specious. This can be alleviated if K', as defined by Eq(71), is used in place of K, i.e., $\xi_1 = K'\sqrt{4D^{\alpha}t}$. Then Eq(87) takes the form:

$$K' = \frac{1}{\sqrt{\pi}} Q^{\alpha} \left(\frac{c^{\alpha\infty} - c^{\alpha\beta}}{c^{\beta\alpha} - c^{\alpha\beta}}\right) \frac{\exp(-(K')^2)}{1 - \mathrm{erf}(K')}. \tag{92}$$

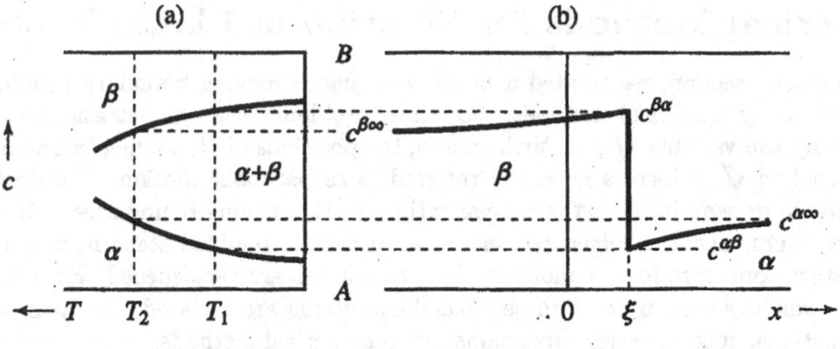

Figure 9: Concentration profile resulting from quenching a system of infinite extent, initially equilibrated in the $\alpha + \beta$ field of the phase diagram at temperature T_2, to a lower temperature, T_1. Growth of the β phase has taken place by boundary motion from its initial position, $x = 0$.

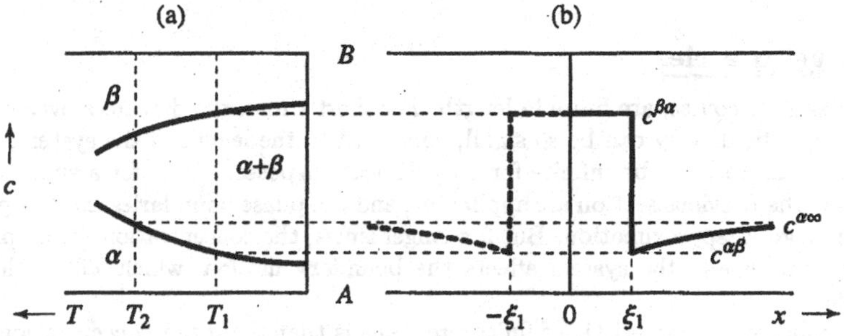

Figure 10: Concentration profile resulting from quenching a system of semi-infinite extent (concentration $c^{\alpha\infty}$) initially equilibrated in the α field of the phase diagram at temperature T_2 to a lower temperature, T_1. The β phase is presumed to nucleate, and growth of a platelet of β occurs via the concentration gradients established in the α phase.

In terms of K', Eq(72) for the concentration profile (for $x > 0$) becomes simply:

$$c^{\alpha} = c^{\alpha\infty} - (c^{\alpha\infty} - c^{\alpha\beta}) \left[\frac{1 - \mathrm{erf}(x/\sqrt{4D^{\alpha}t})}{1 - \mathrm{erf}(K')} \right]. \tag{93}$$

The interested reader can doubtless construct many other examples of quenching in planar systems that correspond formally to solutions already developed in the previous paragraphs on infinite extent planar problems and semi-infinite extent planar problems. Solubilities that decrease instead of increase with temperature will also lead to changes in detail. Insofar as growth of a planar precipitate is concerned, the situation already discussed in reference to Figure 10 is, perhaps, the most important [4, 29, 30]. Note well, however, that the width of the precipitate plate is actually $2\xi_1$. The two- and three-dimensional analogs of the problem depicted by Figure 10 will be discussed in connection with the section **Analytical Solutions to Non-Planar Problems.** The corresponding concentration profiles will resemble Figure 10, but the detailed shapes will differ.

Numerical Methods for Solution of Planar Problems

In the previous section, we treated a number of planar moving boundary problems that can be solved by analytical methods. For those problems, the concentrations were functions of only the variable $x/\sqrt{t}$. Furthermore, the positions of all moving boundaries were proportional to $\sqrt{t}$, which is generally referred to as parabolic motion. The system was either infinite or semi-infinite, the concentrations at the moving boundaries were assumed to be constants, and the initial concentration profiles resembled steps from a region of one constant concentration to another. Such conditions are very special. For other problems, for which one or more of these special conditions are violated, one must resort to approximations, such as series expansions, or to numerical methods.

In this section, we give a brief description of numerical methods for planar moving boundary problems. Our aim is not to be comprehensive, but only to indicate how such problems can be solved, and to point out some general features of their solutions. Rapid advances in computational hardware and algorithms have made this a rapidly expanding field, details of which are beyond the scope of the present article.

Finite geometries

Real systems, of course, are finite in length. For short times, the distances over which the concentration fields vary can be so small, compared to the length of the system, that the system appears to be to be infinite for all practical purposes. Thus, for a short time, the methods of the previous section are applicable, and the phase boundaries execute parabolic motion to a good approximation. But for longer times, the concentration fields spread out and the finite size of the system affects the boundary motion, which will no longer be parabolic.

An important characteristic of finite problems is that the mean concentration[12] of solute in the system, $\bar{c}$, can be defined. Then if there are no external fluxes, the average composition of the system is a constant, and the "endpoint" of the diffusion problem is dictated by the phase diagram.

We illustrate this for a finite $\beta\alpha$ problem corresponding to the phase diagram in Figure 1, and depicted in Figure 11. The initial length of the α phase is ℓ_0^α and that of the β phase is ℓ_0^β. Thus the mean concentration in the system is

$$\bar{c} = \frac{\ell_0^\beta c^{\beta 0} + \ell_0^\alpha c^{\alpha 0}}{\ell_0^\beta + \ell_0^\alpha}. \tag{94}$$

If $\bar{V}_A^\alpha = \bar{V}_A^\beta$ and $\bar{V}_B^\alpha = \bar{V}_B^\beta$, then the total volume of the system will remain constant as diffusion and phase transformation take place, as shown in section **Formalism**. Under these circumstances, $\bar{c}$ will remain constant, and dictate the final equilibrium state of the system. This state might be a single homogeneous phase, α or β, or a two-phase system having a definite proportion of α with concentration $c^{\alpha\beta}$ and β with concentration $c^{\beta\alpha}$, as given by the lever rule.[13] If both α and β are present in the final state of the system, the interface will eventually stop moving. For example, the interface could move to expand the

[12] As in the section **Analytical Solutions to Planar Problems**, we use c to denote the concentration of B atoms, with Greek superscripts denoting phases, as needed.

[13] More generally, it is the average *composition* of the system, say $\bar{X}_B$, that is conserved, irrespective of the relationship of partial molar volumes.

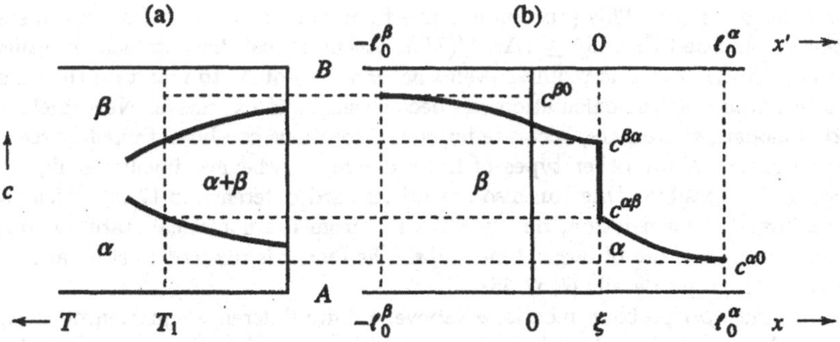

Figure 11: Concentration profile for a system of finite extent with initial concentrations $c^{\alpha 0}$ and $c^{\beta 0}$ in the α and β fields of the phase diagram. The initial length of the β phase is ℓ_0^β and that of the α phase is ℓ_0^α.

β phase, initially with parabolic motion, and then slow down and stop. It is significant, however, that even the sign of the boundary velocity can change as time goes on. Thus, if the final state of the system has an amount of β phase less than its initial amount, the boundary might still move initially to expand the β phase and then turn around and move in the opposite direction until it eventually stops. The direction of its initial velocity is determined by diffusion kinetics, as if the system were infinite, whereas eventually the finite nature of the system causes concentration gradients to adjust, and the interface to move toward its position in the equilibrium state, irrespective of kinetics. This phenomenon of "interface turnaround" has been demonstrated by numerical techniques and experimentally for finite couples of α and β brass [13].

Finite difference techniques

One way of solving planar moving boundary problems numerically is to use finite difference techniques to replace the differential equation for diffusion by a difference equation, and to approximate the concentration gradients that appear in the flux condition for boundary motion by finite differences. Thus, the differential equation Eq(65) can be approximated by the difference equation [7]

$$\frac{c_{i+1,j} - c_{i,j}}{\Delta t} = D\frac{c_{i,j+1} - 2c_{i,j} + c_{i,j-1}}{(\Delta x)^2} \tag{95}$$

where the index i denotes a time coordinate, which has been discretized in units of Δt, and the index j denotes a space coordinate, which has been discretized in units of Δx. In Eq(95), the time derivative $\partial c/\partial t$ at the spatial grid point j has been approximated by a first forward difference between c evaluated at time $(i+1)\Delta t$ and at time $i\Delta t$. The expression on the right hand side of Eq(95) is known as a second central difference and is an approximation to $\partial^2 c/\partial x^2$. It is based on values of c at the current time i but at three consecutive grid points, $j+1$, j, and $j-1$, the central one corresponding to j and the other two located at distances Δx ahead and behind. For each j, Eq(95) can be solved for $c_{i+1,j}$ in terms of known values of c at time step i, resulting in a knowledge of the entire concentration field at time step $i+1$. This procedure is free from cumulative error (numerical stability condition[14]) provided that $\Delta t \leq (\Delta x)^2/(2D)$. Accuracy can be obtained by using very small values of Δx, but this requires even smaller values of Δt to maintain the numerical stability condition, so the calculation can become slow and expensive. Nevertheless, with the aid of modern computers, accurate numerical solutions can be obtained at reasonable times and costs. Many other types of finite difference schemes, involving higher order differences, are possible. One can also use a backward difference in time, which requires a matrix inversion for solution, but which is free from the numerical stability condition, and thus permits the use of larger time steps. The reader is referred to standard texts on numerical analysis for details [7, 52, 56–59].

For the finite $\beta\alpha$ problem mentioned above, a finite difference equation of the form of Eq(95) would have to be solved in each phase with appropriate boundary conditions. At the fixed ends of the specimen, imposing boundary conditions presents no problems. For example, if the concentration is specified at a fixed end, one simply uses its value in the difference equation to update the concentration at its neighboring point in time. Moreover, if there is no flux of solute at a fixed end, say the extreme right end that corresponds to gridpoint N, one can simply invent a fictitious grid point $N+1$ located a distance Δx outside the specimen with concentration always equal to the concentration at the actual grid point $N-1$, located a distance Δx inside the specimen. In other words, $c_{i,N+1} = c_{i,N-1}$ for all i. This will guarantee a zero concentration gradient, and hence a zero flux, at the fixed boundary for all time, while allowing $c_{i,N}$ to change there according to Eq(95).

Handling conditions at the moving boundary is, however, more difficult. We illustrate this for the simple case of $\bar{V}_A^\alpha = \bar{V}_A^\beta$ and $\bar{V}_B^\alpha = \bar{V}_B^\beta$, which precludes any changes in the total volume of the system as well as relative motion of the bulk phases. Thus $\epsilon^\alpha = \epsilon^\beta = 0$ in Eq(50) and Eq(51) and the simplified flux condition Eq(53) applies. If the interface corresponds to gridpoint J, a finite difference form of Eq(53) is

$$(c_{i,J}^\beta - c_{i,J}^\alpha)\left(\frac{\xi_{i+1}-\xi_i}{\Delta t}\right) = -D_\beta\left(\frac{c_{i,J}^\beta - c_{i,J-1}^\beta}{\Delta x}\right) + D_\alpha\left(\frac{c_{i,J+1}^\alpha - c_{i,J}^\alpha}{\Delta x}\right). \tag{96}$$

This equation is based on a first forward difference in time, a first backward difference in space for β, and a first forward difference in space for α.[15] The interface concentrations $c_{i,J}^\beta$ and $c_{i,J}^\alpha$ can be taken to be the local equilibrium concentrations $c^{\beta\alpha}$ and $c^{\alpha\beta}$, respectively, or can even be variables determined from some other condition. Eq(96) can be then be solved to determine ξ_{i+1}, the new position of the interface. Unfortunately, the new interface position is unlikely to correspond to one of the spatial grid points. Its position would therefore have to be determined by interpolation, with appropriate modification of Eq(96) for subsequent time steps. Although this approach can be made to work, interpolation tends to introduce errors in calculating the boundary motion, which is usually the quantity of most interest.

[14] This is a sufficient condition for numerical stability. For the diffusion equation in n spatial dimensions, it is $\Delta t \leq (\Delta x)^2/(2nD)$.

[15] More accurate approximations for the spatial derivatives would be the second backward difference $(3c_{i,J}^\beta + c_{i,J-2}^\beta - 4c_{i,J-1}^\beta)/(2\Delta x)$ and the second forward difference $(4c_{i,J+1}^\alpha - c_{i,J+2}^\alpha - 3c_{i,J}^\alpha)/(2\Delta x)$. These expressions account for curvature of the concentration profiles near the boundary.

Expanding and contracting coordinate systems

An alternative technique is to first make a transformation that maps each phase onto a domain of fixed length, and then to apply finite difference techniques. Then the boundary will always coincide precisely with a fixed grid point in the mapped domain. We illustrate this for the simple case of $\bar{V}_A^\alpha = \bar{V}_A^\beta$ and $\bar{V}_B^\alpha = \bar{V}_B^\beta$ discussed above. Initially, the β phase occupies $-\ell_0^\beta < x < 0$ and the α phase occupies $0 < x < \ell_{\alpha 0}$. We furthermore assume that the fluxes of both A and B atoms are zero at $x = -\ell_0^\beta$ and $x = \ell_0^\alpha$, so that these boundaries do not move. At some later time, suppose that the $\alpha\beta$ boundary is located at $x = \xi(t)$. Then we introduce expanding and contracting coordinates x' in these phases as follows:

$$x' = \frac{x-\xi}{\ell_0^\beta+\xi}\ell_0^\beta \quad \text{for } x \text{ in } \beta \tag{97}$$

$$x' = \frac{x-\xi}{\ell_0^\alpha-\xi}\ell_0^\alpha \quad \text{for } x \text{ in } \alpha. \tag{98}$$

Thus x' ranges from $-\ell_0^\beta$ to 0 in β and from 0 to ℓ_0^α in α, and the boundary remains at $x' = 0$. Then $\partial c^\beta/\partial x = (\partial c^\beta/\partial x')(\partial x'/\partial x) = (\partial c^\beta/\partial x')[\ell_0^\beta/(\ell_{\beta 0}+\xi)]$ and similarly $\partial c^\alpha/\partial x = (\partial c^\alpha/\partial x')[\ell_0^\alpha/(\ell_0^\alpha-\xi)]$. The spatial derivatives in x and x' therefore differ by time-dependent expansion or contraction factors that are ratios of the original lengths of the respective phases to their current lengths. The corresponding time derivatives are

$$\begin{aligned}\left(\frac{\partial c^\beta}{\partial t}\right)_x &= \left(\frac{\partial c^\beta}{\partial t}\right)_{x'} - \left(\frac{\partial c^\beta}{\partial x}\right)_t \left(\frac{\partial x}{\partial t}\right)_{x'} \\ &= \left(\frac{\partial c^\beta}{\partial t}\right)_{x'} - \left(\frac{\partial c^\beta}{\partial x'}\right)_t \frac{\ell_{\beta 0}}{(\ell_0^\beta+\xi)} \frac{(\ell_0^\beta+x')}{\ell_{\beta 0}} \frac{d\xi}{dt}\end{aligned}$$

and similarly

$$\left(\frac{\partial c^\alpha}{\partial t}\right)_x = \left(\frac{\partial c^\alpha}{\partial t}\right)_{x'} - \left(\frac{\partial c^\alpha}{\partial x'}\right)_t \frac{\ell_{\alpha 0}}{(\ell_0^\alpha-\xi)} \frac{(\ell_0^\alpha-x')}{\ell_{\alpha 0}} \frac{d\xi}{dt}.$$

The diffusion equations in β and α therefore become

$$\left(\frac{\partial c^\beta}{\partial t}\right)_{x'} = \left(\frac{\partial c^\beta}{\partial x'}\right)_t \frac{\ell_{\beta 0}}{(\ell_0^\beta+\xi)} \frac{(\ell_0^\beta+x')}{\ell_{\beta 0}} \frac{d\xi}{dt} + D^\beta \left(\frac{\partial^2 c^\beta}{\partial x'^2}\right)_t \left[\frac{\ell_0^\beta}{\ell_0^\beta+\xi}\right]^2 \tag{99}$$

$$\left(\frac{\partial c^\alpha}{\partial t}\right)_{x'} = \left(\frac{\partial c^\alpha}{\partial x'}\right)_t \frac{\ell_{\alpha 0}}{(\ell_0^\alpha-\xi)} \frac{(\ell_0^\alpha-x')}{\ell_{\alpha 0}} \frac{d\xi}{dt} + D^\alpha \left(\frac{\partial^2 c^\alpha}{\partial x'^2}\right)_t \left[\frac{\ell_0^\alpha}{\ell_0^\alpha-\xi}\right]^2. \tag{100}$$

The interface flux condition becomes

$$(c^{\beta\alpha}-c^{\alpha\beta})\frac{\partial\xi}{\partial t} = -D_\beta \left(\frac{\partial c^\beta}{\partial x'}\right)_t \frac{\ell_{\beta 0}}{(\ell_0^\beta+\xi)} + D_\alpha \left(\frac{\partial c^\alpha}{\partial x'}\right)_t \frac{\ell_{\alpha 0}}{(\ell_0^\alpha-\xi)} \tag{101}$$

where the spatial derivatives are to be evaluated at $x' = 0$.

We can now proceed to convert Eqs(99-101) to finite difference equations by discretizing t and x'. In β, we divide ℓ_0^β into N^α equal intervals of size Δx_0^β and in α, we divide ℓ_0^α into N^β equal intervals of size Δx_0^α. This allows us to write the finite difference approximations to the first and second spatial derivatives with respect to x' in a manner similar to that done above for x. But for convenience of notation, we define the variable grid spacings

$$\Delta x_i^\beta = \Delta x_0^\alpha [(\ell_0^\beta + \xi_i)/\ell_0^\beta] = (\ell_0^\beta + \xi_i)/N^\beta \tag{102}$$
$$\Delta x_i^\alpha = \Delta x_0^\beta [(\ell_0^\alpha - \xi_i)/\ell_0^\alpha] = (\ell_0^\alpha - \xi_i)/N^\alpha \tag{103}$$

that incorporate the expansion and contraction factors, and are therefore dependent on the time index i through the interface position ξ_i. Then the finite difference forms of Eqs(99-101) are

$$\frac{c_{i+1,j}^\beta - c_{i,j}^\beta}{\Delta t} = \frac{c_{i,j+1}^\beta - c_{i,j-1}^\beta}{2\Delta x_i^\beta}\left(\frac{N^\beta + j}{N^\beta}\right)\left(\frac{\xi_{i+1} - \xi_i}{\Delta t}\right) + D^\beta \frac{c_{i,j+1}^\beta - 2c_{i,j}^\beta + c_{i,j-1}^\beta}{(\Delta x_i^\beta)^2} \tag{104}$$
$$\frac{c_{i+1,j}^\alpha - c_{i,j}^\alpha}{\Delta t} = \frac{c_{i,j+1}^\alpha - c_{i,j-1}^\alpha}{2\Delta x_i^\alpha}\left(\frac{N^\alpha - j}{N^\alpha}\right)\left(\frac{\xi_{i+1} - \xi_i}{\Delta t}\right) + D^\alpha \frac{c_{i,j+1}^\alpha - 2c_{i,j}^\alpha + c_{i,j-1}^\alpha}{(\Delta x_i^\alpha)^2} \tag{105}$$

and

$$(c_{i,0}^\beta - c_{i,0}^\alpha)\left(\frac{\xi_{i+1} - \xi_i}{\Delta t}\right) = -D_\beta\left(\frac{c_{i,0}^\beta - c_{i,-1}^\beta}{\Delta x_i^\beta}\right) + D_\alpha\left(\frac{c_{i,1}^\alpha - c_{i,0}^\alpha}{\Delta x_i^\alpha}\right). \tag{106}$$

From the initial concentration profile, Eq(106) can be solved to determine $\xi_{i+1} - \xi_i$ and substituted into Eq(104) and Eq(105), which can then be solved to determine $c_{i+1,j}^\beta$ and $c_{i+1,j}^\alpha$. This process can then be repeated. The factors $[(N^\beta + j)/N^\beta]$ and $[(N^\alpha - j)/N^\alpha]$ determine the fraction of the boundary velocity that corresponds to the jth grid point in each phase (in the β phase, $x' = j\Delta x_0^\beta$ so $j \leq 0$ there). These factors are equal to zero at the fixed boundaries and equal to one at the moving boundary. The terms containing these factors can be regarded as "corrections" needed to deal with moving grid points [56], but similar terms also appear in the continuum equations, Eq(99) and Eq(100). Note also that the variable grid spacings Δx_i^β and Δx_i^α must be determined for each time step i. If these become too large or too small, one could lose accuracy or violate the numerical stability condition, respectively. With these precautions in mind, however, such finite difference equations can be solved quite efficiently on modern computers.

We have already remarked that the boundary concentrations $c_{i,0}^\beta$ and $c_{i,0}^\alpha$ can be taken as variables rather than the local equilibrium values $c^{\beta\alpha}$ and $c^{\alpha\beta}$. One possibility would be to allow for departures from local equilibrium of the forms $c_{i,0}^\beta - c^{\beta\alpha} = \mu^\beta d\xi/dt$ and $c_{i,0}^\alpha - c^{\alpha\beta} = -\mu^\alpha d\xi/dt$ where μ^β and μ^α are linear kinetic coefficients. These linear kinetic laws can then be substituted into Eq(106) with $d\xi/dt$ approximated by $(\xi_{i+1} - \xi_i)/\Delta t$. Another complication that can be handled numerically is that of more general initial conditions. For example, a concentration profile obtained experimentally from an electron microprobe analysis may be digitized to provide an initial condition. Furthermore, one can treat problems involving several regions of one phase embedded in another, such as occur during the homogenization of compacted blends of powders, the growth and dissolution of second phase particles, the degradation of hot pressed composite materials, or the formation of second phases at the interface between bonded constituents [8, 9, 11–13, 17, 18].

Concentration-dependent interdiffusion coefficients

Throughout the preceding analysis, we have assumed that the interdiffusion coefficients were independent of concentration. This is a valid assumption in many problems where the overall accuracy of all the concentration parameters (solubilities) and diffusion coefficients does not warrant a more complex analysis. In many instances, however, the dependence of the interdiffusion coefficients on concentration is large. Two approaches to such a situation are valid. Either a mean interdiffusion coefficient for the phase of interest can be used, or known data for the dependence on concentration can be treated by employing numerical techniques [60].

If the diffusivity is no longer constant, Eq(65) must be replaced by

$$\frac{\partial c}{\partial t} = \frac{\partial}{\partial x}\left(D\frac{\partial c}{\partial x}\right) \tag{107}$$

in each phase. Its finite difference form, corresponding to Eq(95), is

$$\frac{c_{i+1,j} - c_{i,j}}{\Delta t} = \left(\frac{D_{i,j+1} - D_{i,j-1}}{2\Delta x}\right)\left(\frac{c_{i,j+1} - c_{i,j-1}}{2\Delta x}\right) + D_{i,j}\frac{c_{i,j+1} - 2c_{i,j} + c_{i,j-1}}{(\Delta x)^2}. \tag{108}$$

Eq(108) can also be written in the form

$$\frac{c_{i+1,j} - c_{i,j}}{\Delta t} = \left(\frac{D(c_{i,j+1}) - D(c_{i,j-1})}{c_{i,j+1} - c_{i,j-1}}\right)\left(\frac{c_{i,j+1} - c_{i,j-1}}{2\Delta x}\right)^2 + D(c_{i,j})\frac{c_{i,j+1} - 2c_{i,j} + c_{i,j-1}}{(\Delta x)^2} \tag{109}$$

which is more useful if the function $D(c)$ is known. Eq(96) is still valid with D^β and D^α replaced by $D^\beta_{i,j} = D^\beta(c_{i,j})$ and $D^\alpha_{i,j} = D^\alpha(c_{i,j})$. Finite difference equations for expanding and contracting coordinate systems can also be developed along the lines already indicated.

Analytical Solutions to Non-Planar Problems

In this section, we present analytical solutions to a number of problems for which a moving phase boundary has a non-planar shape and for which transport is governed by diffusion.

Nonplanar models

For nonplanar systems, we present two models, the precipitation model and the solidification model. The former pertains to the isothermal precipitation of a β phase from an α phase in a binary alloy, and was the only case treated in the original article [1]. The latter pertains to the solidification of a pure crystal from its melt. With suitable assumptions, we show that these models are isomorphous. In the examples that follow, we actually use the notation of the solidification model.

Precipitation model We consider the precipitation of a β phase from an α phase in a binary alloy. The β phase is assumed to be a nearly stoichiometric compound or a pure substance, such that for all practical purposes, its concentration is uniform and may be denoted by a single constant value, $c^{\beta\alpha}$. Motion of the $\alpha - \beta$ phase boundary is assumed to be governed by diffusion in the α phase under conditions for which Eq(25) and Eq(38) are applicable, with constant diffusivity. Furthermore, as was discussed toward the end of section **Formalism**, it will be assumed that $\bar{V}_i^\alpha = \bar{V}_i^\beta$ for $i = A, B$ in order to avoid complications associated with deformation or flow. Thus, for constant D^α, the concentration of species B in the α phase is governed by the equation

$$\frac{\partial c^\alpha}{\partial t} = D^\alpha \nabla^2 c^\alpha, \tag{110}$$

subject to the conditions

$$\Delta c U = D^{\alpha} \nabla c^{\alpha} \cdot \hat{\mathbf{n}} \tag{111}$$

and $c^{\alpha} = c^{\alpha\beta}$ at the α-β boundary, and to the condition $c^{\alpha} = c^{\alpha\infty}$ in the α phase infinitely far from the β phase. Here, U is the local normal growth speed, $\hat{\mathbf{n}}$ is the local unit normal at the α-β boundary (directed toward the α phase) and $\Delta c = c^{\beta\alpha} - c^{\alpha\beta}$. Eq(111) is the three dimensional generalization of Eq(53) when there is no diffusion in the β phase.

Solidification model A very similar problem arises for the solidification (crystallization) of an *isothermal* solid, at its melting point T_M, from a pure supercooled liquid (melt). Provided that there is no convection, the temperature field T in the liquid satisfies a diffusion equation of the form

$$\frac{\partial T}{\partial t} = \kappa \nabla^2 T, \tag{112}$$

where κ is the thermal diffusivity, assumed constant. Furthermore, we assume that the density of the system is uniform and that the solid-liquid interface is at temperature T_M (taken to be constant, and therefore ignoring corrections due to interface curvature). Then the boundary conditions at that interface are

$$L_o U = -k \nabla T \cdot \hat{\mathbf{n}} \tag{113}$$

and $T = T_M$, where L_o is the (assumed constant) latent heat per unit volume, k is the (assumed constant) thermal conductivity of the liquid, U is the local normal speed of solidification, and $\hat{\mathbf{n}}$ is the unit normal to the interface, pointing from solid to liquid. In the liquid, infinitely far from the solid, the temperature is taken to be $T = T_{\infty}$, also a constant.

Isomorphous relationship The fact that the precipitation model and the solidification model are isomorphous can be demonstrated by introducing dimensionless variables as follows:

dimensionless length $(\tilde{x}, \tilde{y}, \tilde{z}) = (x, y, z)/\mathcal{L}$ and $\tilde{\nabla} = \mathcal{L}\nabla$ where $\mathcal{L}$ is some constant length.

dimensionless time $\tilde{t} = t(D^{\alpha}/\mathcal{L}^2)$ for precipitation and $\tilde{t} = t(\kappa/\mathcal{L}^2)$ for solidification.

dimensionless growth speed $\tilde{U} = U(\mathcal{L}/D^{\alpha})$ for precipitation and $\tilde{U} = U(\mathcal{L}/\kappa)$ for solidification.

dimensionless field $\psi = (c^{\alpha\beta} - c^{\alpha})/\Delta c$ for precipitation and $\psi = c_o(T - T_M)/L_o$ for solidification, where c_o is the (assumed constant) heat capacity of the liquid per unit volume.

dimensionless driving force $S = (c^{\alpha\infty} - c^{\alpha\beta})/\Delta c$ for the dimensionless supersaturation and $S = c_o(T_M - T_{\infty})/L_o$ for the dimensionless supercooling.

Then if we note that $\kappa = k/c_o$, the precipitation model and the solidification model reduce to the common problem

$$\frac{\partial \psi}{\partial \tilde{t}} = \tilde{\nabla}^2 \psi, \tag{114}$$

subject to the conditions

$$\tilde{U} = -\tilde{\nabla}\psi \cdot \hat{\mathbf{n}} \tag{115}$$

and $\psi = 0$ at the moving boundary, and to the far-field condition $\psi = -S$.

Similarity solutions and the method of Ivantsov

Analytical solutions to non-planar moving boundary problems are not readily obtainable for arbitrary shapes; instead, a small number of solutions for very simple shapes have been discovered. In every case, these solutions correspond to the discovery of a specific function, say $\xi(x, y, z, t)$, such that T depends on x, y, z, and t *only* through ξ. Eq(112) can then be converted into an *ordinary* differential equation in ξ and the boundary condition Eq (113) will also be expressible with ξ as the only variable. These solutions are called *similarity solutions* because all of the isotherms belong to a similar class of shapes corresponding to constant values of the *similarity variable* $\xi(x, y, z, t)$.

An elegant method of finding allowable functions $\xi(x, y, z, t)$ was invented by Ivantsov [61, 62]. We give a brief description of the method of Ivantsov, as elaborated by Horvay and Cahn [63] and Canright and Davis [64], in order to motivate the choice of similarity variables. The method consists of generalizing the flux boundary condition, Eq(113), to a first order *nonlinear* partial differential equation that can be solved to yield isotherms that belong to some shape class, and that evolve in time in a definite way; this results in identification of a potential similarity variable. Then one seeks compatible solutions of the linear partial differential equation, Eq(112), that depend only on that variable. If this procedure is successful, the result is an *ordinary* differential equation for $T(\xi)$.

To carry out this process, one notes that the total derivative of $T(x, y, z, t) = T_M$ gives $\mathbf{v} \cdot \nabla T + \partial T/\partial t = 0$ where the vector $\mathbf{v} = (dx/dt,\ dy/dt,\ dz/dt)$. The normal to an isothermal surface can be expressed in the form $\hat{\mathbf{n}} = \nabla T/|\nabla T|$. Thus the normal growth speed $U = \mathbf{v} \cdot \hat{\mathbf{n}} = -(1/|\nabla T|)(\partial T/\partial t)$. Substitution into Eq(113) gives

$$\frac{L_o}{k}\frac{\partial T}{\partial t} = |\nabla T|^2. \tag{116}$$

Ivantsov generalized Eq(116) by replacing L_o/k by a function $f(T)$ of temperature to obtain

$$f(T)\frac{\partial T}{\partial t} = |\nabla T|^2 \tag{117}$$

which is assumed to hold *everywhere*, and yields Eq(116), provided that $f(T_M) = k/L_o$.

To solve Eq(117), we make the transformation

$$u(T) = -\int^{T} \frac{dT'}{f(T')} \tag{118}$$

to obtain

$$\frac{\partial u}{\partial t} + \left(\frac{\partial u}{\partial x}\right)^2 + \left(\frac{\partial u}{\partial y}\right)^2 + \left(\frac{\partial u}{\partial z}\right)^2 = 0. \tag{119}$$

Elementary solutions to the first order nonlinear Eq(119) can be found by substituting a *linear* function of the form $u = Ax + By + Cz + Dt + E$, which solves the equation, provided that the constants are related by $D = -(A^2 + B^2 + C^2)$, and E is arbitrary. This solution can be rewritten[16] in the elegant form

$$\mathcal{F}(C_1, C_2, C_3, C_4, C_5) \equiv \frac{C_1^2 + C_2^2 + C_3^2}{C_4} u(T) + C_1 x + C_2 y + C_3 z + C_4 t + C_5 = 0 \quad (120)$$

where all five constants C_i are arbitrary. More general solutions to Eq(119) can then be obtained by finding surfaces that are tangent to infinitely many of the linear functions represented by Eq(120), a process known as *envelope formation*. To do this, one needs to specify functional relationships among the C_i and then eliminate the remaining ones by means of the additional equations $\partial \mathcal{F}/\partial C_i = 0$.

Provided that $\mathcal{F}$ is still a homogeneous function of degree one in the remaining C_i, this process is tractable.[17]

Ellipsoids We choose [63,64] $C_5 = \alpha_1 C_1^2 + \alpha_2 C_2^2 + \alpha_3 C_3^2)/C_4$, which results in

$$\mathcal{F}(C_1, C_2, C_3, C_4) \equiv \frac{C_1^2(u+\alpha_1) + C_2^2(u+\alpha_2) + C_3^2(u+\alpha_3)}{C_4} + C_1 x + C_2 y + C_3 z + C_4 t. \quad (121)$$

Then differentiation of $\mathcal{F}(C_1, C_2, C_3, C_4) = 0$ with respect to C_1, C_2, C_3 and C_4, respectively, gives $2C_1/C_4 = -x/(u+\alpha_1)$, $2C_2/C_4 = -y/(u+\alpha_2)$, $2C_3/C_4 = -z/(u+\alpha_3)$, and $[C_1^2(u+\alpha_1) + C_2^2(u+\alpha_2) + C_3^2(u+\alpha_3)]/C_4^2 = t$. Eliminating the C_i from these expressions gives

$$\frac{x^2}{u+\alpha_1} + \frac{y^2}{u+\alpha_2} + \frac{z^2}{u+\alpha_3} = 4t. \quad (122)$$

For a range of u such that the quantities $u + \alpha_i$ are all positive, the isotherms are ellipsoids with linear dimensions that grow as $\sqrt{t}$. One of these isotherms, corresponding to a special value of u, will correspond to the solid-liquid interface, which will be an ellipsoid with solid inside and liquid outside. Spherical interfaces correspond to the special case $\alpha_1 = \alpha_2 = \alpha_3$.

Hyperboloids For a range of u such that $u + \alpha_1$ is negative, $u + \alpha_2$ is negative and $u + \alpha_3$ is positive, Eq(122) represents an elliptical hyperboloid of two sheets, one opening toward negative z and the other opening toward positive z. The tips of these hyperboloids are located at $x = y = 0$ and $z = \pm\sqrt{4(u+\alpha_3)t}$. A special value of u will correspond to the solid-liquid interface, so the corresponding hyperboloid, which will contain the solid phase, will *recede* from the plane $z = 0$, which describes *melting*. For a range of u such that $u + \alpha_1$ is positive, $u + \alpha_2$ is positive and $u + \alpha_3$ is negative, Eq(122) represents an elliptical hyperboloid of one sheet, of infinite extent along the z axis. The trace of this hyperboloid in the plane $z = 0$ is an ellipse, with linear dimensions that grow like $\sqrt{t}$. Again, a special value of u will correspond to the solid-liquid interface; the solid will be outside this interface, and will be melting.

[16]The transformation is $A = -C_1C_4/Q$, $B = -C_2C_4/Q$, $C = -C_3C_4/Q$, $D = -C_4^2/Q$, and $E = -C_5C_4/Q$, where $Q = C_1^2 + C_2^2 + C_3^2$.

[17]The function $\mathcal{F}$ is homogeneous of degree one in the variables C_i, so by Euler's theorem of homogeneous functions, $\mathcal{F} = \sum_i C_i \partial\mathcal{F}/\partial C_i$. Thus if $\partial\mathcal{F}/\partial C_i$ vanishes for each i, $\mathcal{F}$ also vanishes. We can therefore work only with the partial derivatives of $\mathcal{F} = 0$, which are relatively simple functions.

Paraboloids We choose [63,64] $C_5 = (\alpha_1 C_1^2 + \alpha_2 C_2^2)/C_4$ and $C_4 = -VC_3$, which results in

$$\mathcal{F}(C_1, C_2, C_3) \equiv -\frac{C_1^2(u+\alpha_1) + C_2^2(u+\alpha_2)}{VC_3} + C_1 x + C_2 y + C_3(z - Vt - \frac{u}{V}). \quad (123)$$

Then differentiation of $\mathcal{F}(C_1, C_2, C_3) = 0$ with respect to C_1, C_2 and C_3 gives, respectively, $2C_1/C_3 = Vx/(u+\alpha_1)$, $2C_2/C_3 = Vy/(u+\alpha_2)$, and $[C_1^2(u+\alpha_1) + C_2^2(u+\alpha_2)]/(VC_3^2) + z - Vt - u/V = 0$. Eliminating the C_i from these expressions gives

$$-\frac{Vx^2}{4(u+\alpha_1)} - \frac{Vy^2}{4(u+\alpha_2)} + \frac{u}{V} = z - Vt, \quad (124)$$

which is an elliptical paraboloid that extends an infinite distance along the negative z axis and translates uniformly in the positive z direction with velocity V.

This process of envelope formation results in ellipsoids, hyperboloids and paraboloids of various aspect ratios and orientations. The resulting shapes are *quadric* surfaces, i.e., surfaces represented by polynomials of degree two in x, y and z. For such surfaces, it turns out that the assumption that T depends only on u results in conversion of Eq(112) to an ordinary differential equation, resulting in success of the method of Ivantsov. To the best of our knowledge, exact solutions to the solidification problem are only possible for these quadric surfaces.

Once we recognize the existence of these exact solutions, we can obtain them by a more straightforward method due to Ham [65], who also considered similarity variables that depend on time as $\sqrt{t}$ or as $\sqrt{\tau - t}$ where τ is a constant. In some cases, this allows one to treat both freezing and melting problems for an interface of a given shape class (e.g., hyperboloids). Such generalizations are not trivial because Eq(112) is parabolic, so it is not invariant under time reversal. In the following sections, we use the methods of Ivantsov and Ham, and a somewhat different notation, to catalog these solutions.

Uniformly translating solutions

We proceed to obtain solutions for uniformly translating elliptical paraboloids of the type represented by Eq(124) that was obtained by the method of Ivantsov. We represent these by the equation[18]

$$\frac{(Vx/2\kappa)^2}{\xi} + \frac{(Vy/2\kappa)^2}{\xi + B} = \xi - 2\,(V/2\kappa)\,(z - Vt) \quad (125)$$

where V is the constant velocity of translation. The similarity variable is now represented by ξ, which is dimensionless. We choose the dimensionless parameter B to be positive so that the y axis is the major axis of the elliptical cross section in a plane perpendicular to the z axis.

We first observe that ξ depends on t only through the variable $z - Vt$. Thus if T depends only on ξ, we have $\partial T/\partial t = -V \partial T/\partial z$, so T obeys the steady-state equation

$$\frac{\partial^2 T}{\partial x^2} + \frac{\partial^2 T}{\partial y^2} + \frac{\partial^2 T}{\partial z^2} + \frac{V}{\kappa}\frac{\partial T}{\partial z} = 0, \quad (126)$$

[18] The correspondence in notation is $\xi = u/\kappa$, $\alpha_1 = 0$ and $B = \alpha_2/\kappa$.

which also holds in a moving reference frame, with z replaced by $z' = z - Vt$ and T now regarded as a function of only x, y and z'. The boundary condition, Eq(116), becomes

$$-V\frac{\partial T}{\partial z} = \frac{k}{L_o}|\nabla T|^2. \tag{127}$$

We next introduce dimensionless variables $(X, Y, Z) = (x, y, z - Vt)/(2\kappa/V)$ in terms of which Eqs(125-127) become

$$F(X, Y, Z) := \frac{X^2}{\xi} + \frac{Y^2}{\xi + B} + 2Z - \xi = 0, \tag{128}$$

$$T_{XX} + T_{YY} + T_{ZZ} + 2T_Z = 0, \tag{129}$$

and

$$-2(L_o/c_o)T_Z = (T_X)^2 + (T_Y)^2 + (T_Z)^2, \tag{130}$$

where subscripts denote partial differentiation. Then, since T depends only on ξ, Eq(129) and Eq(130) become

$$(\xi_X^2 + \xi_Y^2 + \xi_Z^2)T_{\xi,\xi} + (\xi_{XX} + \xi_{YY} + \xi_{ZZ} + 2\xi_Z)T_\xi = 0 \tag{131}$$

and

$$-2(L_o/c_o)\xi_Z = (\xi_X^2 + \xi_Y^2 + \xi_Z^2)T_\xi. \tag{132}$$

We proceed by implicit differentiation[19] to recast the partial differential equation (PDE) Eq(131) into an ordinary differential equation (ODE) of the form

$$\frac{d^2T}{d\xi^2} + \left[\frac{1}{2\xi} + \frac{1}{2(B+\xi)} + 1\right]\frac{dT}{d\xi} = 0. \tag{133}$$

Similarly the boundary condition, Eq(132), becomes

$$\left(\frac{L_o}{c_o}\right) = -\left(\frac{dT}{d\xi}\right)_{\xi=\xi_o} \tag{134}$$

where $\xi = \xi_o$ denotes the value of ξ corresponding to the solid-liquid interface.

Eq(133) can be solved by finding an integrating factor[20] which allows it to be written in the form

$$\left[[\xi(\xi + B)]^{1/2} \exp(\xi)\ T_\xi\right]_\xi = 0. \tag{135}$$

Its solution that equals T_M at $\xi = \xi_o$ and T_∞ at $\xi = \infty$ is

$$T = T_\infty + (T_M - T_\infty)\frac{G_p(\xi)}{G_p(\xi_o)} \tag{136}$$

where

$$G_p(\xi) := \int_\xi^\infty \frac{\exp(-w)}{[w(w+B)]^{1/2}}dw. \tag{137}$$

[19]This process is straightforward but tedious. For example, $F_X + F_\xi \xi_X = 0$, and $F_{XX} + 2F_{X\xi}\xi_X + F_{\xi\xi}(\xi_X)^2 + F_\xi \xi_{XX} = 0$, with similar expressions for Y and Z. Thus $\xi_Z = -2/F_\xi$, $\xi_X^2 + \xi_Y^2 + \xi_Z^2 = -4/F_\xi$, and $\xi_{XX} + \xi_{YY} + \xi_{ZZ} = -(2/F_\xi)[1/\xi + 1/(\xi + B)]$.

[20]The integrating factor is $\exp\left(\int^\xi [1/2u + 1/2(u + B) + 1]\, du\right) = [\xi(\xi + B)]^{1/2} \exp(\xi)$.

Satisfaction of Eq(134) then requires ξ_o to obey the transcendental equation

$$S = [\xi_o(\xi_o + B)]^{1/2} \exp(\xi_o)\, G_p(\xi_o) \tag{138}$$

where $S = c_o(T_M - T_\infty)/L_o$ is the dimensionless supercooling, identified previously as a dimensionless driving force.

At the tip of the paraboloid ($x = y = 0$), the radius of curvature in the $x - z$ plane is

$$\rho_x = \left(\frac{\partial^2 z}{\partial x^2}\right)^{-1}_{\xi=\xi_o} = (2\kappa/V)\,\xi_o. \tag{139}$$

Similarly, in the $y - z$ plane

$$\rho_y = \left(\frac{\partial^2 z}{\partial y^2}\right)^{-1}_{\xi=\xi_o} = (2\kappa/V)\,(\xi_o + B). \tag{140}$$

Therefore, B is a shape parameter that describes the degree of eccentricity. From Eq(139), we note that

$$\xi_o = \frac{V\rho_x}{2\kappa} = P \tag{141}$$

where P is known as the Peclet number, referred to the smaller tip radius ρ_x.

The two limiting cases $B = 0$ and $B = \infty$ yield Ivantsov's results for the circular paraboloid and the parabolic cylinder [61,62]. For $B = 0$, $G_p(\xi)$ becomes the exponential integral

$$E_1(\xi) := \int_\xi^\infty \frac{\exp(-w)}{w}\, dw, \tag{142}$$

so Eq(138) becomes

$$S = P \exp(P) E_1(P). \tag{143}$$

For $B = \infty$, Eq(138) becomes

$$S = \sqrt{\pi P} \exp(P) \operatorname{erfc}(\sqrt{P}). \tag{144}$$

where

$$\operatorname{erfc}(u) := \frac{2}{\sqrt{\pi}} \int_u^\infty \exp(-y^2)\, dy \tag{145}$$

is the complementary error function.

In the general case of an elliptical paraboloid, we note that $\rho_y/\rho_x = (P + B)/P$, so if B were held constant and $P = \xi_o$ were to change with S according to Eq(138), the shape of the paraboloid would also change with S. In order to study the growth of a paraboloid of constant *shape* as a function of S, we write[21] $B = rP$ where r is a constant, resulting in $\rho_y/\rho_x = 1 + r$. Then Eq(138) can be rewritten in the form

$$S = (1 + r)^{1/2} P \exp(P) \int_1^\infty \frac{\exp(-Pu)}{[(u + r)u]^{1/2}}\, du. \tag{146}$$

Plots of S as a function of P for several values of r are shown in Figure(12). Note that there are no solutions for $S > 1$, a situation known as hypercooling [66]. As shown previosly,

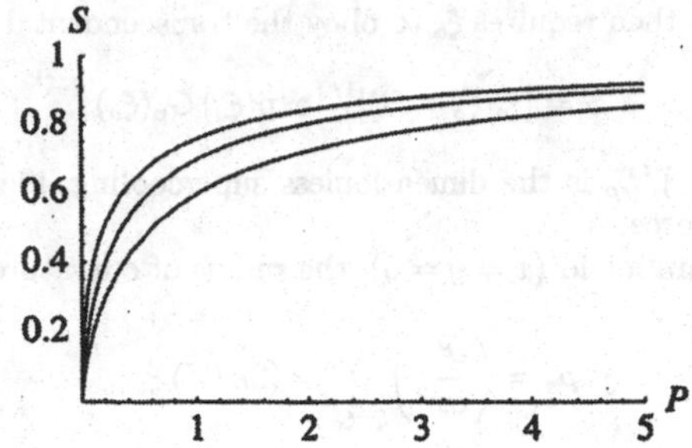

Figure 12: Plots of S as a function of P according to Eq(146) for an elliptical paraboloid for $r = 0$, 3 and ∞, where $\rho_y/\rho_x = 1 + r$. The bottom curve is for $r = 0$ and corresponds to a paraboloid of revolution, Eq(143). The top curve is for $r = \infty$ and corresponds to a parabolic cylinder, Eq(144).

this solidification problem is isomorphous to a precipitation problem, so the results apply with a concomitant change of notation. [22]

Solutions with similarity variable depending on $\sqrt{t}$

In this subsection, we explore solutions for cases in which the moving interfaces and isotherms are described by the quadric surfaces

$$\frac{x^2}{\xi^2 - a^2} + \frac{y^2}{\xi^2 - b^2} + \frac{z^2}{\xi^2} = 4\kappa t, \tag{147}$$

where ξ is a dimensionless generalized coordinate and a and b are fixed parameters.[23] As detailed in Table 1, this equation can be used to describe a variety of quadric surfaces by adjusting the values of a and b and the range of ξ. In order to simplify the notation, we introduce the dimensionless coordinates

$$(X, Y, Z) = \frac{(x, y, z)}{\sqrt{4\kappa t}} \tag{148}$$

such that Eq(147) is transformed into the time-independent form

$$F := \frac{X^2}{\xi^2 - a^2} + \frac{Y^2}{\xi^2 - b^2} + \frac{Z^2}{\xi^2} - 1 = 0. \tag{149}$$

The governing PDE, Eq(112), and the boundary condition, Eq(116), become

$$T_{XX} + T_{YY} + T_{ZZ} + 2(XT_X + YT_Y + ZT_Z) = 0 \tag{150}$$

[21] Horvay and Cahn [63] use a related notation $\mathcal{A} = 1/\sqrt{1+r}$, where $\mathcal{A}$ is the aspect ratio of an elliptical cross section in a plane perpendicular to the z axis.

[22] The prescription for change of notation is $\kappa \to D^\alpha$, $S \to (c^{\alpha\infty} - c^{\alpha\beta})/\Delta c$, the supersaturation, so Eq(141) becomes $P = V\rho_x/(2D^\alpha)$ and Eq(146) or any of its special cases apply directly. For the fields, $(T - T_\infty) \to (c^{\alpha\infty} - c^\alpha)$ and $(T_M - T_\infty) \to (c^{\alpha\infty} - c^{\alpha\beta})$, so Eq(136) becomes $c^\alpha = c^{\alpha\infty} - (c^{\alpha\infty} - c^{\alpha\beta})G_p(\xi)/G_p(\xi_o)$, and the concentration profile resembles the right hand side of Figure 10.

[23] The correspondence to Eq(122) is $\xi^2 = u/\kappa$, $a^2 = -\alpha_1/\kappa$, $b^2 = -\alpha_2/\kappa$, and $\alpha_3 = 0$.

Quadric Surface	Parameters			$f(w)$
	a	b	ξ	
Ellipsoid	a	b	$\xi > b > a$	$[(w^2 - a^2)(w^2 - b^2)]^{1/2}$
Prolate spheroid	b	b	$\xi > b$	$(w^2 - b^2)$
Oblate spheroid	0	b	$\xi > b$	$[w^2(w^2 - b^2)]^{1/2}$
Sphere	0	0	$\xi > 0$	w^2
Elliptic hyperboloid	a	b	$\xi < a < b$	$[(a^2 - w^2)(b^2 - w^2)]^{1/2}$ (two sheets)
	a	b	$a < \xi < b$	$[(w^2 - a^2)(b^2 - w^2)]^{1/2}$ (one sheet)
Hyperboloid of revolution	b	b	$\xi < b$	$(b^2 - w^2)$ (two sheets)
	0	b	$\xi < b$	$[w^2(b^2 - w^2)]^{1/2}$ (one sheet)
Elliptic cylinder	a	∞	$\xi > a$	$(w^2 - a^2)^{1/2}$
Circular cylinder	0	∞	$\xi > 0$	w
Hyperbolic cylinder	a	∞	$\xi < a$	$(a^2 - w^2)^{1/2}$
Plane	∞	∞	finite	1

Table 1: Classification of quadric surfaces given by Eq(149).

and

$$-2(L_o/c_o)(XT_X + YT_Y + ZT_Z) = T_X^2 + T_Y^2 + T_Z^2. \tag{151}$$

For a temperature field that depends only on ξ, Eq(150) and Eq(151) become

$$(\xi_X^2 + \xi_Y^2 + \xi_Z^2)T_{\xi,\xi} + [\xi_{XX} + \xi_{YY} + \xi_{ZZ} + 2(X\xi_X + Y\xi_Y + Z\xi_Z)]T_\xi = 0 \tag{152}$$

and

$$-2(L_o/c_o)(X\xi_X + Y\xi_Y + Z\xi_Z) = (\xi_X^2 + \xi_Y^2 + \xi_Z^2)T_\xi. \tag{153}$$

As in the case of the paraboloid, we proceed by implicit differentiation[24] to recast Eq(152) and Eq(153) in the forms

$$\frac{d^2T}{d\xi^2} + \left[\frac{\xi}{(\xi^2 - a^2)} + \frac{\xi}{(\xi^2 - b^2)} + 2\xi\right]\frac{dT}{d\xi} = 0 \tag{154}$$

and

$$2\left(\frac{L_o}{c_o}\right)\xi_o = -\left(\frac{dT}{d\xi}\right)_{\xi=\xi_o} \tag{155}$$

where $\xi = \xi_o$ again denotes the value of ξ corresponding to the solid-liquid interface.

Eq(154) can be solved by finding an integrating factor[25] which allows it to be written in the form

$$\left[f(\xi)\exp(\xi^2)\, T_\xi\right]_\xi = 0, \tag{156}$$

[24]We proceed as for the paraboloid, except F is now given by Eq(149). We find $X\xi_X + Y\xi_Y + Z\xi_Z = -2/F_\xi$, $\xi_X^2 + \xi_Y^2 + \xi_Z^2 = -2/(\xi F_\xi)$, and $\xi_{XX} + \xi_{YY} + \xi_{ZZ} = -(2/F_\xi)[1/(\xi^2 - a^2) + 1/(\xi^2 - b^2)]$.

[25]The integrating factor is $\exp\left(\int^\xi [u/(u^2 - a^2) + u/(u^2 - b^2) + 2u]\, du\right)$ which can be evaluated to yield $[(\xi^2 - a^2)(\xi^2 - b^2)]^{1/2}\exp(\xi^2)$. Depending on the range of ξ^2 relative to a^2 and b^2, the factors in the square root can be reversed to make the result real, as detailed in Table 1.

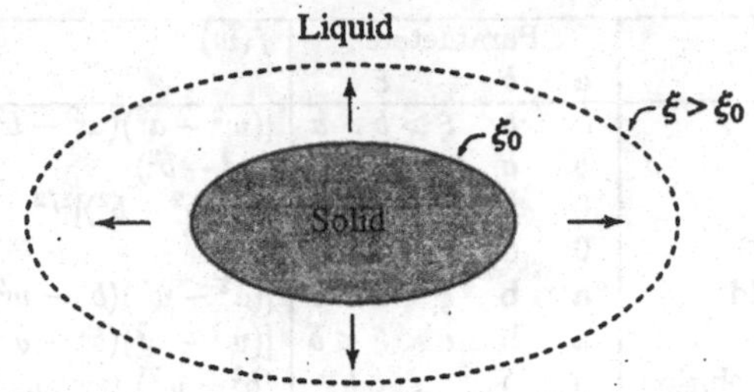

Figure 13: Cross section of a growing ellipsoidal solid, $\xi_o > b > a > 0$. The solid-liquid interface is located at the melting point isotherm, which corresponds to ξ_o. The dashed curve represents an isotherm in the supercooled melt.

where $f(\xi)$ is given in Table 1. Its solutions are therefore of the form

$$T(\xi) = \alpha \int^{\xi} \frac{\exp(-w^2)}{f(w)} dw + \beta \tag{157}$$

where α and β are to be determined by the boundary conditions.

<u>**Growing solid ellipsoid**</u> We first discuss the ellipsoidal case for which $\xi \geq b \geq a \geq 0$. For solidification toward an infinite supercooled melt, the temperature is given by

$$T = T_\infty + (T_M - T_\infty)\frac{G_e(\xi)}{G_e(\xi_o)}, \quad \xi \geq \xi_o \tag{158}$$

where

$$G_e(\xi) := \frac{2}{\sqrt{\pi}} \int_\xi^\infty \frac{\exp(-w^2)}{f_e(w)} dw \tag{159}$$

with $f_e(w) = [(w^2 - a^2)(w^2 - b^2)]^{1/2}$. The value of ξ_o is determined by the condition of energy conservation, Eq(155), which leads to

$$S = \sqrt{\pi}\xi_o \exp(\xi_0^2) f_e(\xi_o) G_e(\xi_o). \tag{160}$$

As illustrated in Figure 13, the solid-liquid interface is an ellipsoid which is initially a point, the origin, and grows with principal axes $(4\kappa t)^{1/2}(\xi_o^2 - a^2)^{1/2}$, $(4\kappa t)^{1/2}(\xi_o^2 - b^2)^{1/2}$ and $(4\kappa t)^{1/2}\xi_o$ along the x, y, z axes respectively. The ratios of these principal axes are independent of time, so the ellipsoid grows with constant shape and with any linear dimension proportional to $t^{1/2}$. The initial condition for any of these solutions may be found by noting that except for the point $x = y = z = 0$, Eq(147) requires $\xi \to \infty$ as $t \to 0$, and since $G_e(\infty) = 0$, Eq(158) yields $T = T_\infty$.

In order to study an ellipsoid of fixed shape as a function of the dimensionless supercooling S, we can write $a = r_a\xi_o$ and $b = r_b\xi_o$ where r_a and r_b are constants, in which case Eq(160) takes the form

$$S = 2\xi_o^2 \exp(\xi_o^2) \left[(1 - r_a^2)(1 - r_b^2)\right]^{1/2} \int_1^\infty \frac{\exp(-\xi_o^2 u^2)}{[(u^2 - r_a^2)(u^2 - r_b^2)]^{1/2}} du. \tag{161}$$

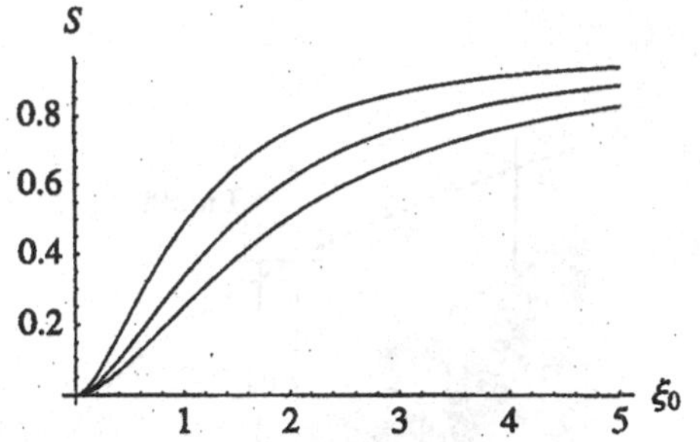

Figure 14: Plots of S as a function of ξ_o according to Eq(161) for a growing sphere and several prolate spheroids. The curves are for $a = b = r\xi_o$. The top curve is for $r = 0$ and corresponds to a sphere, Eq(162), the middle curve is for $r = 0.8$ and the bottom curve is for $r = 0.9$.

For small S, ξ_o will also be small; we can approximate the exponentials by unity, and note that ξ_0 will be proportional to $\sqrt{S}$ with a proportionality constant that depends on shape.

Special cases of the ellipse are the prolate spheroid ($0 \neq a = b < \xi$) with z being the axis of rotational symmetry, the oblate spheroid ($a = 0,\ 0 \neq b < \xi$) with y being the axis of rotational symmetry, and the sphere ($a = b = 0$). Plots of S versus ξ_o are shown in Figure 14 for the sphere and several prolate spheroids. For the sphere ($r_a = r_b = 0$), $G_e(\xi) = (2/\xi\sqrt{\pi})\exp(-\xi^2) - 2\,\mathrm{erfc}(\xi)$, and Eq(161) becomes

$$S = 2\xi_o^2\left[1 - \sqrt{\pi}\xi_o \exp(\xi_o^2)\,\mathrm{erfc}(\xi_o)\right] \tag{162}$$

which is well known [52,67]. For small S, one has $\xi_o \approx \sqrt{S/2}$.

Other special cases are the elliptic cylinder ($0 \neq a < \xi,\ b = \infty$) and the circular cylinder ($0 = a < \xi,\ b = \infty$), with y being the axis of rotational symmetry for the latter. For the circular cylinder, the function G_e becomes $E_1(\xi^2)/\sqrt{\pi}$ and the value of ξ_o is determined by

$$S = \xi_o^2 \exp(\xi_o^2) E_1(\xi_o^2). \tag{163}$$

Eq(163) has a remarkable resemblance to Eq(143) for the circular paraboloid, with ξ_o^2 now replacing the Peclet number P. This can be understood by two observations: First, the Peclet number for the circular cylinder is actually ξ_o^2. This follows because the radius of that cylinder is $R = \sqrt{x^2+z^2} = \sqrt{4\kappa t}\,\xi_o$ and its growth speed is $V = dR/dt = \sqrt{\kappa/t}\,\xi_o$, so $P = VR/2\kappa = \xi_o^2$. Thus for the circular cylinder, $\sqrt{x^2+z^2} = \sqrt{4\kappa t}\,\sqrt{P}$. Second, the *lateral* dimension of the circular paraboloid, at a fixed value of $z = \kappa\xi_o/V = \rho_x/2$, which is the position of the tip of the paraboloid at $t = 0$, actually increases as $\sqrt{t}$. But for the paraboloid, $\xi_o = P$. Thus from Eq(125) with $B = 0$, $\xi = \xi_o$ and $z = \kappa\xi_o/V$, we have $\sqrt{x^2+y^2} = \sqrt{4\kappa t}\,\sqrt{\xi_o} = \sqrt{4\kappa t}\,\sqrt{P}$, exactly as for the cylinder. In other words, from the moment that the tip of the circular paraboloid reaches a given value of z, the paraboloid thickens laterally at that fixed z in proportion to $\sqrt{t}$, with exactly the same growth rate as would an infinite circular cylinder that started to grow at that same moment. This is illustrated in Figure 15.

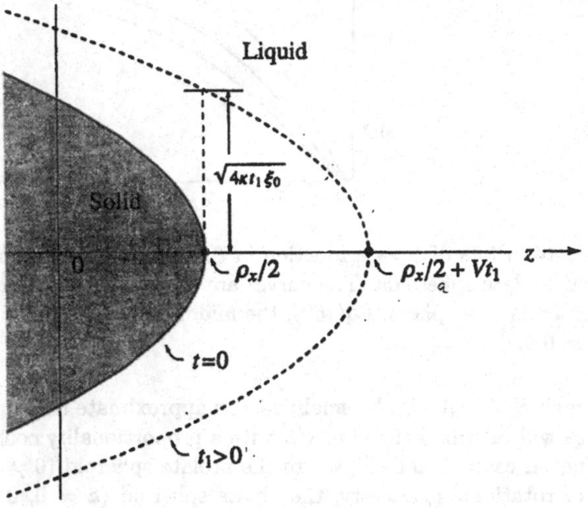

Figure 15: Cross section of a uniformly translating paraboloid of revolution. The lateral distance between the paraboloid and an observer located at $z = \kappa\xi_o/V = \rho_x/2$ increases in exactly the same way as would the radius of an infinite circular cylinder.

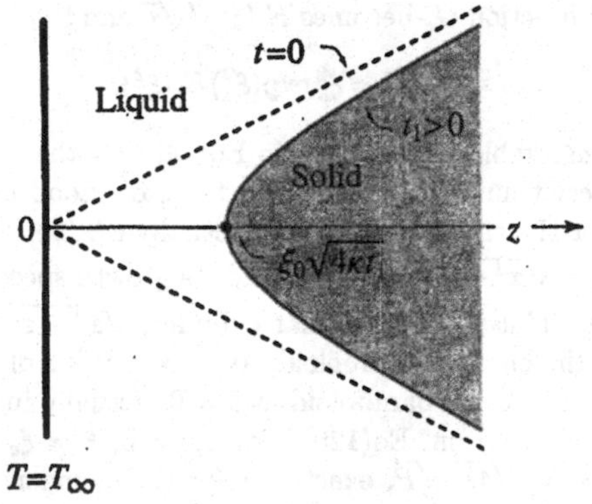

Figure 16: Cross section of a melting hyperboloidal solid. The hyperboloid is initially a cone with a tip that moves according according to $z = \xi_o\sqrt{4\kappa t}$.

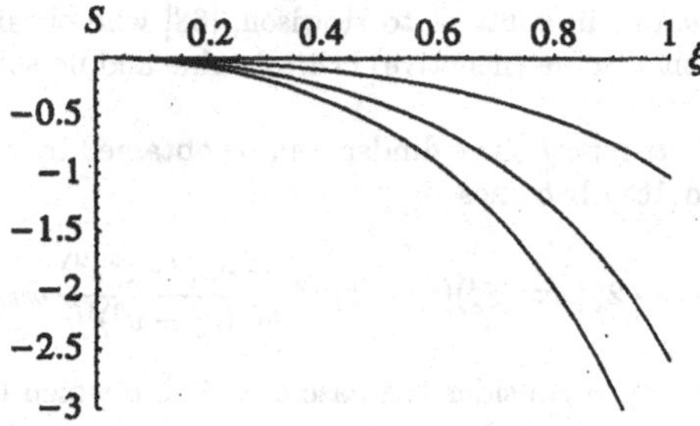

Figure 17: Plots of S as a function of ξ_o according to Eq(168) for a melting circular hyperboloid. The curves are for $r_a = r_b = 1.1$, 1.5 and 11 from top to bottom.

Melting solid hyperboloid If we choose $\xi \le \xi_o < a \le b$, the interface is an elliptical hyperboloid of two sheets

$$-\frac{x^2}{(a^2-\xi_o^2)} - \frac{y^2}{(b^2-\xi_o^2)} + \frac{z^2}{\xi_o^2} = 4\kappa t \qquad (164)$$

which has a hyperbolic trace on the $x - z$ plane with foci at $x = y = 0$, $z = \pm a(4\kappa t)^{1/2}$ and asymptotes the lines $y = 0$, $x = \pm\left[(a/\xi_o)^2 - 1\right]^{1/2} z$. It has a different hyperbolic trace on the $y - z$ plane with foci at $x = y = 0$, $z = \pm b(4\kappa t)^{1/2}$ and asymptotes the lines $y = 0$, $x = \pm\left[(b/\xi_o)^2 - 1\right]^{1/2} z$. Because of the symmetry for $z \to -z$, one can consider only the half-space problem in which the temperature is specified to be T_∞ in the plane $z = 0$, which corresponds to $\xi = 0$. As illustrated in Figure 16, the solid is initially a cone with apex at the origin. The solid melts away from the origin as a hyperboloid with a focus that increases in proportion to $t^{1/2}$.

The temperature field is therefore

$$T(\xi) = T_\infty + (T_M - T_\infty)\frac{G_h(\xi)}{G_h(\xi_o)}, \quad 0 < \xi < \xi_o \qquad (165)$$

where

$$G_h(\xi) := \frac{2}{\sqrt{\pi}}\int_0^\xi \frac{\exp(-w^2)}{f(w)}dw \qquad (166)$$

with $f(w) = [(a^2 - w^2)(b^2 - w^2)]^{1/2}$. The value of ξ_o is now determined by

$$S = -\sqrt{\pi}\xi_o \exp(\xi_o^2) f(\xi_o) G_h(\xi_o). \qquad (167)$$

The only solutions are for $S < 0$, which implies that $T_\infty > T_M$. The tip of the hyperboloid in the right half plane moves as $z = \xi_o\sqrt{4\kappa t}$, so the hyperboloid is melting.

We can study a hyperboloid of fixed shape as a function of S, by writing $a = r_a\xi_o$ and $b = r_b\xi_o$ where $r_a > 1$ and $r_b > 1$ are constants. Then Eq(167) takes the form

$$S = -2\xi_o^2 \exp(\xi_o^2)\left[(r_a^2 - 1)(r_b^2 - 1)\right]^{1/2}\int_0^1 \frac{\exp(-\xi_o^2 u^2)}{[(r_a^2 - u^2)(r_b^2 - u^2)]^{1/2}} du \qquad (168)$$

which should be compared with Eq(161). Figure 17 shows a plot of S versus ξ_o for several circular hyperboloids, with $r_a = r_b = r$. By using a and b proportional to ξ_o, we obtain only one solution for each S, in contrast to Howison, [68] who obtained two solutions for each value of S greater than some (negative) critical value and no solutions for S less than the critical value.

Another special case, a hyperbolic cylinder, can be obtained by taking the formal limit $b \rightarrow \infty$. In this case, Eq(168) becomes

$$S = -2\xi_o^2 \exp(\xi_o^2)(r_a^2 - 1)^{1/2} \int_0^1 \frac{\exp(-\xi_o^2 u^2)}{(r_a^2 - u^2)^{1/2}} du. \tag{169}$$

In principle, we could also consider the case $a < \xi < b$ which corresponds to hyperboloids of one sheet,

$$\frac{x^2}{(\xi^2 - a^2)} - \frac{y^2}{(b^2 - \xi^2)} + \frac{z^2}{\xi^2} = 4\kappa t, \tag{170}$$

with the solid-liquid interface corresponding to ξ_o. For general a and b, it would be necessary to specify the temperature T_∞ on some other moving hyperboloid, which is not very interesting physically. One possibly interesting solution would be for $a = 0$ and finite b, which would allow T_∞ to be specified at a very small value $\xi = \epsilon\xi_o$. Such a surface is essentially a thin hot wire along the y axis, but the radius of such a wire would have to increase as $\sqrt{t}$, which is not of practical interest.

Growing/melting planar interface The final special case of Eq(147) that we discuss corresponds formally to the limit $a = b = \infty$, which results in a planar interface located at

$$z = \sqrt{4\kappa t}\,\xi_o. \tag{171}$$

Such an interface is located initially at $z = 0$ and moves toward the right if we assume $\xi_o > 0$. Growth occurs if the solid is on the left and supercooled liquid extends to infinity on the right, in which case the problem can be thought of as a limiting case of a growing ellipsoid. Melting occurs if the solid is on the right and a temperature in excess of the melting point is specified in the plane $z = 0$; this can be thought of as the limiting case of a melting hyperboloid.

The corresponding temperature is

$$T = T_\infty + (T_M - T_\infty)\frac{G_{pl}(\xi)}{G_{pl}(\xi_o)} \tag{172}$$

where $G_{pl}(\xi) = \text{erfc}(\xi)$ for growth and $G_{pl}(\xi) = \text{erf}(\xi)$ for melting. Thus

$$S = \sqrt{\pi}\xi_o \exp(\xi_o^2)\text{erfc}(\xi_o) \tag{173}$$

for growth, and

$$S = -\sqrt{\pi}\xi_o \exp(\xi_o^2)\text{erf}(\xi_o) \tag{174}$$

for melting. Plots of S versus ξ for these cases are shown in Figure 18. Note that Eq(173) has the same form as Eq(144) for the parabolic cylinder, except that the effective Peclet number is now $\xi_o^2 = z(t)\,(dz/dt)/(2\kappa)$. In other words, from the moment that the tip of the parabolic cylinder reaches a given value of z, the paraboloic cylinder thickens laterally at that fixed z in proportion to $\sqrt{t}$, with exactly the same growth rate as would an infinite plane that started to grow at that same moment.

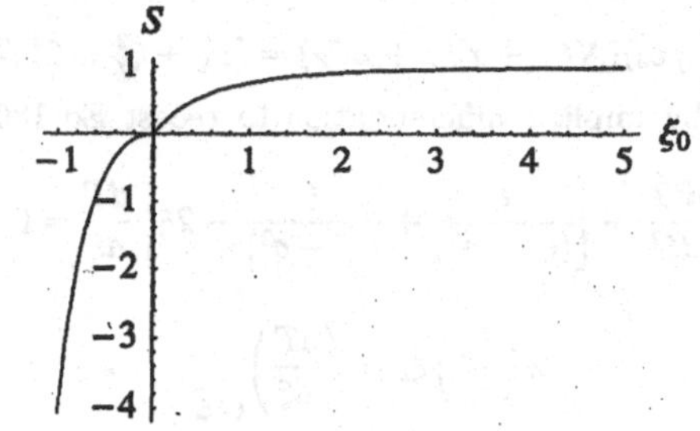

Figure 18: Plot of S as a function of ξ_o for the planar case. $S > 0$ corresponds to growth, Eq(173) and $S < 0$ to melting, Eq(174).

We can also examine the possibility of $\xi_o < 0$ in which case an interface located initially at $z = 0$ moves toward the left. Eq(173) now admits a new solution for $S < 0$, which corresponds to melting away from a hot liquid that extends to $z = +\infty$. But Eq(174) is unchanged because the product $\xi_o \text{erfc}(\xi_o)$ remains positive, so it continues to represent melting, although now toward the left of a hot plane located at $z = 0$.

Solutions with similarity variable depending on $\sqrt{\tau - t}$

In this section, we investigate similarity solutions for which the quadric surfaces depend on time as [65]

$$\frac{x^2}{\xi^2 - a^2} + \frac{y^2}{\xi^2 - b^2} + \frac{z^2}{\xi^2} = 4\kappa(\tau - t) \tag{175}$$

where τ is a constant. We introduce the dimensionless coordinates

$$(X, Y, Z) = \frac{(x, y, z)}{\sqrt{4\kappa(\tau - t)}} \tag{176}$$

such that Eq(175) is transformed into

$$F := \frac{X^2}{\xi^2 - a^2} + \frac{Y^2}{\xi^2 - b^2} + \frac{Z^2}{\xi^2} - 1 = 0. \tag{177}$$

Eq(176) is exactly the same as Eq(149) and therefore also represents the quadric surfaces detailed in Table 1. But because (X, Y, Z) now depend on time as $\sqrt{\tau - t}$ rather than $\sqrt{t}$, the governing PDE, Eq(112), and the boundary condition, Eq(116), become

$$T_{XX} + T_{YY} + T_{ZZ} - 2(XT_X + YT_Y + ZT_Z) = 0 \tag{178}$$

and

$$2(L_o/c_o)(XT_X + YT_Y + ZT_Z) = T_X^2 + T_Y^2 + T_Z^2. \tag{179}$$

Note the sign changes in comparison to Eq(150) and Eq(151). For a temperature field that depends only on ξ, Eq(178) and Eq(179) become

$$(\xi_X^2 + \xi_Y^2 + \xi_Z^2)T_{\xi,\xi} + [\xi_{XX} + \xi_{YY} + \xi_{ZZ} - 2(X\xi_X + Y\xi_Y + Z\xi_Z)]T_\xi = 0 \tag{180}$$

and

$$2(L_o/c_o)(X\xi_X + Y\xi_Y + Z\xi_Z) = (\xi_X^2 + \xi_Y^2 + \xi_Z^2)T_\xi. \tag{181}$$

Again, we proceed by implicit differentiation to recast Eq(180) and Eq(181) in the forms

$$\frac{d^2T}{d\xi^2} + \left[\frac{\xi}{(\xi^2 - a^2)} + \frac{\xi}{(\xi^2 - b^2)} - 2\xi\right]\frac{dT}{d\xi} = 0 \tag{182}$$

and

$$2\left(\frac{L_o}{c_o}\right)\xi_o = \left(\frac{dT}{d\xi}\right)_{\xi=\xi_o} \tag{183}$$

where $\xi = \xi_o$ denotes the value of ξ corresponding to the solid-liquid interface. Eq(182) can be solved by finding an integrating factor[26] which allows it to be written in the form

$$\left[f(\xi)\exp(-\xi^2)\ T_\xi\right]_\xi = 0, \tag{184}$$

where $f(\xi)$ is given in Table 1. Its solutions are therefore of the form

$$T(\xi) = \alpha \int^\xi \frac{\exp(w^2)}{f(w)}\,dw + \beta \tag{185}$$

where α and β are to be determined by the boundary conditions. The sign change in Eq(183) relative to Eq(155) will lead to an interchange of growth and melting. Note, however, that the integrand in Eq(185) contains a factor $\exp(w^2)$ instead of $\exp(-w^2)$ which occurs in Eq(157). Since this factor diverges as $w \to \infty$, we must now confine our solutions to a finite domain of ξ. Thus, unfortunately, the problem of a melting ellipsoid in an infinite domain becomes impossible to treat by this method.

Growing hyperboloids On the other hand, the problem of a hyperboloidal solid growing toward a cold plane is possible to treat. For $0 \le \xi \le \xi_o < a \le b$, Eq(175) becomes

$$-\frac{x^2}{a^2 - \xi^2} - \frac{y^2}{b^2 - \xi^2} + \frac{z^2}{\xi^2} = 4\kappa(\tau - t) \tag{186}$$

which describes an elliptical hyperboloid of two sheets. By symmetry, it also describes the half-plane ($z \ge 0$) problem, depicted in Figure 19, in which a hyperboloid grows toward a cold plane located at $z = 0$, which corresponds to $\xi = 0$. Such a hyperboloid grows toward the plane for only a finite time τ, when it becomes a cone traveling at infinite speed.

The corresponding temperature field is

$$T = T_\infty + (T_M - T_\infty)\frac{G_{gh}(\xi)}{G_{gh}(\xi_o)}, \quad \xi_o > \xi \ge 0 \tag{187}$$

where

$$G_{gh}(\xi) := \frac{2}{\sqrt{\pi}} \int_0^\xi \frac{\exp(w^2)}{f(w)}dw \tag{188}$$

[26]The integrating factor is now $\exp\left(\int^\xi [u/(u^2 - a^2) + u/(u^2 - b^2) - 2u]\,du\right)$ which can be evaluated to yield $\left[(\xi^2 - a^2)(\xi^2 - b^2)\right]^{1/2}\exp(-\xi^2)$.

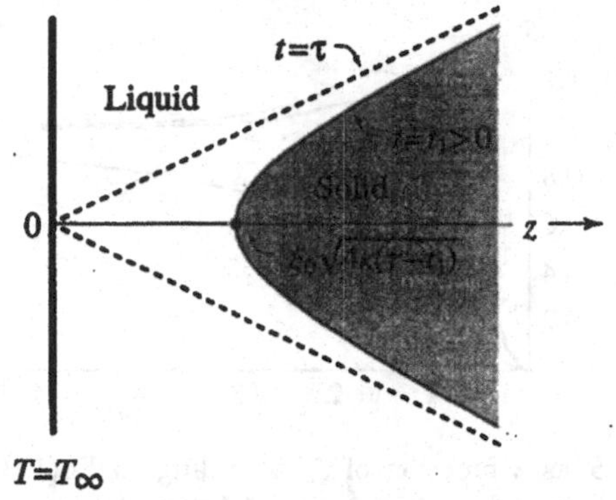

Figure 19: Cross section of a hyperboloidal solid growing toward a cold plane $z = 0$. The tip of the hyperboloid is located at $z = \xi_o\sqrt{4\kappa(\tau - t)}$. The hyperboloid becomes a cone at $t = \tau$.

and $f(w) = [(a^2 - w^2)(b^2 - w^2)]^{1/2}$. The flux condition, Eq(183), requires ξ_o to satisfy

$$S = \sqrt{\pi}\xi_o \exp(-\xi_o^2) f(\xi_o) G_{gh}(\xi_o). \tag{189}$$

The tip of the hyperboloid ($x = y = 0$) is located at

$$z = \xi_o\sqrt{4\kappa(\tau - t)} \tag{190}$$

and its *growth* velocity, $V = -dz/dt$, is

$$V = \xi_o\sqrt{\kappa/(\tau - t)} = 2\kappa\xi_o^2/z. \tag{191}$$

Unlike the paraboloid growing into an infinite supercooled melt at constant velocity, the hyperboloid speeds up as it approaches the plane. However, if it is far from the plane, in the sense that $\tau >> t$, its growth velocity is nearly constant. At the tip of the hyperboloid ($x = y = 0$), the radius of curvature in the x - z plane is

$$\rho_{xz} = \left(\frac{\partial^2 z}{\partial x^2}\right)^{-1}_{\xi=\xi_o} = \frac{a^2 - \xi_o^2}{\xi_o}\sqrt{4\kappa(\tau - t)} = \frac{a^2 - \xi_o^2}{\xi_o^2} z, \tag{192}$$

so the tip radius decreases with time. But the corresponding Peclet number is $\rho_{xz}V/2\kappa = a^2 - \xi_o^2$, which is constant in time. Similarly, $\rho_{yz} = (b^2 - \xi_o^2)z/\xi_o^2$ and $\rho_{yz}V/2\kappa = b^2 - \xi_o^2$.

To study a hyperboloid with fixed *asymptotes* as a function of S, we write $a = r_a\xi_o$ and $b = r_b\xi_o$, where $r_a > 1$ and $r_b > 1$. Eq(189) then takes the form

$$S = 2\xi_o^2 \exp(-\xi_o^2)\left[(r_a^2 - 1)(r_b^2 - 1)\right]^{1/2} \int_0^1 \frac{\exp(\xi_o^2 u^2)}{[(r_a^2 - u^2)(r_b^2 - u^2)]^{1/2}} du. \tag{193}$$

Plots of S versus ξ_o for several values of $r = r_a = r_b$ are shown in Figure 20. There is one solution for $S < 1$, two solutions for $1 < S < S_{crit}$ and no solutions for $S > S_{crit}$. Since ξ_o determines the temperature field via Eq(187) and the exact shape of the hyperboloid via Eq(186), the multiple solutions correspond to different initial conditions. For small S, ξ_o will also be small; then for $r_a = r_b = r$ we can evaluate the integral approximately to obtain $S \approx \xi_o^2[(r^2 - 1)/r]\ln[(r + 1)/(r - 1)]$.

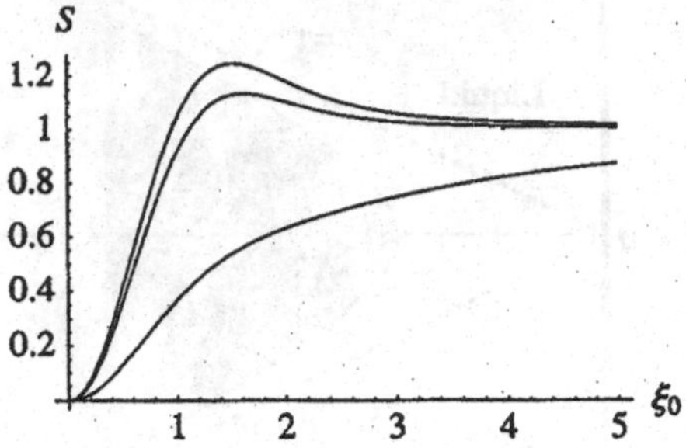

Figure 20: Plots of S as a function of ξ_o according to Eq(193) for a growing circular hyperboloid. The curves are for $r_a = r_b = r = 1.1$, 2 and 4 from bottom to top.

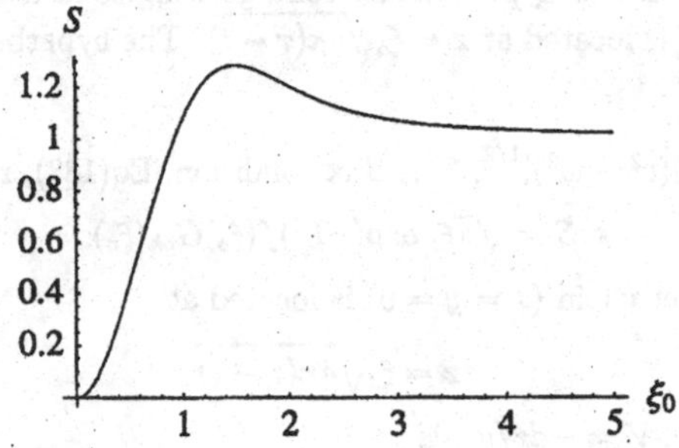

Figure 21: Plot of S as a function of ξ_o for a planar solid growing toward a cold plane.

Planar solid growing toward a cold plane We finally discuss the limiting case of $a = b = \infty$. The solid-liquid interface is a plane described by $z = \sqrt{4\kappa(\tau - t)}\xi_o$, which moves toward the origin and coincides with the plane $z = 0$ at $t = \tau$. In this case,

$$S = 2\xi_o \exp(-\xi_o^2) \int_0^{\xi_o} \exp(w^2)dw. \tag{194}$$

As shown in Figure 21, there is one solution for ξ_o if $S < 1$ but two solutions if $1 < S < S_{crit}$, where S_{crit} is the critical value beyond which no solution exists. The value of $S_{crit} \simeq 1.285$ corresponds to $\xi_o \simeq 1.502$. In case there are two solutions, they correspond to different initial conditions. Since $-z(dz/dt) = 2\kappa\xi_o^2$, the solution with the larger initial value of z has a smaller initial growth speed.

Discussion and Limitations

We conclude with a brief discussion of the assumptions and limitations of the solutions to moving phase boundary problems that have been described in the previous sections.

First of all, it is important to recognize that moving phase boundary problems are a subset of a much broader class of problems, known as free boundary problems. All of the solutions discussed in the previous sections pertain to boundaries that maintain definite shapes throughout their motion. On the other hand, a completely free boundary, described implicitly by an equation of the form [69]

$$f(x, y, z, t) = 0, \tag{195}$$

can change its shape as well as just move with time. This leads to questions of morphological stability [70] of the solutions for definite shapes. Specifically, even though perfectly rigorous mathematical solutions can be found for moving boundaries of fixed shapes, there is no guarantee that such solutions will describe a physically realizable situation. Indeed, other solutions for slightly or even greatly distorted shapes might be preferred by the system in the following sense: Given the option of taking on a new shape, the boundary might continue to deviate from its original shape. In such a case, the original shape is said to be unstable.

This instability phenomenon has been quantified by Mullins and Sekerka [70] who find that instability of an originally spherical particle will occur whenever the dimensions of the particle become about an order of magnitude larger than the nucleation radius. This criterion is established by study of the solutions corresponding to boundaries whose shapes are only slightly different from a boundary of spherical shape. The new shapes are called perturbed shapes and are imagined to arise by adding small perturbations to the spherical boundary. The condition that these perturbations grow spontaneously is the condition for instability. This phenomenon is very general and has been studied for a wide variety of shapes, including planes, cylinders, spheres and paraboloids. Details may be found in several review articles [71–76]. Nevertheless, the solutions for boundaries of fixed shape are at least a starting point for instability calculations. Perhaps even for slightly unstable shapes, such solutions characterize approximately the average motion of the phase boundary; however, if conditions for instability are severe, phenomena such as dendritic growth can result.

We also emphasize that the explicit solutions discussed in previous sections are based on the assumption that the concentrations (or for the non-planar problems the temperature) at the moving phase boundaries are constants. This is known as the assumption of local equilibrium. For non-planar boundaries, even the assumption of local equilibrium leads to boundary concentrations or temperatures that depend on local boundary curvature, the so-called Gibbs-Thomson effect. For large bodies having shapes with small curvatures, these effects are negligible, but they are essential to phenomena such as nucleation or morphological stability.

In some situations, particularly for rapid boundary motion, kinetic processes local to the moving boundary will either be rate limiting or will lead to some retardation of the boundary motion, corresponding formally to variable concentrations (or temperature) at the moving phase boundaries [77–80]. An even more complicating feature is the possible crystalline anisotropy of these local boundary processes; this renders the problem almost intractable analytically, except for the special case treated by Frank [81] in which the rate of boundary advance is presumed to depend only on local boundary orientation. Although it is possible to obtain an analytical solution for parabolic growth of a sphere, including an isotropic Gibbs-Thomson effect and interface kinetics, this solution corresponds to peculiar initial conditions [82].

It is also noteworthy that solutions for the growth of a phase by phase boundary motion cannot be applied to the problem of dissolution of that phase by letting the time variable decrease instead of increase. This arises because dissipative equations are not invariant under a change of the algebraic sign of the time; in particular, the diffusion equation contains $\partial c/\partial t$ and a different equation would result if t were replaced by $-t$. Therefore, treatment of dissolution problems necessitates the finding of new solutions to the diffusion equation, perhaps by numerical techniques or by analytical approximations such as those discussed by Aaron [16].

In general, free boundary problems are characterized by almost insurmountable mathematical complexity because they call for solutions to differential equations that must satisfy boundary conditions on a surface whose position and shape are, themselves, part of the solution [83]. Even numerical finite difference techniques are beset with difficulties because the boundary will not, in general, coincide with the grid points of a two- or three-dimensional network. One must therefore resort to interpolation or to sophisticated mapping techniques that result in the boundary coinciding with one of the coordinates [84]. Another approach, sometimes called the boundary integral technique, is to use Green's functions to express formally the solutions to the differential equations in terms of integrals over sources located on the unknown boundary. This leads to integral equations that must be solved to determine the boundary shape. A detailed description of this technique for solidification problems can be found in a doctoral thesis by Nash [85]. Examples of complex shapes that can be calculated by using the boundary integral technique for slow solidification (quasi-steady state approximation) can be found in a paper by Brush and Sekerka [86].

More recently, phase field models have been developed to approximate the classical sharp interface problem. In phase field models, the sharp interface between phases is replaced by a diffuse interface by inventing an auxiliary variable, ϕ, to identify the phase. For the solidification of a pure material, for example, we can choose $\phi = 0$ in the solid phase and $\phi = 1$ in the liquid phase. In the vicinity of the solid-liquid interface, ϕ changes rapidly but continuously from zero to one over a short distance, the thickness of the diffuse interface. One then proceeds to formulate dynamical equations for ϕ and T by employing the principle of positive local entropy production [87]. This results in a pair of *coupled* partial differential equations that can be solved numerically in the entire spatial domain, without explicitly tracking the moving phase boundary. The boundary can then be located at any time by determining the locus of $\phi = 1/2$ from the solution. If the diffuse interface is very thin, one obtains a good approximation to the solution of the corresponding sharp interface problem, which even incorporates dependence of the melting point on interface curvature (Gibbs-Thomson effect) and departures from local equilibrium at the interface (interface kinetics). Phase field models therefore enable treatment, in principle, of boundaries having very complicated shapes, but their use is still limited by the speed and memory limitations of supercomputers, even for solidification of pure materials [88–96].

Naturally, real precipitation phenomena involve growth at many nucleation sites and overlap of diffusion fields associated with various growth centers eventually results. As soon as there is substantial overlap, the preceding solutions must be modified accordingly. This problem has been treated for a regular array of growth centers by Ham and has been treated statistically by Johnson and Mehl and many followers [97–102]. Moreover, many real problems involve multicomponent systems (ternary or higher order), which necessitate the specification of more than one diffusivity [37, 103–109] as well as other complications such as multiple similarity solutions [110].

The careful reader will doubtless be able to think of many other situations for which the solutions obtained in this article are not strictly valid; the above examples are, therefore, intended to be representative. Nevertheless, it is believed that these solutions constitute a core of understanding about which more refined knowledge revolves; herein lies their great pedagogical value.

Acknowledgments

The authors gratefully acknowledge the continuing financial support of the Division of Materials Research of the National Science Foundation, most recently under grant DMR-9211276, which supported the doctoral work of Shun-Lien Wang, and the current grant DMR-9634056. We would also like to acknowledge the contributions of Richard W. Heckel and Christian L. Jeanfils, who were coauthors of the article originally published in the first edition of this volume. The underlying framework that they helped to provide has been an excellent basis for this revised article.

References

[1] Hubert I. Aaronson, ed., Lectures on the Theory of Phase Transformations, (New York, NY: American Institute of Mining, Metallurgical and Petroleum Engineers, Inc., 1977), 117

[2] Shun-Lien Wang, Computation of Dendritic Growth at Large Supercoolings by Using the Phase Field Model, Doctoral Thesis (Pittsburgh, PA: Carnegie Mellon University, 1995) Appendix A

[3] Paul G. Shewmon, Diffusion in Solids, (New York, NY: McGraw-Hill, 1963)

[4] J.S. Kirkaldy in Decomposition of Austenite by Diffusional Processes, ed. V. F. Zackay and H.I. Aaronson (New York, NY: Wiley-Interscience, 1962), 39

[5] H.I. Aaronson in Decomposition of Austenite by Diffusional Processes, ed. V. F. Zackay and H.I. Aaronson (New York, NY: Wiley-Interscience, 1962), 387

[6] W. Jost, Diffusion in Solids, Liquids, and Gases, (New York, NY: Academic Press 1960)

[7] J. Crank, The Mathematics of Diffusion, (London, UK: Oxford Press, 1960)

[8] R.A. Tanzilli and R. W. Heckel, Trans. AIME, 242 (1968), 2313

[9] R.D. Lanam and R.W. Heckel, Met. Trans., 2 (1971), 2255

[10] G.R. Purdy and J.S. Kirkaldy, Trans. AIME, 227 (1963), 1255

[11] R.A. Tanzilli and R.W. Heckel, Met. Trans., 2 (1971), 1779

[12] R.D. Lanam and R.W. Heckel, Met. Trans., 6A (1975), 421

[13] R.W. Heckel et al., Met. Trans., 3 (1972), 2565

[14] D.L. Baty, R.A. Tanzilli, and R.W. Heckel, Met. Trans., 1 (1970), 1651

[15] R.A. Tanzilli and R.W. Heckel, Met. Trans., 1 (1970), 1863

[16] H.B. Aaron, Metal Science J., 2 (1968), 192

[17] R.A. Tanzilli and R.W. Heckel, Met. Trans., 6A (1975), 329

[18] R.D. Lanam et al., Met. Trans., 6A (1975), 337

[19] A.J. Hickl and R.W. Heckel, Met. Trans., 6A (1975), 431

[20] N. Birks, Decarburization, ISI Publication 133 (London,UK: The Iron and Steel Institute, 1970), 1

[21] A. Pattnaik and A. Lawley, Met. Trans., 5 (1974), 111

[22] Thaddeus B. Massalski, ed., Binary Alloy Phase Diagrams, (Metals Park, OH: ASM International 1990) Volumes 1-3

[23] E.A. Brandes and G.B. Brook, eds., Smithells Metals Reference Book, Seventh Edition, (Oxford, UK: Butterworth - Heinemann Ltd. 1992)

[24] Diffusion in Metals Data Center, Metallurgy Division, National Institute of Standards and Technology, Gaithersburg, MD 20899

[25] Daniel B. Batrymowitz, John R. Manning and Michael E. Read, Diffusion Rate Data for Mass Transport Phenomena for Copper Systems, (Washington, DC: National Bureau of Standards 1977)

[26] Daniel B. Batrymowitz, Diffusion Rate Data for Mass Transport Phenomena for Copper Systems Part II, (Washington, DC: National Bureau of Standards 1981)

[27] V.N. Svechnikov, ed., Diffusion Processes in Metals, (Kiev, USSR: Academy of Sciences of the Ukranian SSR, 1966)

[28] M.A. Krishtal, Diffusion Processes in Iron Alloys, (Moscow, USSR: Metallurgizdat, 1963)

[29] H.I. Aaronson, C. Laird, and K.R. Kinsman in Phase Transformations, (Metals Park, OH: American Society for Metals, 1970), 318

[30] A.E. Nielsen, Kinetics of Precipitation, (New York, NY: Pergamon-Macmillan, 1964)

[31] J.B. Clark, Trans. AIME., 227 (1963), 1250

[32] J.R. Manning, Diffusion Kinetics for Atoms in Crystals, (Princeton, NJ: D. Van Nostrand, 1968)

[33] A. Le Claire in Progress in Metal Physics, ed. B. Chalmers (London, UK: Pergamon Press, 1953), 265

[34] L.A. Girifalco, Atomic Migration in Crystals, (New York, NY: Blaisdell Publishing Co., 1964)

[35] Atom Movements, (Metals Park, OH: American Society for Metals, 1951)

[36] R.B. Bird, W. E. Stewart, and E. N. Lightfoot, Transport Phenomena, (New York, NY: John Wiley & Sons Inc., 1960)

[37] R.F. Sekerka and W.W. Mullins, J. Chem. Phys., 73 (1980), 1413

[38] W.W. Mullins and R.F. Sekerka, Scripta Met., 15 (1981), 29

[39] C.H.P. Lupis, Chemical Thermodynamics of Materials (New York, NY: North Holland, 1983), 51

[40] L.S. Darken and R.W. Gurry, Physical Chemistry of Metals, (New York, NY: McGraw-Hill, 1953)

[41] J.G. Kirkwood and L. Oppenheim, Chemical Thermodynamics, (New York, NY: McGraw-Hill, 1961)

[42] C.S. Barrett and T.B. Massalski, Structure of Metals, (New York, NY: McGraw-Hill, 1966), 360

[43] H.W. King in Alloying Behavior and Effects in Concentrated Solid Solutions, ed. T. B. Massalski (New York, NY: Gordon and Breach, 1965), 85

[44] A. Fick, Annalen der Physik, 94 (1855), 59

[45] L. Boltzmann, Annalen der Physik und Chemie, 53 (1894), 959

[46] C. Matano, Japan. J. Physics, 8 (1933), 109

[47] S. Prager, J. Chem. Phys., 21 (1953), 1344

[48] Donald D. Fitts, Nonequilibrium Thermodynamics, (New York, NY: McGraw-Hill, 1962)

[49] P.S. Perera and R.F. Sekerka, Physics of Fluids, 9 (1997) 376

[50] L.S. Darken, Trans. AIME, 175 (1948), 184

[51] P.V. Danckwerts, Trans. Faraday Soc., 46 (1950), 701

[52] H.S. Carslaw and J.C. Jaeger, Conduction of Heat in Solids, (London, UK: Oxford University Press, 1959)

[53] M. Abramowitz and I.A. Stegun, Handbook of Mathematical Functions, (Washington, DC: U. S. Department of Commerce, National Bureau of Standards, 1964)

[54] R.W. Balluffi and L.L. Seigle, J. Appl. Phys., 25 (1954), 607

[55] H. Fara and R.W. Balluffi, J. Appl. Phys., 27 (1956), 964

[56] W.D. Murray and F. Landis, Trans. ASME, 81 (1959), 106

[57] L.W. Ehrlich, J. Assoc. Comp. Machinery, 5 (1958), 161

[58] M.L. James, G.M. Smith, and J.C. Wolford, Analog and Digital Computer Methods in Engineering Analysis, (Scranton, PA: International Textbook, 1964)

[59] John C. Strikwerda, Finite Difference Schemes and Partial Differential Equations, (Pacific Grove, CA: Wadsworth and Brooks/Cole 1989)

[60] C. Atkinson, Acta Met., 16 (1968), 1019

[61] G. P. Ivantsov, Dokladi Akademii Nauk SSSR, 83 (1952), 573

[62] G. P. Ivantsov, Dokladi Akademii Nauk, SSSR, 58 (1947) 567 [English Translation by G. Horvay, General Electric Research Report No. 60-RL-(2511M), Schenectady, NY, 1960]

[63] G. Horvay and J. W. Cahn, Acta Met., 9 (1961), 695

[64] D. Canright and S. H. Davis, Met. Trans., 20A (1989), 225

[65] F. S. Ham, Quart. J. Appl. Math., 17 (1959), 137

[66] M.E. Glicksman and R.J. Schaefer, J. Crystal Growth, 1 (1967), 297

[67] F. C. Frank, Proc. Roy. Soc. A, 201 (1950), 586

[68] S. D. Howison, IMA J. Appl. Math., 40 (1988), 147

[69] R.F. Sekerka and M.E. Glicksman in Problems in Materials Science, ed. H. S. Merchant (New York, NY: Gordon and Breach, 1972), 169

[70] W.W. Mullins and R.F. Sekerka, J. Appl. Phys., 3A (1963), 323

[71] R.F. Sekerka in Crystal Growth: An Introduction, ed. P. Hartman (Amsterdam, Netherlands: North Holland, 1973), 403

[72] R.F. Sekerka, J. Crystal Growth, 3/4 (1968), 71

[73] A.A. Chernov, J. Crystal Growth, 24/25 (1974), 11

[74] R.T. Delves in Crystal Growth, Vol. 1, ed. B.R. Pamplin (New York, NY: Pergamon Press, 1974), 90

[75] J.S. Langer, Rev. Mod. Phys, 52 (1980), 1

[76] S.R. Coriell and G.B. McFadden in Handbook of Crystal Growth 1a Fundamentals, Transport and Stability, ed. D. T. J. Hurle (Amsterdam, Netherlands: North-Holland, 1993), 785

[77] J.S. Langer and R.F. Sekerka, Acta Met., 23 (1975), 1225

[78] M. Hillert in The Mechanism of Phase Transformations in Crystalline Solids, (London, UK: Institute of Metals, 1969), 231

[79] F.V. Nolfi, P.G. Shewmon, and J.S. Foster, Met. Trans., 1 (1970), 2291

[80] H.L. Frisch and F.C. Collins, J. Chem. Phys., 20 (1952), 1997 [See also: Erratum, J. Chem. Phys., 21 (1953), 1116]

[81] F.C. Frank in Growth and Perfection in Crystals, ed. R.H. Doremus, B.W. Roberts, and D. Turnbull (New York, NY: John Wiley & Sons Inc., 1958), 411

[82] R.F. Sekerka et al., J. Crystal Growth, 87 (1988) 415

[83] A.A. Chernov, Soviet Physics Crystallography, 8 (1964), 901

[84] W. Oldfield in Solidification of Metals, (London, UK: Iron and Steel Institute, 1967), 70

[85] Gerald E. Nash, A Self-Consistent Theory of Steady State Dendritic Growth, Doctoral Thesis (Washington, DC: George Washington University, 1975)

[86] L.N. Brush and R.F. Sekerka, J. Crystal Growth, 96 (1989), 419

[87] S.L. Wang et al., Physica D, 66 (1993), 189

[88] R. Kobayashi, Bull. Japan. Soc. Ind. Appl. Math., 1 (1991), 22

[89] R. Kobayashi, Physica D, 63 (1993), 410

[90] A.A. Wheeler, B.T. Murray and R.J. Schaefer, Physica D, 66 (1993), 243

[91] B. T. Murray et al. in Heat Transfer in Melting, Solidification and Crystal Growth, ed. I.S. Habib and S. Thynell (New York, NY: ASME, 1993), 67

[92] Shun-Lien Wang and Robert F. Sekerka, Phys. Rev. E, 53 (1996), 53

[93] Shun-Lien Wang and Robert F. Sekerka, J. Computational Phys., 127 (1996), 110

[94] B.T. Murray, A.A. Wheeler and M.E. Glicksman, J. Crystal Growth, 154 (1995), 386

[95] A. Karma and W. J. Rappel, Phys. Rev. Lett., 77 (1996), 4050

[96] A. Karma and W. J. Rappel, Phys. Rev. E, 53 (1996), R3017

[97] F.S. Ham, J. Phys. Chem. Solids, 6 (1958), 335

[98] W.A. Johnson and R.F. Mehl, Trans. AIME, 135 (1939), 416

[99] M. Avrami, J. Chem. Phys., 7 (1939) 1103; 8 (1940), 212; 1 (1941), 177

[100] C. Wert and C. Zener, J. Appl. Phys., 21 (1950), 5

[101] J. W. Christian, The Theory of Transformations in Metals and Alloys, (New York, NY: Pergamon Press, 1965), 471

[102] John W. Cahn, Mat. Res. Soc. Proc. Vol. 398 (Materials Research Society, 1996), 425

[103] L. Onsager, Phys. Rev., 37 (1931), 405; 38 (1931), 2265

[104] M. Hillert, Met. Trans., 6A (1975), 5

[105] L.S. Darken, Trans. AIME, 180 (1949), 430

[106] J.S. Kirkaldy, Can. J. Phys., 35 (1957), 435

[107] H. Fujita and L.J. Gostling, J. Am. Chem. Soc., 78 (1956), 1099

[108] J.S. Kirkaldy, Can. J. Phys., 37 (1959), 3034

[109] J.S. Kirkaldy and D.G. Fedak, Trans. AIME, 224 (1962), 490

[110] S.R. Coriell et al., J. Crystal Growth, 191 (1998), 573